PRACTICAL MACROMOLECULAR ORGANIC CHEMISTRY

MMI PRESS POLYMER MONOGRAPH SERIES
Edited by Hans-Georg Elias

This series will present international accounts of research and developments in the specialized areas of macromolecular chemistry.

Volume 1 Thermodynamics of Polymer Solutions
By Michio Kurata
Translated from the Japanese by Hiroshi Fujita

Volume 2 Practical Macromolecular Organic Chemistry
By D. Braun, H. Cherdron, and W. Kern
Translated from the German by Kenneth J. Ivin

Additional volumes in preparation:

^{13}C-NMR Spectroscopy: A Working Manual with Exercises
By E. Breitmeier and G. Bauer
Translated from the German by Bruce K. Cassels

Polymer Analytics
By M. Hoffmann, H. Kromer, and R. Kuhn
Translated from the German by Horst G. Stahlberg

ISSN: 0275-7265

The publisher will accept continuation orders for this series, which may be cancelled at any time and which provide for automatic billing and shipping of each title in the series upon publication. Please write for details.

PRACTICAL MACROMOLECULAR ORGANIC CHEMISTRY

Dietrich Braun
Institut für Makromolekulare Chemie
der Technischen Hochschule Darmstadt
und
Deutsches Kunststoff-Institut, Darmstadt

Harald Cherdron
Hoechst AG Kunststoff-Forschung
Frankfurt, Main

Werner Kern
Organisch-Chemisches Institut
der Universität Mainz

Third Revised and Enlarged Edition
Translated from the German
by
Kenneth J. Ivin
Department of Chemistry
The Queen's University of Belfast.

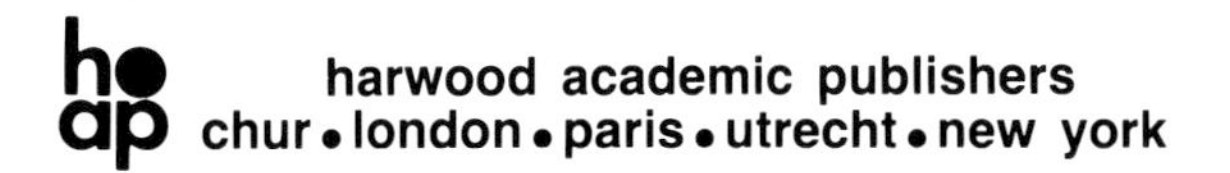
harwood academic publishers
chur • london • paris • utrecht • new york

Published under license by:

Harwood Academic Publishers

Poststrasse 22
7000 Chur
Switzerland

42 William IV Street
London, WC2N 4DE
England

58, rue Lhomond
75005 Paris
France

P.O. Box 15053
3501 BB Utrecht
The Netherlands

P.O. Box 786
Cooper Station
New York, NY 10276
United States of America

Originally published in German in 1966, revised edition published in 1971 and 1979, as Praktikum der Makromolekularen Organischen Chemie by Dr. Alfred Hüthig Verlag GmbH

Library of Congress Cataloging in Publication Data

Braun, Dietrich.
Practical macromolecular organic chemistry.

(MMI Press polymer monograph series; v. 2)
Translation of the 3rd ed. of Praktikum der makromolekularen organischen Chemie.
Bibliography: p.
Includes index.
1. Polymers and polymerization—Laboratory manuals.
I. Cherdron, Harald. II. Kern, Werner. III. Title.
IV. Series.
QD385.B7313 1983 661'.8 83-12985
ISBN 3-7186-0059-5

 Printed in Great Britain by Bell and Bain Ltd., Glasgow.

CONTENTS

PREFACE TO THE THIRD EDITION

The entire text of this new third edition has been revised but the original concept of using simple practical examples to illustrate the preparation and investigation of polymers, which characterized the two previous editions of Praktikum der Makromolekularen Organischen Chemie (1971, 1966), has been retained.

Three of the old examples have been deleted and nine new ones inserted, e.g. metathesis polymerization, polyimides, microencapsulation, and butadiene-styrene block copolymers. This third edition also contains, in response to requests from colleagues, more experiments on physical properties of macromolecules. However, only those experiments which can be performed with the least amount of time and apparatus have been included.

We are grateful to colleagues in numerous academic and industrial laboratories, both at home and abroad, whose valuable suggestions and criticisms have helped to improve this work.

We particularly thank Dr. H. Bartl (Leverkusen), Dr. L. Bohn (Hoechst), Prof. O. Fuchs (Hofheim), Prof. H. Höcker (Bayreuth), Dr. M. Hoffmann (Leverkusen), Prof. J. P. Kennedy (Akron), Prof. H. J. Klein (Braunschweig), Dr. H. J. Leugering (Hoechst), Prof. O. F. Olaj (Wien), Dr. H. Pohlemann (Ludwigshafen), Prof. R. C. Schulz (Mainz), Prof. O. Vogl (Amherst), Dr. P. Wittmer (Ludwigshafen) and Prof. B. A. Wolf (Mainz).

We also thank our colleagues in the German Plastics Institute (Darmstadt), who played a key role in the careful screening of the newly adopted examples, and the "Macromolecular Chemistry" group of the German Chemical Society for their financial support.

We are indebted to Dr. J. H. Wendorff and Dr. G. Disselhoff (Darmstadt) for critical scrutiny of several chapters and for assistance in examining and supplementing the experimental instructions and the literature references.

Dietrich Braun
Harald Cherdron
and
Werner Kern

PREFACE TO THE SECOND EDITION

The good response to the first edition of Praktikum der Makromolekularen Organischen Chemie (1971) and the appearance of several foreign-language editions has been gratifying.

This new second edition has been carefully and extensively revised. The sections on "Characterization of Polymers" and "Theory of Condensation Polymerization" has been expanded. Eleven new examples have been introduced, including foamable polystyrene, poly(phenylene oxide), polysiloxane and gel permeation chromatography. The method of numbering the examples has also been changed from that used in the first edition.

Numerous suggestions were made that we should include experiments on physical properties of macromolecules in this edition. After careful consideration we decided against their inclusion so as not to change the character of the "Praktikum" as a mainly preparative introduction to macromolecular chemistry. Nevertheless, Chapter 2 has been expanded in various places.

We wish to thank Dr. G. Henrici-Olivé and Dr. S. Olivé (Zürich) for valuable comments on the section "Kinetics of Radical Polymerization"; and Prof. Dr. O. Fuchs, Dr. L. Bohn and Dr. H. J. Leugering (Farbwerke Hoechst) for advice and experimental results relating to the section on "Characterization of Polymers."

We again offer our grateful thanks to Miss R. Weis of the Institute of Organic Chemistry at the University of Mainz for her careful reading of the proofs, and to the publishers for their dedicated assistance.

Dietrich Braun
Harald Cherdron
and
Werner Kern

PREFACE TO THE FIRST EDITION

The field of macromolecular chemistry has undergone dramatic and continuous growth in the four decades since its foundation by Hermann Staudinger, and the flood of publications has become so great that it is hard to keep abreast of developments.

For those who teach the subject of macromolecular chemistry it is essential to be able to provide students with laboratory work which will bring alive the theoretical concepts taught in lectures. This book is designed to help in this respect. It describes general working methods and the most important procedures for the preparation and investigation of polymers, together with specific examples.

All the examples are carefully related both to the preparation of macromolecular substances and, as far as possible, to the underlying theory. In addition to preparations there are examples of kinetics of polymerization, and the demanding and unusual laboratory techniques required for the preparation, work-up, purification and characterization of high-molecular-weight substances. Examples of the applications of polymers are indicated wherever appropriate, e.g. emulsions, foamed materials, resins, films and ion-exchangers.

We have not included the physical chemistry and physics of macromolecules or the large area of natural macromolecular compounds. The text is limited essentially to the synthetic organic chemistry of macromolecules and its place within the compass of organic chemistry.

We hope this work will find acceptance in university teaching laboratories, industrial laboratories, and wherever students wish to become conversant with the concepts of macromolecular chemistry.

We are grateful to Prof. Dr. O. Horn (Farbwerke Hoechst), former Chairman of the Plastics and Rubber Group of the German Chemical Society, and the Group Committee for their encouragement and financial support which enabled our coworkers to thoroughly test the experimental instructions (88 examples).

The examples, carefully adapted to the requirements of students and the resources of teaching establishments, were worked out from our own experience and from tested laboratory instructions that were placed at our disposal by numerous firms in the chemical industry. We are particularly indebted to Prof. Dr. O. Bayer (Farbenfabriken Bayer), Prof. Dr. B. Blaser (Henkel), Dr. E. Heisenberg (Vereinigte Glanzstoff-Fabriken), Prof. Dr. H. Hellman

(Chemische Werke Hüls), Prof. Dr. O. Horn (Farbwerke Hoechst), Prof. Dr. F. Korte (Shell Grundlagenforschung, Birlinghoven), Dr. S. Nitzsche (Wacker-Chemie), Dr. R. O. Sauer (Chemische Werke Albert), Prof. Dr. A. Steinhofer (Badische Anilin und Soda-Fabrik), Dr. E. Trommsdorff (Röhm and Haas), and their coworkers for valuable assistance.

For the experimental testing of the examples we thank W. Neumann (Darmstadt), R. Kern and A. Sanner (Mainz). To W. Neumann we owe further thanks for critical evaluation of the experience gained from practical courses on macromolecular chemistry taken by students of the Technische Hochschule Darmstadt in the Deutsche Kunststoff-Institut. Many examples were also tested in the practical organic chemistry course of the University of Mainz under the guidance of Dr. L. Dulog, to whom our thanks are also due.

We wish to thank Miss R. Weis and Dr. E. Radlmann of the Institute of Organic Chemistry at the University of Mainz for their assistance in proof-reading, and the publishers for their unfailing guidance.

Dietrich Braun
Harald Cherdron
and
Werner Kern

TRANSLATOR'S NOTE

At the request of the authors, a number of minor changes have been made during translation in order to bring the nomenclature into line with IUPAC recommendations. In making these changes I have been greatly assisted by Dr. H. Reimlinger, whose knowledge of such matters is beyond compare.

Some of the more important changes may be mentioned here. The term "polyaddition" has been replaced by "stepwise addition polymerization" or just "addition polymerization" where the context is clear. The term "macromolecular compound" has been deliberately avoided in favour of "macromolecular substance". The abbreviation CRU has been liberally used for "constitutional repeating unit." A word like "butyl" without a prefix always means "*n*-butyl".

It is of course difficult to be entirely consistent in matters of nomenclature without becoming pedantic, and I take full credit/blame for any irritating inconsistencies which the reader may find.

K. J. Ivin

1. Introduction

Macromolecular organic chemistry is concerned with substances that are distinguished by especially high molecular weight. A sharp boundary cannot, however, be drawn between low-molecular-weight substances and macromolecular substances; rather there is a gradual transition between them. However, one can say that macromolecules consist of a minimum of several hundred atoms; accordingly the lower limit for the molecular weight of macromolecular substances can arbitrarily be taken as around 10^3 g mol^{-1}.

Pure, low-molecular-weight compounds consist, by definition, of molecules of identical structure and size; but in the case of macromolecular compounds this is generally not the case. Some naturally occurring high-molecular-weight compounds (e.g. certain nucleic acids) possess a uniform structure and constant molecular size; but most polymer samples consist of mixtures of macromolecules with different molecular weights, though with the same or similar structural units. These and other differences from low-molecular-weight compounds (e.g. structural- and stereo-isomerism), together with the high molecular weights, give rise to a range of properties and characteristics that must be taken into account by the preparative chemist when synthesizing and analyzing macromolecular substances.

1.1. SYNTHESIS OF MACROMOLECULAR SUBSTANCES

In a chemical reaction between two molecules, the constitution of the reaction product can be unequivocally deduced if the starting materials possess functional groups which react selectively under the conditions chosen. If an organic compound contains one reactive group that can give rise to one linkage in the intended reaction, it is called monofunctional; for two, three or more groups it is called bi-, tri- or poly-functional respectively. However, this statement concerning the functionality of a compound is only significant in relation to a specific reaction. For example, the primary amino group is monofunctional with respect to the formation of Schiff's bases, but up to trifunctional when reacted with alkyl halides. Simple unsaturated compounds, epoxides and cyclic esters are monofunctional in their addition reactions with monofunctional compounds, but bifunctional in polymerizations (see Section 3.2.4).

Compounds suitable for the formation of macromolecules must be at least bifunctional with respect to the desired polymerization; they are termed monomers. Linear macromolecules result from the coupling of bifunctional molecules with each other or with other bifunctional molecules; in contrast, branched or crosslinked polymers are formed when tri- or poly-functional compounds are involved.

Stepwise polymerizations such as condensation polymerization (see Section 4.1) and certain types of addition polymerization (see Section 4.2) yield molecules that still possess free reactive functional groups which can react further. On the other hand, monofunctional compounds always give low-molecular-weight products even when reacted with polyfunctional compounds. Not all reactions of bifunctional monomers lead to macromolecular substances. For example, ethylene oxide can be dimerized to 1,4-dioxane, and formaldehyde undergoes trimerization to 1,3,5-trioxane. Moreover, many condensation and addition reactions of bifunctional compounds yield low-molecular-weight cyclic oligomers as well as the desired linear polymer. Thus, intramolecular cyclization sometimes competes with the intermolecular coupling; under suitable conditions it can become the main reaction, e.g. in the synthesis of large rings according to the Ruggli-Ziegler dilution principle.

Starting from suitable monomers, one can make macromolecules by various polymerization reactions that differ in kinetics and mechanism. They may be divided according to their mechanism into chain reactions and stepwise reactions. In chain reactions initiation generates active centres which generally propagate the reaction very rapidly via macroradicals or macroions; chain termination yields inactive macromolecules. Most polymerizations belong to this class. In contrast, the formation of macromolecules by stepwise reactions involves the stepwise coupling of monomer, oligomer or polymer molecules with reactive end groups, without any true initiation or termination reactions.

Addition polymerizations of single compounds involve reactions in which monomeric compounds, containing reactive double bonds or rings, are transformed into macromolecules, either spontaneously or under the action of catalysts, by successive addition of monomer molecules to the reactive chain ends. The characteristic feature of such polymerizations lies in their behaviour as chain reactions, the mechanism of which may be radical, cationic or anionic. Some polymerizations that are initiated by organometallic mixed catalysts do not quite fit into this scheme; although they are chain reactions, the nature of the active chain end depends on various factors and in many cases is not fully clarified.

The formation of relatively low-molecular-weight members of a homologous polymeric series (dimers, trimers etc., generally called oligomers) is termed oligomerization, regardless of the reaction mechanism. Oligomers and polymers with two identical, generally reactive, end groups are said to be telechelic.

The ability of a monomer to polymerize radically or ionically depends, above all, on the polarizability of the double bond (or other bonds in cyclic monomers), on the structure, position and number of substituents, on the initiator, and on the external conditions such as temperature, pressure etc. While the tendency of a monomer towards radical polymerization can generally be established by means of a few experiments, this is usually more difficult with ionic polymerization, especially if one wants to obtain high molecular weights or particular polymeric structures. In Table 1.1 a selection of monomers of different classes is set out and their polymerizability indicated. Some monomers can be copolymerized, yet do not homopolymerize or do so only with difficulty; these are noted in the final column.

Condensation polymerization or polycondensation (for details see Section 4.1) refers to the formation of macromolecules by coupling of bi- or higher-functional molecules with the elimination of low-molecular-weight compounds, such as water or alcohol, derived from the reacting groups. This is a typical stepwise reaction and differs fundamentally from radical or ionic polymerization. The resulting high-molecular-weight products are called condensation polymers. Table 1.2 shows some monomeric compounds that are polycondensible.

Some types of addition polymerization (for details see Section 4.2) also consist of the coupling of bi- or poly-functional reaction partners, but without elimination of low-molecular-weight compounds. Frequently a hydrogen atom migrates at each addition step. As in polycondensation, the macromolecules are built up stepwise; this type of addition polymerization is thus mechanistically related to polycondensation. Addition polymerizations of this type are less common than either addition polymerizations involving single compounds or condensation polymerizations; some addition polymerizations of this stepwise type are summarized in Table 1.3.

The most important differences between the two types of polymerizations are listed below for comparison:

Chain reactions	*Stepwise reactions*
Reaction generally requires initiators or catalysts	Reaction often proceeds without the need for catalysts
Only active species (e.g. macroradicals or macro-ions) can add further monomer molecules in the propagation process	Both monomer and polymer molecules with suitable functional end groups can react
Activation energy of chain initiation is very much greater than that of progagation	Activation energy is about the same for each reaction step
Monomer concentration decreases with reaction time	Monomer molecules disappear very quickly; more than 99% of monomer molecules have already reacted when the degree of polymerization is 10
Macromolecules are formed from the beginning of the reaction	Monomer molecules first give oligomers; high polymer is formed only towards the end of reaction
The average molecular weight of the polymer normally changes little with reaction time	The average molecular weight increases steadily with reaction time; long reaction times are usually necessary to produce high molecular weights

Numerous macromolecular substances can also be obtained by treatment of reactive polymers with low-molecular-weight compounds. Such reactions can be used not only to modify known polymers but sometimes provide the only route to a particular polymer (see Chapter 5).

1.2. STRUCTURE AND NOMENCLATURE OF MACROMOLECULAR SUBSTANCES

Macromolecules may consist of one or several types of monomeric units and may accordingly be described as homopolymers or copolymers respectively. Linear macromolecules can also be classified according to their chain structure; isochains contain only carbon atoms in the main chain (e.g. polymers of vinyl compounds), while heterochains contain more

TABLE 1.1

Monomers that form Macromolecules by Addition Polymerization of a Single Compound

Monomer	Polymerizability radical (a)	cationic (b)	anionic[1] (c)	Constitutional repeating unit (CRU)	Preparation of the polymers[b,c]
Monomers with C=C *double bonds*					
1. *Olefins*					
Ethylene	+	+	+	$—CH_2—CH_2—$	a) *Houben-Weyl 14/1* (1961) 592. b) *ibid*, p. 588 c) Example 3–31
Propene	–	+	+	$—CH_2—CH(CH_3)—$	Example 3–32
2-Methylpropene (isobutene)	–	+	–	$—CH_2—C(CH_3)_2—$	Example 3–23
Styrene	+	+	+	$—CH_2—CH(C_6H_5)—$	a) Example 3–01 *et seq*. b) Example 3–25 c) Example 3–28
α-Methylstyrene	+	+	+	$—CH_2—C(CH_3)(C_6H_5)—$	a) *Houben-Weyl 14/1* (1961) 817 (copolymerization) b) Example 3–26 c) Example 3–27
Tetrafluoroethylene	+	–	–	$—CF_2—CF_2—$	*Houben-Weyl 14/1* (1961) 844.

TABLE 1.1 (*Continued*)

Monomers that form Macromolecules by Addition Polymerization of a Single Compound

Monomer	Polymerizability radical (a)	cationic (b)	anionic[1] (c)	Constitutional repeating unit (CRU)	Preparation of the polymers[2,3]
2. *Dienes*					
Butadiene	+	+	+	see p. 181	a) *Houben-Weyl 14/1* (1961) 674 cf. Example 3–47 b) J. Polym Sci. *13* (1954) 325 c) Example 3–34 Macromol. Synth. I, 125, 129
2-Methyl-1,3-butadiene (isoprene)	+	+	+	see p. 16	a) Example 3–21 b) J. Polym. Sci. *13* (1954) 325 c) Example 3–30 Macromol. Synth. I, 137, 141
2-Chloro-1,3-butadiene (chloroprene)	+	–	–	$-CH_2-C(Cl)=CH-CH_2-$ (*trans*-1,4)	*Houben-Weyl 14/1* (1961) 745
3. *Vinyl compounds*					
Vinyl chloride	+	–	+	$-CH_2-CH(Cl)-$	a) cf. Example 3–49 b) J. Polym. Sci. Part C*4* (1964) 299
Vinyl acetate	+	–	–	$-CH_2-CH(OCOCH_3)-$	Example 3–06 *et seq*.
Isobutyl vinyl ether	+	+	–	$-CH_2-CH(O-CH_2CH(CH_3)_2)-$	a) *Houben-Weyl 14/1* (1961) 956 (copolymerization) b) Example 3–24
Vinylsulfonic acid	+	–	–	$-CH_2-CH(SO_3H)-$	*Houben-Weyl 14/1* (1961) 1099.
Methyl vinyl ketone	+	+	–	$-CH_2-CH(COCH_3)-$	a) *Houben-Weyl 14/1* (1961) 1093 b) *ibid*, p. 1095
1-Vinylpyrrolidone	+	+	–	$-CH_2-CH-$	a) *Houben-Weyl 14/1* (1961) 1113

4. *Acrylic compounds*					
Acrylic acid	+	–	–	$-CH_2-CH(COOH)-$	*Houben-Weyl 14/1* (1961) 1023
Methacrylic acid	+	–	–	$-CH_2-C(CH_3)(COOH)-$	a) Example 3–10
Methyl acrylate	+	–	+	$-CH_2-CH(COOCH_3)-$	a) *Sorenson-Campbell*, p. 176 c) *Houben-Weyl 14/1* (1961) 1065
Methyl methacrylate	+	–	+	$-CH_2-C(CH_3)(COOCH_3)-$	a) Example 3–05 et seq. b) Example 3–43 (copolymerization) c) Example 3–29
Acrylonitrile	+	–	+	$-CH_2-CH(CN)-$	a) Example 3–20 c) *Houben-Weyl 14/1* (1961) 977
Acrylamide	+	–	+	a) $-CH_2-CH(CONH_2)-$	a) Example 3–18
				c) $-CH_2-CH_2-C(=O)-N(H)-$	c) *Sorenson-Campbell*, p. 89; Macromol. Synth. I, 95.
Acrolein	+	+	+	a) and b) $-CH_2-CH(CHO)-$	a) *Houben-Weyl 14/1* (1961) 1084; Macromol. Synth. I, 227. b) Makromol. Chem. *17* (1955) 62
				c) $-C(H)(CH=CH_2)-O-$	c) Makromol. Chem. *60* (1963) 139; Macromol. Synth. I, 231

TABLE 1.1 (*Continued*)

Monomers that form Macromolecules by Addition Polymerization of a Single Compound

Monomer	Polymerizability radical (a)	cationic (b)	anionic[a] (c)	Constitutional repeating unit (CRU)	Preparation of the polymers[b,c]
5. *Allyl compounds*					
Allyl alcohol	+	–	–	$-CH_2-CH(CH_2OH)-$	*Houben-Weyl 14/1* (1961) 1138
Allyl chloride	+	+	–	$-CH_2-CH(CH_2Cl)-$	a) *Houben-Weyl 14/1* (1961) 1139 b) *ibid*, p. 1140
Allyl acetate	+	–	–	$-CH_2-CH(CH_2OCOCH_3)-$	*ibid*, p. 1141
Monomers with C=O *bonds*					
Formaldehyde	–	+	+	$-CH_2-O-$	c) Example 3–36
Acetaldehyde	–	+	+	$-CH(CH_3)-O-$	b) Macromol. Synth. I, 283
Monomers with N=C *bonds*					
Butyl isocyanate	–	–	+	$-N(C_4H_9)-C(=O)-$	Example 3–38
Cyclic monomers that polymerize by ring-opening					
1. *Cyclic ethers*					
Ethylene oxide	–	+	+	$-CH_2-CH_2-O-$	c) *Sorenson-Campbell*, p. 241; Macromol. Synth. I, 289
Propylene oxide	–	+	+	$-CH(CH_3)-CH_2-O-$	b) *ibid*, p. 242
3,3-Bis(chloromethyl) -oxetane	–	+	+	$-CH_2-C(CH_2Cl)_2-CH_2-O-$	b) *Sorenson-Campbell*, p. 246 c) Makromol. Chem. *53* (1962) 203
				[illegible]	Example 3–39

Cyclic siloxanes (e.g. 2,4,6-hexamethyl-cyclotrisiloxane)	–	+	+	$-Si(CH_3)_2-O-$	c) Example 4–18
1-Chloro-2,3-epoxypropane (epichlorohydrin)	–	+	+	$-CH_2-CH(CH_2Cl)-O-$	a) Macromol. Synth. I, 433
2. *Cyclic acetals*					
1,3,5-Trioxane	–	+	–	$-CH_2-O-CH_2-O-CH_2-O-$	Example 3–40
3. *Cyclic amides*					
β-Propiolactam	–	–	+	$-C(=O)-(CH_2)_2-N(H)-$	*Houben-Weyl 14/2* (1963) 118
γ-Butyrolactam	–	–	+	$-C(=O)-(CH_2)_3-N(H)-$	*ibid*, p. 119
ε-Caprolactam	–	+	+	$-C(=O)-(CH_2)_5-N(H)-$	c) Example 3–42
4. *Cyclic esters*					
β-Propiolactone	–	+	+	$-C(=O)-(CH_2)_2-O-$	b) Makromol. Chem. *56* (1962) 179
δ-Valerolactone	–	+	+	$-C(=O)-(CH_2)_4-O-$	b) Makromol. Chem. *56* (1962) 179
ε-Caprolactone	–	+	+	$-C(=O)-(CH_2)_5-O-$	b) Makromol. Chem. *56* (1962) 179 Macromol. Synth. I, 327

[a] Including organometallic mixed catalysts.
[b] Where there is no corresponding Example in Chapter 3, suitable references have been selected from the literature.
[c] The examples from *Sorenson-Campbell (W. R. Sorenson, T. W. Campbell*, "Preparative Methods of Polymer Chemistry", Interscience Publishers, New York, 1961; German translation, *Th. Lyssy*, Verlag Chemie, Weinheim) are taken from the 1st German edition. Macromol. Synth. I: "Macromolecular Synthesis", Coll. Vol. I (1977). *Houben-Weyl*: "Methoden der Organischen Chemie"; Ed. Eu Müller, Thieme Verlag, Stuttgart. *Cf.* Section 1.5, Literature on macromolecular substances, p. 31.

TABLE 1.2

Monomers that form Macromolecules by Polycondensation

Monomer 1	Monomer 2	Polymer	Constitutional repeating unit (CRU)	Preparation of the polymers[a,b]
ω-Hydroxy-carboxylic acid	—	Linear polyesters	$-C(=O)-(CH_2)_x-O-$	*Houben-Weyl 14/2* (1963) 6
Diols	Dicarboxylic acids or their derivatives	Linear polyesters	$-C(=O)-(CH_2)_x-C(=O)-O-(CH_2)_y-O-$	Examples 4–01, 4–02, 4–03, 4–04
Triols or polyols	Di- or poly-carboxylic acids or their derivatives	Branched or crosslinked polyesters resp.	—	Example 4–06
ω-Amino carboxylic acids	—	Linear polyamides	$-C(=O)-(CH_2)_x-N(H)-$	Example 4–08
Diamines	Dicarboxylic acids or their derivatives	Linear polyamides	$-C(=O)-(CH_2)_x-C(=O)-N(H)-(CH_2)_y-N(H)-$	Example 4–09, 4–10
Diamines	Phosgene	Linear polyureas	$-N(H)(CH_2)_x-N(H)-C(=O)-$	*Houben-Weyl 14/2* (1963) 165
Dicarboxylic acids	—	Polyanhydrides	$-C(=O)-(CH_2)_x-C(=O)-O-$	*Sorenson-Campbell*, p. 138

Potassium hydrogen-phosphate	—	Polyphosphate	$-\overset{\overset{\displaystyle O}{\|}}{\underset{\underset{\displaystyle OK}{\|}}{P}}-O-$	*Sorenson-Campbell*, p. 141
Dimethyl-silanediol	—	Polysiloxane	$-O-\overset{\overset{\displaystyle CH_3}{\|}}{\underset{\underset{\displaystyle CH_3}{\|}}{Si}}-$	*Sorenson-Campbell*, p. 250 cf. also Example 4–18
α,ω-Dihalo-alkanes	Sodium polysulfide	Poly(thioalkylenes)	$-(CH_2)_x-S_y-$	Example 4–17
Phenol	Formaldehyde	Phenol/formaldehyde condensation polymer	—	Example 4–13, 4–14
Urea	Formaldehyde	Urea/formaldehyde condensation polymer	—	Example 4–15
Melamine	Formaldehyde	Melamine/formalde-hyde condensation polymer	—	Example 4–16

[a] Where there is no corresponding Example in Chapter 4, suitable references have been selected from the literature.
[b] See footnote c of Table 1.1.

TABLE 1.3

Monomers that form Macromolecules by Stepwise Addition Polymerization involving Two Monomers

Monomer 1	Monomer 2	Polymer	Constitutional repeating unit (CRU)	Preparation of the polymers[a,b]
Diols	Diisocyanates	Polyurethanes	$-\underset{\underset{O}{\Vert}}{C}-\underset{\underset{H}{\vert}}{N}-(CH_2)_x-\underset{\underset{H}{\vert}}{N}-\underset{\underset{O}{\Vert}}{C}-O-(CH_2)_y-O-$	Example 4–22
Diamines	Diisocyanates	Polyureas	$-\underset{\underset{O}{\Vert}}{C}-\underset{\underset{H}{\vert}}{N}-(CH_2)_x-\underset{\underset{H}{\vert}}{N}-\underset{\underset{O}{\Vert}}{C}-\underset{\underset{H}{\vert}}{N}-(CH_2)_y-\underset{\underset{H}{\vert}}{N}-$	*Sorenson–Campbell*, p. 93
Di- or poly-epoxides	Amines, anhydrides	Epoxy resins	—	Example 4–26
Non-conjugated dienes	Dithiols	Polythioethers	$-S-(CH_2)_x-S-(CH_2)_y-$	*Sorenson–Campbell*, p. 126

[a] Where there is no corresponding Example in Chapter 4, suitable references have been selected from the literature.
[b] See footnote c of Table 1.1.

than one type of element, (e.g. polyacetals contain oxygen atoms as well as carbon, and polyamides contain nitrogen atoms as well as carbon[1]).

Basic definitions of terms relating to polymers have been put forward in recent years by the Commission on Nomenclature of Macromolecules of the International Union of Pure and Applied Chemistry (IUPAC)[2], superseding the older nomenclature.[3,4] The most important new definitions are indicated below.

A *polymer* is defined as a substance consisting of molecules that are characterized by multiple repetition of one or more species of atoms, or groups or atoms (constitutional units), linked to each other in amounts sufficient to provide a set of properties that do not vary markedly with the addition or removal of one or a few of the constitutional units. The polymer chain

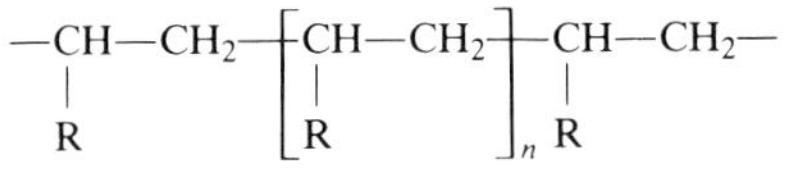

$$\text{—}\underset{\displaystyle R}{\underset{|}{CH}}\text{—}CH_2\text{—}\left[\underset{\displaystyle R}{\underset{|}{CH}}\text{—}CH_2\right]_n\text{—}\underset{\displaystyle R}{\underset{|}{CH}}\text{—}CH_2\text{—}$$

contains the following constitutional units:

$$\underset{\text{(a)}}{\text{—}\underset{\displaystyle R}{\underset{|}{CH}}\text{—}CH_2\text{—}} \qquad \underset{\text{(b)}}{\text{—}CH_2\text{—}\underset{\displaystyle R}{\underset{|}{CH}}\text{—}} \qquad \text{—}CH_2\text{—} \qquad \text{—}\underset{\displaystyle R}{\underset{|}{CH}}\text{—} \qquad \text{etc.}$$

Of these only (a) and (b) are the smallest constitutional units by which the regular polymer chain may be completely described and are examples of *constitutional repeating units* (CRU's).

That part of a macromolecule which corresponds to the largest unit in the chain or to one of the molecules from which the macromolecule was built (or can be regarded as being built) is designated as a *monomeric unit*. In the case of addition polymerization of a single monomer the CRU is the same as the monomeric unit:

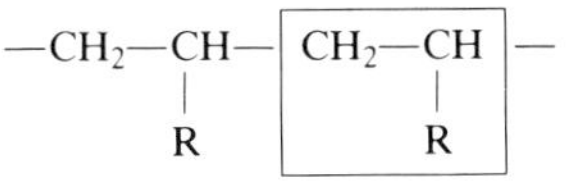

$$\text{—}CH_2\text{—}\underset{\displaystyle R}{\underset{|}{CH}}\text{—}\boxed{CH_2\text{—}\underset{\displaystyle R}{\underset{|}{CH}}}\text{—}$$

CRU identical with monomeric unit

---

[1] Besides this chemical classification, it is also possible to divide polymers according to physical or technological properties, e.g. into thermoplastics, elastomers and thermosetting resins.

[2] IUPAC, Pure Appl. Chem. *40* (1974) 477

[3] *J.W. Breitenbach* et al., Makromol. Chem. *38* (1960) 1.

[4] *M.L. Huggins, G. Natta, V. Desreux* and *H. Mark*, Makromol. Chem. *82* (1965) 1.

A CRU can, however, contain several monomeric units; this is the case for many condensation polymers and for alternating copolymers:

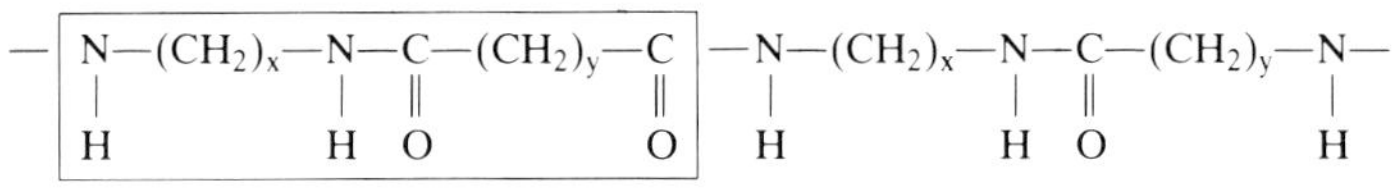

$$-\boxed{\underset{\displaystyle\mathrm{H}}{\underset{|}{\mathrm{N}}}-(\mathrm{CH_2})_x-\underset{\displaystyle\mathrm{H}}{\underset{|}{\mathrm{N}}}-\underset{\displaystyle\mathrm{O}}{\underset{\|}{\mathrm{C}}}-(\mathrm{CH_2})_y-\underset{\displaystyle\mathrm{O}}{\underset{\|}{\mathrm{C}}}}-\underset{\displaystyle\mathrm{H}}{\underset{|}{\mathrm{N}}}-(\mathrm{CH_2})_x-\underset{\displaystyle\mathrm{H}}{\underset{|}{\mathrm{N}}}-\underset{\displaystyle\mathrm{O}}{\underset{\|}{\mathrm{C}}}-(\mathrm{CH_2})_y-\underset{\displaystyle\mathrm{H}}{\underset{|}{\mathrm{N}}}-$$

CRU consisting of two different monomeric units

A random copolymer with the partial structure:

$$-\underset{\displaystyle\mathrm{Y}}{\underset{|}{\mathrm{CH}}}-\mathrm{CH_2}-\underset{\displaystyle\mathrm{R}}{\underset{|}{\mathrm{CH}}}-\mathrm{CH_2}-\underset{\displaystyle\mathrm{Y}}{\underset{|}{\mathrm{CH}}}-\mathrm{CH_2}-\underset{\displaystyle\mathrm{Y}}{\underset{|}{\mathrm{CH}}}-\mathrm{CH_2}-\underset{\displaystyle\mathrm{R}}{\underset{|}{\mathrm{CH}}}-\mathrm{CH_2}-$$

is made up of more than one type of monomeric unit, but cannot be described in terms of any CRU; such polymers are said to be irregular, e.g. copolymer of styrene and methyl methacrylate.

CRU's are always related to, and may be identical with, the compounds that have been used to synthesize the polymer. Many natural macromolecules are also made up of CRU's. For polymers of vinyl compounds the CRU contains two chain atoms. CRU's can also have three or more chain atoms or even only one:

$$-\mathrm{CH_2}-\mathrm{CH_2}-\mathrm{CH_2}-\boxed{\mathrm{CH_2}}-$$

CRU with one chain atom (polymethylene)

$$-\mathrm{CH_2}-\mathrm{CH_2}-\boxed{\mathrm{CH_2}-\mathrm{CH_2}}-$$

Monomeric unit with two chain atoms (polyethylene)

$$-\mathrm{CH_2}-\mathrm{CH_2}-\mathrm{O}-\boxed{\mathrm{CH_2}-\mathrm{CH_2}-\mathrm{O}}-$$

CRU with three chain atoms [poly(oxyethylene)]

The systematic nomenclature of regular single-strand polymers takes as its starting point the naming of the CRU as a group with two free valencies, conforming as far as possible to the rule of nomenclature of organic chemistry. The name of the polymer is then simply obtained by adding the prefix "poly"; the direction and sequence of CRU's by which the polymer is named is defined by rules whereby the sub-units are arranged in decreasing rank from left to right. For example:

$$\left[O - \langle \bigcirc \rangle \right]_n$$ poly(phenylene oxide) = poly(oxy-1,4-phenylene)

$$\left[NH-CO-(CH_2)_4-CO-NH-(CH_2)_6 \right]_n$$

polyhexamethyleneadipamide (Nylon-6,6) = poly(iminoadipoyliminohexamethylene)

Alongside the systematic, structure-based names, some of the existing source-based semi-systematic or trivial names are still allowed. Brackets are used when the name of the monomer, from which that of the polymer is derived, contains more than one word, for example poly(vinyl chloride).

Macromolecules that consist of the same constitutional units can differ as a result of structural isomerism and stereoisomerism, and may also have different end groups. Linear, branched, and crosslinked macromolecules of the same monomer are structural isomers. Branching is to be expected when trifunctional monomers are used in the synthesis, or when side reactions, such as transfer reactions with the polymer, take place during polymerization (see Section 3.1). Crosslinking can occur when tri- or poly-functional reactants are involved in the formation of the polymer (e.g. polycondensations with triols, or polymerizations of vinyl compounds in the presence of divinyl compounds which react as tetrafunctional agents). Fig. 1.1 shows the schematic structure of polystyrenes that are respectively linear, branched, and crosslinked with 1,4-divinylbenzene (see Example 3–50).

Another type of structural isomerism occurs in the polymers of vinyl or vinylidene compounds. In the addition polymerization of these compounds there are two possible modes of addition of the monomer to the growing chain, and therefore two possible arrangements of the CRU's:

$$-CH_2-\underset{X}{CH} \vdots CH_2-\underset{X}{CH} \vdots CH_2-\underset{X}{CH}-$$

head-tail arrangement

$$-CH_2-\underset{X}{CH} \vdots \underset{X}{CH}-CH_2 \vdots CH_2-\underset{X}{CH}-$$

head-head tail-tail arrangement

Of these the head-tail structure is by far the most predominant; the proportion of head-head structure is small and can only be determined experimentally in special cases (see Example 5–18).

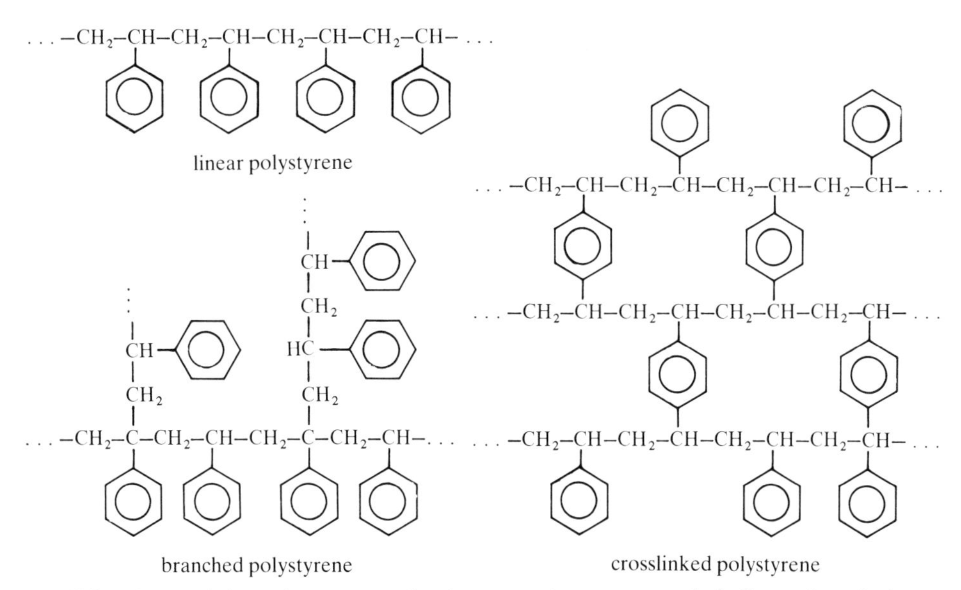

FIGURE 1.1 Schematic structure of polystyrenes that are respectively linear, branched, and crosslinked with 1,4-divinylbenzene.

Further types of isomeric chain structure are formed in the polymerization of conjugated dienes. Addition to the growing chain can occur in the 1,2-position, in the 1,4-position and, in the case of unsymmetrically substituted dienes, also in the 3,4-position (which may also give rise to head-head structures as described above for vinyl compounds):

$CH_2{=}C(CH_3){-}CH{=}CH_2$ →

$\ldots -CH_2-C(CH_3){=}CH-CH_2 \vdots CH_2-C(CH_3){=}CH-CH_2- \ldots$ 1,4-addition

$\ldots -CH_2-C(CH_3)(CH{=}CH_2) \vdots CH_2-C(CH_3)(CH{=}CH_2)- \ldots$ 1,2-addition

$\ldots -CH_2-CH(C(CH_3){=}CH_2) \vdots CH_2-CH(C(CH_3){=}CH_2)- \ldots$ 3,4-addition

When there are double bonds in the main chain, as in 1,4-addition, there is the further possibility of *cis/trans* isomerism. In some polymers all the double bonds have *cis* structure (e.g. natural rubber) while others have an all-*trans* structure (e.g. gutta percha, balata):

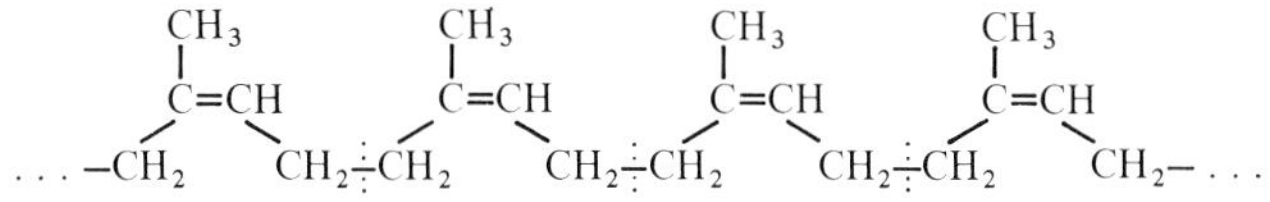

```
        CH3              CH3              CH3              CH3
        |                |                |                |
        C=CH             C=CH             C=CH             C=CH
       /    \           /    \           /    \           /    \
...-CH2      CH2 ⋮ CH2       CH2 ⋮ CH2       CH2 ⋮ CH2       CH2-...
```

cis–1,4–polyisoprene (natural rubber)

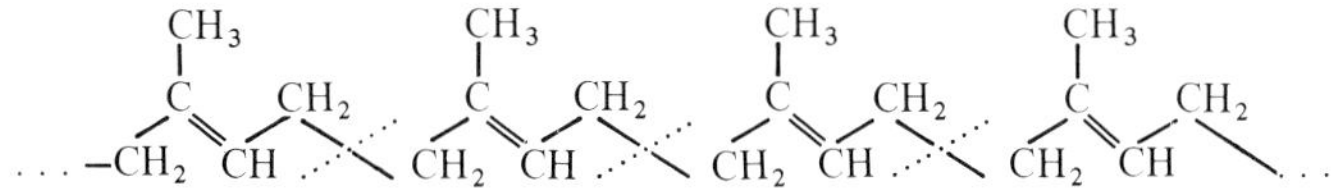

```
        CH3               CH3               CH3               CH3
        |                 |                 |                 |
        C     CH2         C     CH2         C     CH2         C     CH2
       / \\   /   \  ⋰   / \\   /   \  ⋰   / \\   /   \  ⋰   / \\   /   \
...-CH2    CH   ⋰   CH2    CH   ⋰   CH2    CH   ⋰   CH2    CH         ...
```

trans-1,4-polyisoprene (gutta percha)

Synthetic polymeric dienes are generally not so structurally uniform as naturally occurring polyisoprene. However, using special initiator systems and under carefully controlled polymerization conditions (see Example 3–30), it is possible to prepare *cis*-1,4-polyisoprene ("synthetic natural rubber"), even on a large scale.

Linear macromolecules having pendant groups X, for example polymers of vinyl compounds with the CRU —CH_2—CHX—, can exhibit two further kinds of stereoisomerism, namely optical isomerism and tacticity. (In certain cases the latter can also give rise to optical activity).

The type of stereoisomerism designated by Natta[1] as tacticity stems from the possibility of different spatial arrangements of the substituents X. Consider a polymer chain having the CRU —CH_2—CHX—. If one arranges the chain carbon atoms in planar zig-zag form (Fig. 1.2), the substituents X can lie either all above or all below the plane (I); or they may lie alternately above and below the plane (II); or they may be irregularly arranged (III).

Polymers with configuration I are called isotactic. Their chains consist of a regular sequence of CRU's, containing tertiary carbon atoms with the same steric configuration. When the substituted carbon atoms have alternating configuration (II), the polymer is called syndiotactic. If there is no regularity, for example a random distribution of configurations (III), the polymer is said to be atactic. Up to the present time, tactic polymers in the sense of I and II have been prepared mainly from vinyl and acrylic

---

[1] *G. Natta,* Angew. Chem. *68* (1956) 393; *76* (1964) 553; *M Farina, M. Peraldo* and *G. Natta,* Angew. Chem. *77* (1965) 149.

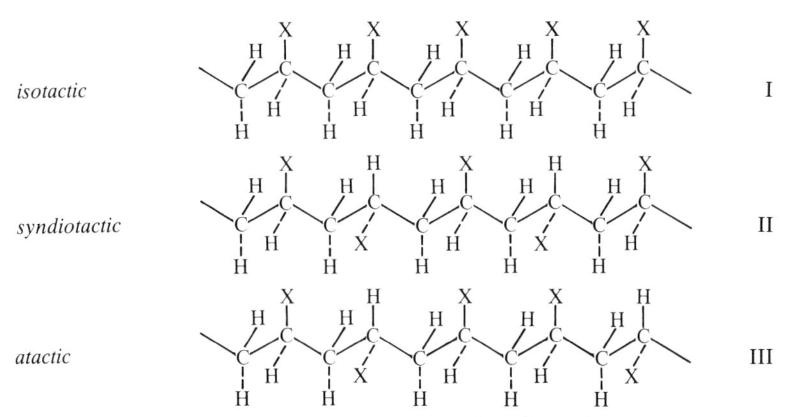

FIGURE 1.2 Tacticity of macromolecules consisting of $—CH_2—CHX—$constitutional repeating units. The main chain bonds lie in a horizontal plane. Other bonds lie either above the plane (full lines) or below the plane (dashed lines).

compounds; they can also be obtained by the polymerization of conjugated dienes through the 1,2- and 3,4-positions. Substituted epoxides can also be polymerized stereospecifically.

Optical isomerism is possible whenever the substituent X of a vinyl compound contains an asymmetric carbon atom; the polymers derived from the enantiomers are optically active[1], but the specific rotation of the polymer is generally significantly different from that of the monomer. In principle, optical isomerism is also possible when there are asymmetric carbon atoms in the main chain.[2]

The same types of structural- and stereo-isomerism can occur in copolymers as in homopolymers. Structural isomers can also result from the various possible distributions of two (or more) types of CRU within the polymer chains. In a linear copolymer of monomers A and B, the CRU's may either alternate, or be distributed at random, or occur in blocks; these are described as alternating, random (statistical) and block copolymers respectively. Finally, there is yet another possibility, namely that B-type chains may be grafted on to a main chain consisting of A-type units or on to a random copolymer; this is known as a graft copolymer. Such structures can also be produced by condensation polymerization.

Stereoblock copolymers represent a special case. They consist of a long tactic block of A units followed by a similar tactic block of B units. Stereoblock polymers are homopolymers consisting, for example, of an isotactic block of A units followed by a syndiotactic block of A units.

---

[1] See *D. Braun* in *Houben-Weyl 14/1* (1961) 119; *R.C. Schulz* and *E. Kaiser*, Adv. Polymer Sci. *4* (1965) 236.

[2] *P. Pino*, Adv. Polymer Sci. *4* (1967) 393.

```
—ABABABABABAB—          alternating copolymer
—AABABBABBABA—          random copolymer
—AAAA—BBBBB—AAAA—       block copolymer
—AAAAAAAAAAAAA—         graft copolymer
   |          |
   B          B
   B          B
   B          B
   B          B
   B          B
              B
```

Whether a copolymerization yields an alternating or random copolymer depends on the composition of the monomer feed, and above all on the type of initiation and on the reactivity ratios (see Section 3.3). Block and graft copolymers can often be tailor-made (see Section 3.3.2).

1.3. SHAPE OF MACROMOLECULES

Isolated macromolecules do not take up a precisely defined three-dimensional shape, rather they assume a statistically most probable form that approximates to the state of maximum possible entropy. From calculations on models, Kuhn[1] showed that this is neither a compact sphere nor an extended stiff chain, but a loose statistical coil; see Fig. 1.3.

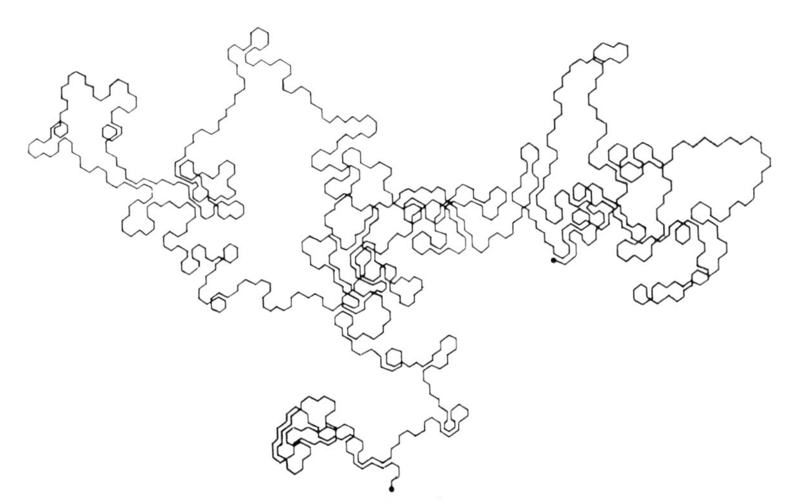

FIGURE 1.3 Representation in two dimensions of a statistically coiled linear macromolecule

[1] *W. Kuhn*, Kolloid-Z *68* (1934) 2.

Real chain molecules are not able to assume the shape of an ideal statistical coil, as would be required by random-flight statistics, since molecular parameters such as fixed bond angles and restricted rotation about the bonds affect the shape of the coil. Branching and incorporation of rigid chain members (aromatic rings, heterocycles) also affect the coil form. The shape of real chain molecules can be calculated by statistical methods when the molecular parameters of the chain molecules are known (theory of rotational isomers).[1]

1.4. STRUCTURE AND PROPERTIES OF MACROMOLECULAR SUBSTANCES[2]

1.4.1 Macromolecules in solution

In certain cases chain molecules in very dilute solution can take up the shape that is predicted by the theory of isolated chain molecules. In general, however, the interactions between the solvent molecules and the macromolecules have a significant effect. In "poor" solvents these interactions are such that the coil dimensions, which are characterized by the radius of gyration, tend towards the case of an isolated chain. This is the result of preferred interaction between like partners (chain molecules with chain molecules, solvent molecules with solvent molecules). The shape of the isolated coil can be realized under certain conditions, i.e. the so-called θ-conditions[3] (in a θ-solvent at its θ-temperature).

In "good" solvents the preferred interactions between unlike partners leads to an expansion of the coil compared with the isolated macromolecule. The chain molecules in these coils are strongly solvated, i.e. the coils contain within them a considerable amount of trapped solvent. The enclosed solvent is in a state of continuous exchange (by diffusion) with the surrounding solvent, but is nevertheless held sufficiently firmly that, in certain situations, it may be regarded as moving with the coil as a whole, for example during sedimentation in the ultracentrifuge. The macromolecular coils are thus comparable with small gel particles, which, like a fully soaked sponge, consist of a framework (the coiled macromolecules) and the embedded solvent. This concept is schematically portrayed in Fig. 1.4.

---

[1] *P.J. Flory*, "Statistical Mechanics of Chain Molecules", Interscience Publishers, New York, London, Sydney, Toronto 1969.

[2] *D.W. van Krevelen*, "Properties of Polymers, their Estimation and Correlation with Chemical Structure", Elsevier Publishers, Amsterdam 1976.

[3] Actually in the θ-state there is compensation of different effects, each of which can cause a deviation from ideal behaviour; for details see textbooks on macromolecular chemistry.

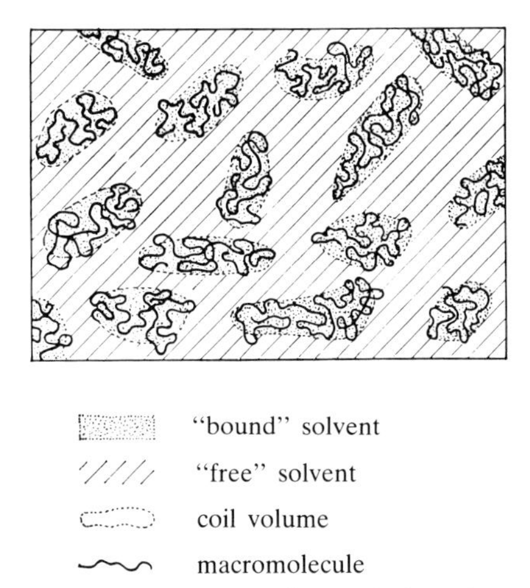

FIGURE 1.4 Schematic representation of a dilute solution of macromolecules.

With increasing concentration of polymer, the coils take up a greater proportion of the total volume of solution until finally, at a certain "critical" concentration there is mutual contact of the coils. At still higher concentration the coils can interpenetrate or, if this is not possible on account of incompatibility effects, the interaction may be confined to the boundary regions.[1] The compatibility of mixtures of two different polymers is, as a rule, governed by thermodynamics; in general, a mixing of two polymer solutions results in phase separation or the appearance of turbidity (cf. Section 2.3.1).

Depending on the molecular weight of the macromolecules, the coil volume in such solutions may be 20 to 1 000 times the chain volume, and may thus consist of more than 99% of solvent. Since the diameter of such gel coils may be between ten and several hundred nm, depending on the molecular weight and solvent, these solutions may be classified as colloids. However, in contrast to the colloidal particles of ordinary dispersions, the colloidal particles in macromolecular solutions are identical with the solvated macromolecular coils; they may therefore be termed "molecular colloids".

The properties of solutions of macromolecular substances depend on the solvent, temperature, and molecular weight of the chain molecules. The molecular weight of polymers may be determined, for example, by measurement of the viscosity of dilute solutions; this method is of great practical importance and is discussed in detail in Section 2.3.2.1.

---

[1] *D. Braun*, Angew. Chem. *88* (1976) 487.

1.4.2. Macromolecules in the molten and glassy states

The shape of macromolecules in the melt is probably almost identical with that in dilute solution under θ-conditions. This is indicated not only by theoretical considerations[1], but also by numerous experimental studies on the structure of polymer melts. In the melt there may be deviations from ideality when there is marked orientation of the chain members. The space occupied by a single chain coil is small so that in the melt the chains must be very intertwined. The resulting entanglements play an important role in the properties of polymer melts.

The individual chain members are not regularly arranged in the melt but are packed in tight disorder; there is, however, a short-range order. In many cases when the melt is cooled, it solidifies without any change in the disorder of the chains; this may happen even with crystallizable polymers. The so-called glassy state which is thus formed can be regarded as an isotropic frozen melt. Solidification under these circumstances occurs over a temperature region whose position is characterized by the glass transition temperature (T_g). The glassy state is a thermodynamically metastable state.

The attainment and maintenance of the fully isotropic glassy state is not an easy matter. Thus flow orientation can be frozen in when cooling the polymer from the melt. Orientation of the molecular coils can also occur when the polymer is already in the solid state; this may result, for example, through the action of tensile or shear forces (stretching). It can also occur in the casting of films from solution, for example during the drying of films, as a result of shrinkage or when the film is pulled away from the support. Orientation causes anisotropy of various physical properties; thus orientation in a transparent polymer (e.g. polystyrene) is readily detected by the use of polarized light. Orientation also manifests itself very clearly in the dependence of mechanical properties on direction (see Section 2.3.12). The complete prevention of orientation is not very easy: the polymer must be kept for some time at temperatures above T_g and then allowed to cool under conditions where no deformation by external forces can occur. In technology, of course, orientation is deliberately produced in films and fibres, for example by uniaxial or biaxial drawing, in order to achieve particular properties.

[1] *P.J. Flory*, "Statistical Mechanics of Chain Molecules", Interscience Publishers, New York, London, Sydney, Toronto 1969.

1.4.3. Chain molecules in the crystalline state

Chain molecules that are sufficiently regular can be crystallized by slow cooling. Ordered regions are formed in which the chain segments are arranged regularly on a three-dimensional lattice over distances that are large compared with atomic dimensions; there exists a long-range order. The ordered regions may be disturbed to some extent by lattice defects.

In most cases, however, polymers do not crystallize completely but give partially crystalline material in which crystalline regions are separated by adjacent amorphous regions. In general, the crystalline regions extend over average distances between 10 and 40 nm and are embedded in an amorphous matrix. The fact that polymers generally crystallize only partially is attributable to the fact that the macromolecules have difficulty in changing their form from the coiled state in the melt to the ordered state that is necessary for their incorporation into a crystal. The proportion of material which is crystalline is termed the degree of crystallinity and is an important parameter of such materials.

Two different models have been developed to describe the structure of partially crystalline polymers, the fringed micelle model and the folded lamella model. According to the fringed micelle model (Fig. 1.5), the macromolecules lie parallel to one another in certain regions like elongated threads (extended chain crystals) thereby providing the order necessary for crystallization. These highly ordered regions, designated as crystallites, are interspersed between irregular arrangements of chain molecules, the amorphous regions. Since the length of macromolecules generally exceeds that of crystallites by a large factor, a given polymer chain contributes to the structure of several crystallites, i.e. traverses several crystalline and amorphous regions. The individual crystallites are thus bound together by amorphous regions.

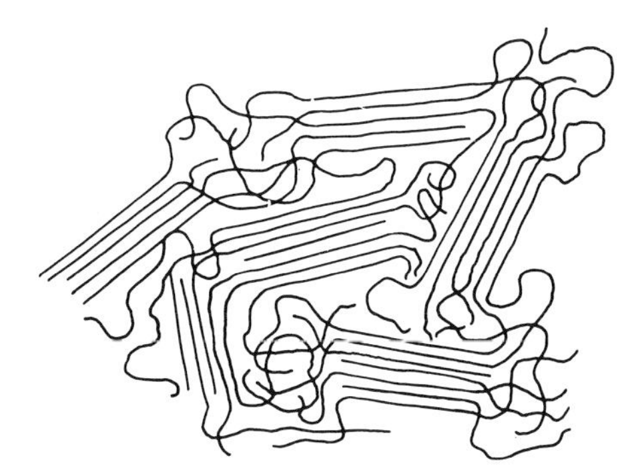

FIGURE 1.5 Fringed micelles (schematic).

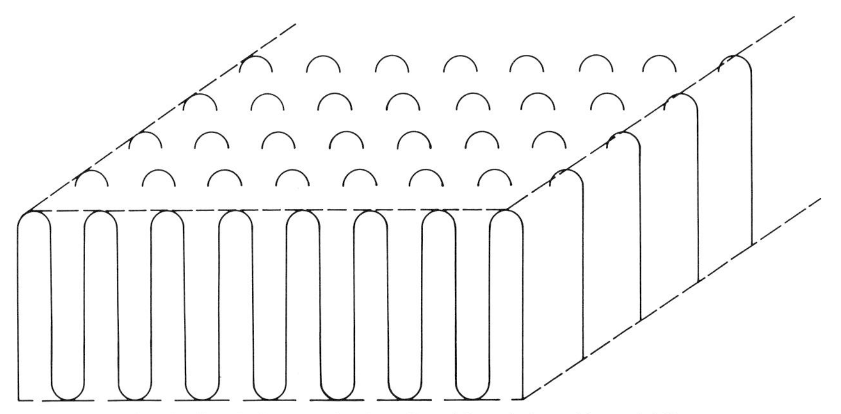

FIGURE 1.6 Idealized picture of a lamella with strictly uniform folding.

The fringed micelle picture is not particularly good for describing synthetic polymers crystallizing under the usual conditions from solution or from the melt. However, the fibrils of many natural substances, such as cellulose and proteins (collagen, silk), consist of bundles of macromolecules in parallel alignment, compatible with the fringed micelle model.

It is more often found that polymers crystallize in such a way that the macromolecules fold with an essentially constant length leading to a lamellar-type crystallite structure (Fig. 1.6). This is especially evident in the case of a single crystal. Such single crystals, which can be grown from some polymers by careful cooling of highly dilute solutions (0.01–0.1%), originate from a nucleus and grow to lamellae through the back-and-forth folding of the macromolecules. Evidence for chain folding comes from X-ray diffraction and electron microscope investigations on single crystals of polyethylene, polypropene and other polymers.[1]

The lamellar thickness of the crystallites is of the order of 10 nm, with the molecular chains aligned perpendicular to the lamellae. In the fully extended form the macromolecules would have a length of 10–1 000 μm (depending on the molecular weight) so that in the crystallites they must be folded. The crystals are generally not completely crystalline since not all the chains will be tightly folded and chain ends may project from the lamellae, giving rise to a disordered portion.

[1] See e.g. *B. Wunderlich*, "Macromolecular Physics", Vol. 1, Academic Press, New York, London 1973, p. 178 et seq.

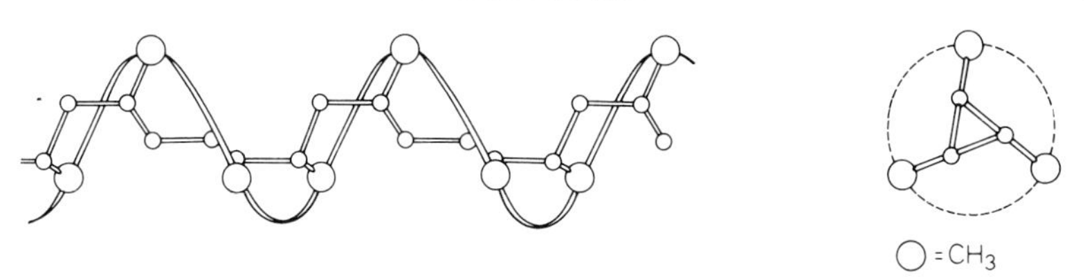

FIGURE 1.7 The three-fold helix of the polypropene macromolecule.

The shape of the macromolecule within a folded lamella is not the same in all polymers. In crystalline polyethylene the chains have a planar zig-zag conformation, but in some polymers the chains prefer a helical shape, as in proteins. The helix may be such that the chain repeats itself in the lattice after every 3,4 or 5 monomeric units; one speaks accordingly of a three-, four-, or five-fold helix (Fig. 1.7).

Finally, it should be mentioned that polymers can exhibit polymorphism, i.e. can crystallize in different types of lattice.[1,2] The different crystal forms generally differ in their physical properties, e.g. crystalline melting point and density.

If the crystallization is carried out in a streaming solution, in which there is present a certain amount of so-called elongational flow, then the morphology of the polymer, e.g. polyethylene, may be completely different from that formed from the undisturbed solution. The elongational flow causes the coiled macromolecules to become extended to some degree so that crystallization from such a solution results in the formation of fibrillar crystals; the stretched molecules stack together side by side in the fibrils, somewhat like a piece of string. At regular distances the fibrils contain laterally arranged lamellar packets. The morphology resembles a shish-kebab and is described by this term.[3,4]

During crystallization from the melt, the situation is very complicated on account of the mutual interaction of different macromolecules. Nevertheless, crystallites are formed according to similar principles. Ordered regions are formed containing lamellae in which there is chain folding. However, the chain folding is not so regular as in single crystals since numerous chain molecules are built into different lamellae. These tie molecules substantially influence the mechanical properties of polymers. Compared with crystallization from solution, the overall shape of the

---

[1] *L. Mandelkern*, "Crystallization of Polymers", McGraw-Hill Company, New York, San Francisco, Toronto, London 1964, p. 138 et seq.

[2] *F. Danusso*, Polymer *8* (1967) 281.

[3] *A.J. Pennings, J.M.M.A. van der Mark and H.C. Booij*, Kolloid-Z. Z. Polym. *236* (1969) 99.

[4] *W. Gordon, H.J. Leugering and H. Cherdron*, Angew. Chem. *90* (1978) 833.

FIGURE 1.8 Spherulites of polypropene.

chains can undergo relatively little change during crystallization from the melt.

On cooling a polymer melt, it generally crystallizes in the form of spherulites, containing both amorphous and crystalline regions. These morphological units are larger than the crystallites and can attain a diameter of several tenths of mm, recognizable under the polarizing microscope by the characteristic "Maltese cross" (Fig. 1.8). The structure of the spherulites is determined by the radial disposition and longitudinal extension of lamellae and amorphous interlayers. The circular banding around the centre point of the spherulites can be attributed to a periodic twisting of the lamellae. The morphology also depends on the temperature at which the crystallization from the melt is carried out. This temperature is always below the melting point of the polymer crystal.[1] Moreover, the number and size of the spherulites depends strongly on external conditions (pressure, orientation) and the number of nuclei. By the use of added nucleating agents, the number of spherulites per unit volume can be substantially raised and their resulting diameter diminished (heterogeneous nucleation[2], see Example 3–32); some of the physical properties are thereby affected, especially optical properties such as transparency.

---

[1] *L. Mandelkern*, "Crystallization of Polymers", McGraw-Hill Company, New York, San Francisco, Toronto, London 1964.

[2] *F.L. Binsbergen*, Polymer *11* (1970) 2531; *F.L. Binsbergen* and *B.G.M. De Lange*, Polymer *11* (1970) 309.

Whether and to what degree a polymer crystallizes depends on various factors, notably on its structure, on the symmetry along the main chain, on the number and length of side chains and on branching (e.g. in high-pressure- and low-pressure-polyethylene). Polar groups that lead to the formation of hydrogen bonds, and thereby an association of chain segments in the molten state (e.g. amide groups in polyamides), substantially promote crystallization. Some polymers, such as polyethylene, can be obtained in the amorphous state only under special conditions; others, such as isotactic polystyrene, can be prepared in both amorphous and crystalline states. Yet others, such as poly(ethylene terephthalate), can be brought into the amorphous state by rapid cooling to a temperature far below the crystallization temperature; they remain stable if kept at low temperature, but begin to crystallize if warmed above the glass transition temperature.

Some polymers do not crystallize immediately but must be induced to crystallize by appropriate treatment such as long heating at a certain temperature (annealing), slow cooling of the molten polymer, or contact with a suitable swelling agent (see isotactic polystyrene, Example 3–28).

1.4.4. Thermal transitions in polymers

When glassy macromolecular substances soften, they do not liquefy abruptly; instead the change from solid to liquid occurs over a finite temperature range. Thus the intermolecular forces are not suddenly overcome by thermal motions. This observation may be attributed to the fact that chain molecules undergo different types of motion. Thus several chain segments may move as a unit within a macromolecule (micro-Brownian motion); on the other hand, the whole macromolecule may move relative to others (macro-Brownian motion). Below the glass transition temperature, T_g, the macro-Brownian motion is completely frozen and the micro-Brownian motion is also largely frozen ("glassy state").[1] Above this temperature, the micro-Brownian motion becomes pronounced so that the material softens and can be strongly deformed by external forces.

---

[1] At the glass transition temperature there is an abrupt change in some of the physical properties, such as relative permittivity, heat capacity, expansion coefficient and elastic modulus. Measurement of the temperature dependence of these quantities can be used for the determination of the glass transition temperature.

Micro-Brownian motion is responsible for softening in all polymers even when they are crosslinked. The softening temperature is often quoted rather than the glass transition temperature T_g, since it is easier to measure; for amorphous polymers, it lies close to T_g. The temperature at which external forces produce a marked deformation depends on the magnitude of the load; hence the conditions under which the softening temperature is determined must be standardized. The fluidity is strongly dependent on the molecular weight so that the softening temperature is often hard to discern with the naked eye because of the high melt viscosity.

For partially crystalline material, the transition from the solid to the molten state occurs more abruptly over a limited temperature range (the melting range), within which the ordered crystalline domains are transformed into the disordered state of the melt.

1.4.5. Viscoelastic properties

It is characteristic of polymers at temperatures above the glass transition temperature (above the melting point for crystalline polymers), that the deformation caused by an external force disappears as soon as the force is removed[1], in other words, the polymer behaves elastically. Such elasticity is not due to deformation of valence angles or bond lengths, but stems from the fact that the molecular coils seek to regain their statistically most probable form (entropic elasticity).[2] The elasticity of polymers above their softening point is, of course, not ideal, i.e. the elastic recovery after deformation is not complete ("residual deformation"). This is because the internal stress of the system, resulting from the deformation of the chain segments, can be relieved by migration of the macromolecules, thereby reducing the restoring force to some extent. This relaxation process occurs more quickly at higher temperatures, since the increased macro-Brownian motion favours migration of molecules. In spite of this, a polymer melt is still elastic, because of the coiling of the macromolecules; the melts thus behave viscoelastically. This behaviour is, of course, only well defined over a certain temperature range: in the immediate vicinity of the softening temperature the polymer chains are still relatively stiff, so that deformation requires considerable force, and recovery occurs very slowly. Well above

[1] See Section 1.4.2.

[2] The processes described here can be mathematically formulated and provide the basis of the theory of rubber elasticity.

the softening temperature the melt deforms easily, but the tendency to flow as a result of increased macro-Brownian motion is still outweighed by the elastic recovery. The temperature range for pronounced elastic behaviour of the polymer melt depends on the chain structure of the polymer (e.g. whether it is branched) and especially on the molecular weight and molecular weight distribution.

The flow behaviour of molten macromolecular substances is generally quite different from that of low-molecular-weight compounds, for example, in the shape of the flow curves and in the occurrence of flow orientation; this again is a consequence of their molecular structure. In ideal liquids (water, glycerol, sulfuric acid etc.), the viscosity is characteristic of the liquid and is independent of the shear rate $\dot{\gamma}$ ("Newtonian flow"). If the shearing force τ is plotted against $\dot{\gamma}$ for a Newtonian liquid, a straight line is obtained with slope η, where η is the Newtonian viscosity. Such a plot is called a flow diagram. Macromolecular melts behave differently in that their melt viscosity depends on both τ and $\dot{\gamma}$, and the lines in the flow diagram are curved. The melt viscosity shows a very sensitive dependence on molecular weight, molecular weight distribution and branching, so that rheological measurements on polymer melts yield not only interesting information on technical processing, but also allow deductions about the structure and size of macromolecules.

If the melt of a macromolecular substance is subjected to the action of external forces, e.g. by kneading or rolling or by extrusion through narrow openings (slits, dies), the closely interwoven coils take up a preferred orientation, which can be frozen into the solid polymer by cooling so rapidly that there is no time for relaxation of the internal forces by migration of molecules ("glassy state"; also see Section 1.4.2).

The glass transition temperature and melting point of a polymer are manifest in its mechanical properties, as, for example, in its tensile strength, elongation, and elastic modulus. The behaviour depends on whether the polymer is amorphous or crystalline. Above the glass transition temperature T_g, the micro-Brownian motion results in slow softening, leading to flow, accompanied by an increase in extensibility and a decrease in rigidity and elastic modulus (see Fig. 1.9). In crystalline polymers, where T_g lies considerably below the melting point (Boyer-Beaman rule, see Section 2.3.4), the rigidity and elastic modulus decrease to some extent at T_g, but fall abruptly only when the crystalline structure begins to collapse at temperatures near the melting point of the crystallites. In all temperature regions the behaviour is viscoelastic.

In highly crosslinked polymers, where the network segments can undergo micro-Brownian but not macro-Brownian motion, softening but not flow or melting is observed. There is, therefore, not much change in the

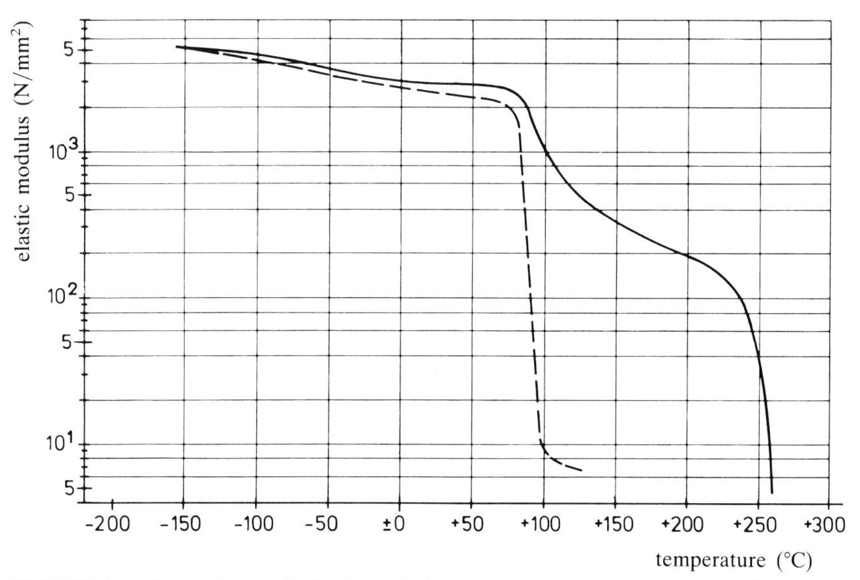

FIGURE 1.9 Dependence of elastic modulus on temperature in amorphous (---) and crystalline (——) poly(ethylene terephthalate).

afore-mentioned mechanical properties on warming; a marked fall does not occur until the decomposition temperature (see Section 1.9) is approached.

1.4.6. Macromolecules in the elastomeric state

As shown in Section 1.4.2, rubber elasticity is a property of all polymers above the glass transition temperature. Amorphous polymers whose glass transition temperatures lie below room temperature are called elastomers; at room temperature they behave in a rubber-like fashion. However, after stretching an uncrosslinked ("unvulcanized") elastomer at 20°C, it does not return completely to its original length. The residual extension is greater, the larger the stretch and the longer the time for which the material is held in the extended state; this is a consequence of the relaxation of the internal tension by migration of the deformed macromolecular coils. If this migration is prevented by crosslinking, e.g. by covalent bonding between chains (vulcanization), the slippage of macromolecules past one another becomes impossible and the elastomer will recover its original form even after high extension for long periods; such materials are known as rubbers.[1] Rubbers are distinguished by very small values of the elastic modulus combined with large and reversible elasticity. The properties of a rubber are determined essentially by the number of crosslinks (degree of crosslinking): weakly crosslinked rubbers are highly elastic and of low modulus; an increase in degree of crosslinking reduces the elasticity and raises the elastic modulus; highly crosslinked rubbers lose their elasticity almost completely (hard rubbers).

As already indicated in connection with the properties of polymer melts, rubber elasticity is a consequence of a particular state of matter. The elasticity of a substance like steel is determined by the tendency of the Helmholtz free energy ΔA ($\Delta A = \Delta U - T\Delta S$) to strive towards a minimum as a consequence of a decrease in internal energy ΔU; but in the case of a rubber it is governed by the increase in entropy ΔS. This "entropic elasticity" also explains the remarkable fact that the tension of a rubber band (at constant length) increases with temperature while that in a steel wire decreases. This is because upon elongation of the rubber band the macromolecules are forced to change from their statistically most probable coil shape to the statistically less probable shape of an extended chain. The higher the temperature, the greater the restoring force, since the free energy change increases with both deformation and temperature ($T\Delta S$).

Orientations in elongated rubbers are sometimes so regular that there is local crystallization of individual chain segments (e.g. in natural rubber). X-ray diffraction patterns of such samples are very similar to those obtained from fibres.

1.5. LITERATURE ON MACROMOLECULAR SUBSTANCES

The following compilation is limited to some selected textbooks and a list of the most important scientific journals that deal with the subject of macromolecular chemistry. Monographs or reviews on individual topics are cited in the appropriate Sections. A comprehensive guide to the literature can also be found in P. Eyerer, "Informationsführer Kunststoffe", 2nd edition, VDI-Verlag GmbH, Düsseldorf 1977.

---

[1] Even in crosslinked elastomers the recovery to the original state after extension is not quite complete. As a consequence, in the stress-strain diagram (see Section 2.3.12.1) the stretching curve (extension at increasing force) does not coincide with the recovery curve (extension at decreasing force). There is thus a hysteresis loop whose breadth is a measure of the residual extension.

The Deutsches Kunststoff-Institut, Darmstadt also issues monthly a rapid literature service "Kunststoff, Kautschuk, Fasern". It reports rapidly on new publications in the realm of chemistry and physics of polymers as well as on their technical applications (about 12 000 references per annum). In the UK a similar service is available from the Rubber & Plastics Research Association, Shrewsbury.

TEXTBOOKS

H. Batzer and F. Lohse, "Einführung in die makromolekulare Chemie", 2nd Edn., Hüthig-Verlag, Heidelberg, New York 1976.

J. M. G. Cowie, "Polymers: Physics and Chemistry of Modern Materials, Intertext Books, Aylesbury 1973.

H. G. Elias, "Makromoleküle", 3. Auflage, Hüthig-Verlag, Heidelberg, New York, 1975.

P. J. Flory, "Principles of Polymer Chemistry", Cornell University Press, Ithaca, New York 1953.

G. Henrici-Olivé and S. Olivé, "Polymerisation: Katalyse – Kinetik – Mechanismen". Verlag Chemie, Weinheim/Bergstrasse 1970.

R. W. Lenz, "Organic Chemistry of Synthetic High Polymers", Interscience Publishers, New York, London, Sydney 1967.

M. Hoffmann, H. Krämer and R. Kuhn, "Polymeranalytik", Thieme Verlag, Stuttgart 1977.

B. Philipp and G. Reinisch, "Grundlagen der makromolekularen Chemie", 2nd Edn., Akademie-Verlag, Berlin 1976.

F. Runge and G. Taeger, "Einführung in die Chemie und Technologie der Kunststoffe", Akademie-Verlag, Berlin 1976.

J. Schurz, "Physikalische Chemie der Hochpolymeren", Springer-Verlag, Berlin, Heidelberg, New York 1974.

H. A. Stuart (Ed.), "Die Physik der Hochpolymeren", Springer-Verlag, Berlin -Göttingen-Heidelberg 1952–1956.

Vol. 1: Die Struktur des freien Moleküls, bzw. H.A. Stuart, Molekülstruktur, 3rd Edn., Springer-Verlag, Berlin, Heidelberg, New York 1967.
Vol. 2: Das Molekül in Lösungen.
Vol. 3: Ordnungszustände und Umwandlungserscheinungen in festen hochpolymeren Stoffen.
Vol. 4: Theorie und molekulare Deutung technologischer Eigenschaften von hochpolymeren Werkstoffen.

J. Ulbricht, "Grundlagen der Synthese von Polymeren", Akademie-Verlag, Berlin 1978.

B. Vollmert, "Polymer Chemistry", Springer-Verlag, Berlin, Heidelberg, New York 1973.

B. Wunderlich, "Macromolecular Physics", Academic Press, New York, London, Vol. I 1973, Vol. II 1976.

HANDBOOKS

J. Brandrup and E. H. Immergut (Eds.), "Polymer Handbook", Interscience Publ., 2nd edn., New York 1975.

Houben-Weyl-Müller, "Methoden der organischen Chemie", Vols. *14/1* (1961) and *14/2* (1963): "Makromolekulare Stoffe", Parts 1 and 2, Georg Thieme Verlag, Stuttgart.

Hummel-Scholl, "Atlas der Kunststoff-Analyse, "Verlag Chemie, Weinheim and Hanser Verlag, München 1968.

H. Mark, N. G. Gaylord and N. M. Bikales (Eds.), "Encyclopedia of Polymer Science and Technology", 16 Vols., Interscience Publishers, New York, London, Sydney 1964 to 1972; Supplementary Vol. 1977.

R. Vieweg (Ed.), "Kunststoff-Handbuch", 11 Vols., Carl Hanser Verlag, München 1963 to 1975.

PRACTICAL BOOKS AND WORKING INSTRUCTIONS

E. M. McCaffery, "Laboratory Preparation for Macromolecular Chemistry", McGraw-Hill, New York 1970.

I. P. Lossew and O. Ja. Fedotowa, "Praktikum der Chemie hochmolekularer Verbindungen, Akademische Verlagsgesellschaft Geest und Portig, Leipzig 1962.

S. H. Pinner, "A Practical Course in Polymer Chemistry", Pergamon Press, Oxford, London, New York, Paris 1961.

W. R. Sorenson and T. W. Campbell, "Präparative Methoden der Polymeren-Chemie", Verlag Chemie, Weinheim/Bergstrasse 1962. (2nd English Edn. = **W. R. Sorenson and T. W. Campbell,** "Preparative Methods of Polymer Chemistry", Interscience Publishers, New York 1968).

E. A. Collins, J. Bares and F. W. Billmeyer jr., "Experiments in Polymer Science", J. Wiley & Sons, New York, London, Sydney, Toronto 1973.

S. R. Sandler and W. Karo, "Polymer Syntheses", Vol. I, 1974, Vol. II, 1977, Academic Press, New York, London.

J. A. Moore (Ed.), Macromolecular Syntheses, Collective Volume I, J. Wiley & Sons, New York, Chichester, Brisbane, Toronto 1978.

JOURNALS[1,2]

Advances in Polymer Science – Fortschritte der Hochpolymeren Forschung	(Fortschr. Hochpolym.-Forsch.)
Applied Polymer Symposia	(Appl. Polym. Symp.)
Biopolymers	(Biopolymers)
British Polymer Journal	(Br. Polym. J.)

[1] The technical journals, with a few exceptions, are not listed.
[2] The abbreviations used by Chemical Abstracts are shown in brackets.

Colloid and Polymer Science, Kolloid-Zeitschrift & Zeitschrift für Polymere	(Kolloid Z.Z. Polym.)
Die Angewandte Makromolekulare Chemie	(Angew. Makromol. Chem.)
Die Makromolekulare Chemie	(Makromol. Chem.)
European Polymer Journal	(Eur. Polym. J.)
Faserforschung und Textiltechnik	(Faserforsch. Textiltech.)
Gummi, Asbest, Kunststoffe	(Gummi, Asbest + Kunstst.)
Journal of Applied Polymer Science	(J. Appl. Polym. Sci.)
Journal of Colloid Science	(J. Colloid Sci.)
Journal of Macromolecular Science	(J. Macromol. Sci.)
Part A Chemistry	(J. Macromol. Sci., Chem.)
Part B Physics	(J. Macromol. Sci., Phys.)
Part C Reviews in Macromolecular Chemistry	(J. Macromol. Sci., Rev. Macromol. Chem.)
D Reviews in Polymer Technology	(J. Macromol. Sci., Rev. Polym. Technol.)
Journal of Polymer Science, Polymer Chemistry Edition	(J. Polym. Sci., Polym. Chem. Ed.)
Journal of Polymer Science, Polymer Physics Edition	(J. Polym. Sci., Polym. Phys. Ed.)
Journal of Polymer Science, Polymer Letters Edition	(J. Polym. Sci., Polym. Lett. Ed.)
Journal of Polymer Science, Polymer Symposia	(J. Polym. Sci., Polym. Symp.)
Journal of Polymer Science, Macromolecular Reviews	(J. Polym. Sci., Macromol. Rev.)
Kautschuk und Gummi, Kunststoffe	(Kautsch., Gummi, Kunstst.)
Kunststoffe	(Kunststoffe)
Macromolecules	(Macromolecules)
Plaste und Kautschuk	(Plaste Kautsch.)
Plastics (London)	(Plastics (London))
Plastics Technology	(Plast. Technol.)
Plastverarbeiter	(Plastverarbeiter)
Polymer	(Polymer)

Polymer Age	(Polym. Age)
Polymer Journal	(Polym. J.)
Polymer Mechanism (English Transl.)	(Polym. Mech. (Engl. Transl.))
Polymer News	(Polym. News)
Polymer-Plastics, Technology and Engineering	(Polym.-Plast. Technol. Eng.)
Polymer Report	(Polym. Rep.)
Polymer Reviews	(Polym. Rev.)
Polymer Science USSR (Engl. Transl.)	(Polym. Sci. USSR (Engl. Transl.))
Polimery (Warsaw)	(Polimery (Warsaw))
Progress in Colloid & Polymer Science – Fortschrittsberichte über Kolloide und Polymere	(Fortschrittsber. Kolloide Polym.)
Progress in Polymer Science	(Prog. Polym. Sci.)
Vysokomolekulyarnye Soedineniya,	(Vysokomol. Soedin., Ser. A)
Seriiya A, Seriiya B	(Vysokomol. Soedin., Ser. B)

2 General Methods of Macromolecular Chemistry

2.1. PREPARATION OF MACROMOLECULAR SUBSTANCES

2.1.1. Working with exclusion of oxygen and moisture[1]

Molecular oxygen has an influence on the course of most polymerizations. In radical polymerizations this may occur through an effect on the initiation or termination reactions; in ionic polymerizations the initiator may be either activated or deactivated by oxygen. Oxygen may also cause oxidative degradation of macromolecules that have already been formed (especially in polycondensations). These effects are often detectable at very low oxygen concentrations and it is therefore advisable to work under nitrogen during the preparation of macromolecular substances; it is often necessary to use especially pure "oxygen-free" nitrogen. Extremely pure nitrogen can be prepared by passing normal cylinder nitrogen either through suitable solutions or over a solid catalyst that reacts with the traces of oxygen. The formerly used copper catalyst of Meyer and Ronge[2] has been superseded by the commercially available "BTS-Kontakt"[3] on account of its effectiveness and simplicity of handling (working temperature 20°C; residual oxygen 10^{-4} to 10^{-5}%).[4] A final purification of nitrogen can also be achieved with the aid of solutions of suitable reagents, e.g. benzene-1,2,3-triol, sodium dithionite, metal ketyls, or organoaluminium

---

[1] See also: *Houben-Weyl 1/2* (1959) 321; *A Weissberger*, "Technique of Organic Chemistry", Interscience Publishers, New York, London *3* (1950) 605.

[2] *F.R. Meyer* and *G. Ronge*, Angew. Chem. *52* (1939) 637.

[3] Manufacturer: BASF, Ludwigshafen, Germany; also see *M. Schütze*, Angew. Chem. *70* (1958) 697.

[4] Gas purification equipment for the laboratory is also available commercially, e.g. from Otto Fritz GmbH, NORMAG, D6328 Hofheim am Taunus, Germany.

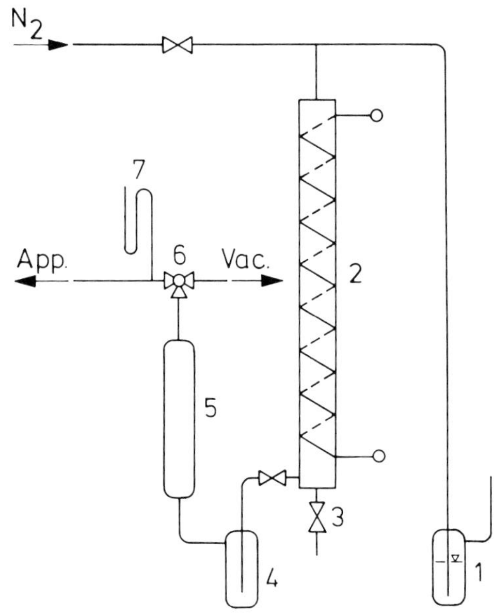

FIGURE 2.1 Apparatus for the purification and drying of nitrogen at room temperature. 1: Hg-valve; 2: column (100 × 5 cm) wrapped with heating spiral (2000W) for reduction of the catalyst (BTS-Kontakt); 3: outlet stopcock for hydrogen and water vapour during the reduction of the catalyst; 4: bubbler (paraffin oil); 5: drying column (for drying agents see Table 2.1); 6: three-way tap for evacuating and filling the apparatus with nitrogen; 7: Hg-manometer.

compounds[1]; however, this method generally offers no advantage over the use of solid catalysts.

Charging of reaction vessels with nitrogen should always be done by repeated evacuation and admission of nitrogen rather than by simply passing nitrogen through the vessel since one can never be sure when all the air has been displaced. Connections should be made with poly(vinyl chloride) tubing (PVC tubing) or glass tubing rather than rubber tubing. Figure 2.1 shows a simple arrangement for both the evacuation and filling of the apparatus through a single connection, and the purification and drying of the nitrogen.

When conducting a reaction or distillation under continuous flow of nitrogen rather than in a closed system, a suitable outlet must be used in order to prevent back-diffusion of oxygen into the apparatus from the

[1] See *Houben-Weyl 1/2* (1959) 321.

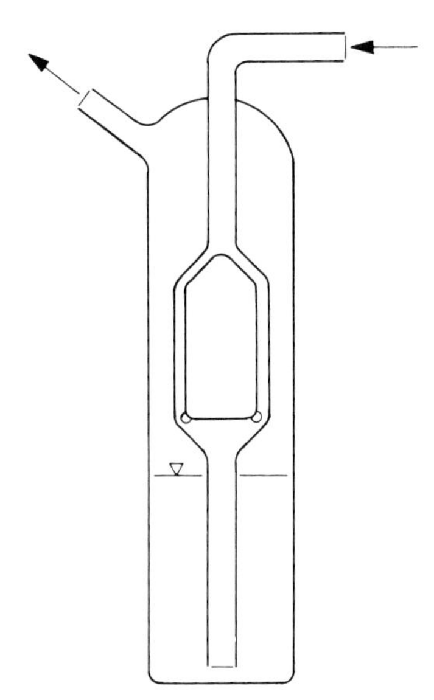

FIGURE 2.2 Liquid-filled shut-off device with ground-in non-return valve.

surrounding air. In the simplest cases, it is sufficient to use a Bunsen valve (a rubber tube with two lengthwise slits, closed at one end with a glass rod), or a mercury or paraffin oil valve (see Figure 2.1). When working with spontaneously inflammable substances (e.g. organometallic compounds) or with reactions where there are rapid changes of pressure, it is advantageous to employ a non-return valve, as shown in Figure 2.2. If the pressure falls, the ground-in float is pushed to the upper end of the chamber by the ascent of the liquid (paraffin oil), thus sealing the apparatus against the outer atmosphere.

It is especially important to exclude oxygen and water in ionic polymerizations; the drying procedures normally used for preparative work are then generally inadequate. Glassware can be freed from water by drying in an oven at 150°C, but better by baking under high vacuum. Gases can be dried by freezing out the moisture or by passage through columns filled with suitable solid reagents (see Table 2.1). Likewise, liquids can be dried by treatment with suitable drying agents (boiling under reflux) or by azeotropic or extractive distillation.[1] Agents for the drying of liquids must, of course, be chemically inert, otherwise undesired side reactions can occur (e.g. styrene polymerizes explosively on contact with concentrated sulfuric acid at room temperature).

[1] *Houben-Weyl 1/1* (1958) 866; *A. Weissberger*, "Technique of Organic Chemistry", 2nd edn., Interscience Publishers, New York *4* (1965) 423.

TABLE 2.1

Drying agents for gases and liquids

Drying agent	Residual moisture (mg H_2O/l gas)
Calcium chloride (according to quality)	0.3 to 1.2
Calcium oxide	0.2
Silica gel	0.006
Calcium sulfate[a]	0.005
Conc. sulfuric acid	0.003
Fused potassium hydroxide	0.002
Molecular sieves[b]	0.001
Cooling to $-75°C$[c]	0.001
Metallic sodium	—
Calcium hydride	—
Magnesium perchlorate	0.0005
Phosphorus pentoxide[d]	0.00002

[a] "Half-roasted gypsum" in granulated form; manufactured by Fluka, Basel.

[b] Zeolites of a certain pore size mixed with clay in bead or rod form, e.g. Bayer-Zeolithe K154, T142 (Bayer AG, Leverkusen) or molecular sieves 3A, 4A, 5A (E. Merck, Darmstadt); see *D.W. Breck et al.*, J. Am. Chem. Soc. *78* (1956) 5963; Russ. Chem. Rev. (Engl. transl.) *9* (1960) 509; *C.K. Hersh*, "Molecular Sieves", Chapman and Hall, London (1961).

[c] It is necessary to ensure that ice crystals are not carried out of the cold trap as an aerosol in the gas stream (filter through wadding or glass wool).

[d] Best used with a solid carrier, e.g. Sicapent (E. Merck, Darmstadt).

2.1.2. Purification and storage of monomers

In all polymerizations the purity of the starting materials is of prime importance. Impurities present to the extent of 10^{-2} to 10^{-4} wt. % often have a considerable influence on the course of the reaction. With unsaturated monomers the following impurities may be encountered:

- by-products formed during their preparation (e.g. ethylbenzene and divinylbenzenes in styrene; acetaldehyde in vinyl acetate);
- added stabilizers (inhibitors);
- autoxidation and decomposition products of the monomers (e.g. peroxides in dienes, benzaldehyde in styrene, hydrogen cyanide in acrylonitrile);

— impurities that derive from the method of storage of monomer (e.g. traces of metal or alkali from the vessels, tap grease etc.);
— dimers, trimers and polymers that are generally soluble in the monomer, but sometimes precipitate out, for example polyacrylonitrile from acrylonitrile.

The purification procedures to be applied depend on the monomer, on the expected impurities and especially on the purpose for which the monomer is to be employed, e.g. whether it is to be used for radical polymerization in aqueous emulsion or for ionic polymerization initiated with sodium naphthalene. It is not possible to devise a general purification scheme; instead the most suitable method must be chosen in each case from those given below. A prerequisite for successful purification is extreme cleanliness of all apparatus (if necessary, treating with hot nitrating acid or chromic/sulfuric acid and repeatedly washing thoroughly with distilled water).

The usual procedures of fractional, azeotropic or extractive distillation under inert gases[1], crystallization, sublimation, and column chromatography, must be carried out very carefully. For liquid, water-insoluble, monomers (e.g. styrene, Example 3–01), it is recommended that phenols or amines, that may be present as stabilizers, should first be removed by shaking with dilute alkali or acid respectively; the relatively high volatility of many of these kinds of stabilizers often makes it difficult to effect their complete removal by distillation. Gaseous monomers (e.g. lower olefins, butadiene, ethylene oxide) can be purified and stored over molecular sieves (see Section 2.1.1 for drying of starting materials). So-called pre-polymerization is frequently used to effect very high purification: monomer which has already been purified by the normal methods is polymerized to about 10–20% conversion by heating or irradiation, or if necessary by addition of initiator, and then separated from the polymer by fractional distillation under nitrogen. Impurities that affect the initiation (e.g. by reaction with the initiator or its fragments) or that react with the growing macro-molecules (causing termination or transfer) are thereby removed.

Measurements of the usual physical constants such as boiling point or refractive index are not sufficiently sensitive to determine the trace amounts of impurities in question. Infrared and ultraviolet spectroscopy, and especially gas chromatography are more suitable for this purpose. The surest criterion for the absence of interfering foreign compounds lies in the

[1] See p. 37

polymerization itself. The purification is repeated until test polymerizations on the course of the reaction under standard conditions are reproducible (conversion-time curve, viscosity number of the polymers).

Storage of monomers and solvents also requires special precautions. The vessels must be so constructed that they permit the removal of their contents under inert gas. Figures 2.3 and 2.4 show graduated storage vessels from which the distilled monomer (or solvent) can be directly transferred to the reaction vessel under nitrogen; the storage vessel shown in Figure 2.5 can be closed with a self-sealing cap (serum cap) through which the contents may be withdrawn by means of a pipette or hypodermic needle (see Section 2.1.3). The constructions shown in Figures 2.4 and 2.5 have the advantage that the contents of the vessel come into very little contact with tap grease.

Most monomers can be stored unchanged under nitrogen only for short times (hours or days), even in the dark at low temperature. For long-term

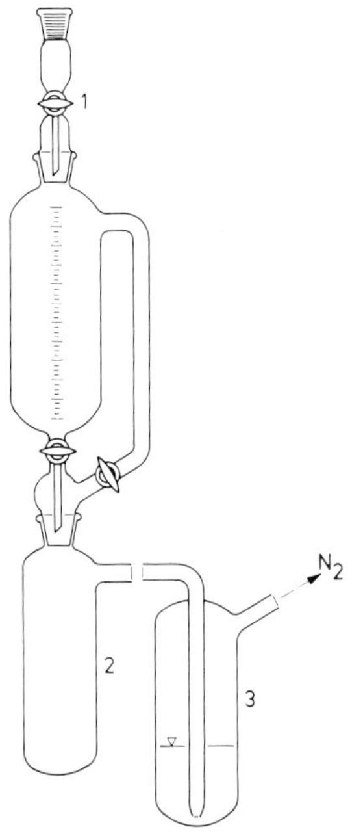

FIGURE 2.3 Graduated receiver for monomers and solvents. 1: Head with dropping tube; 2: receiver for first fraction, with side-arm to mercury seal or for connection to vacuum pump for distillation under reduced pressure; 3: mercury seal for distillation under nitrogen at atmospheric pressure.

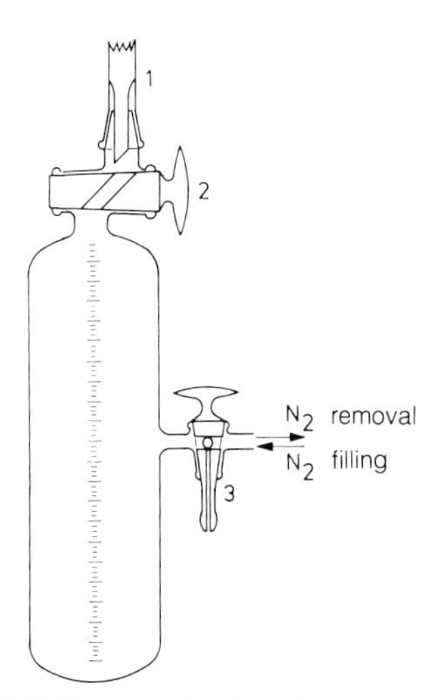

FIGURE 2.4 Graduated reservoir for monomers and solvents (discharged by turning through 180°). 1: Dropping tube; 2: oblique-bore tap; 3: two-way tap.

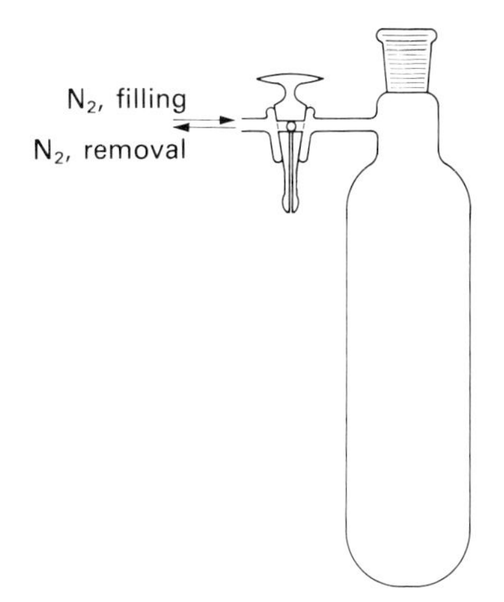

FIGURE 2.5 Reservoir for monomers and solvents.

storage a suitable stabilizer is therefore indispensable. Effective stabilizers (inhibitors) of radical polymerization[1] are quinones, phenols, amines, nitro-compounds and some metals or metal compounds. Sufficient stabilization of many monomers is attained by the addition of 0.1 to 1 wt. % of hydroquinone or 4-*tert*-butylpyrocatechol.

Inhibition of ionic polymerization can be effected by water, alcohols, ethers or amines. However these substances can act in different ways according to their concentration. For example, in polymerizations initiated by Lewis acids (BF_3 with isobutene) or organometallic compounds (aluminium alkyls), water in small concentrations behaves as a cocatalyst (see Section 3.2), but in larger concentrations as an inhibitor (reaction with the initiator or with the ionic propagating species).

2.1.3. Reaction vessels for polymerization reactions

There are described in the literature[2] a multitude of special experimental arrangements that are suitable for particular monomers and types of polymerization, but they may all be regarded as modifications of the apparatus described below.

Addition and condensation polymerizations of solid or liquid monomers are frequently conducted in an autoclave using glass ampoules. These

---

[1] *Houben-Weyl 14/1* (1961) 42; 746.
[2] Summary in: *Houben-Weyl 14/1* (1961) 45.

should he made from thick-walled glass in order to withstand any internal pressure built up during reaction and should be sealed off as strain-free as possible; safety precautions must be taken against possible explosion (wrapping in cloth, insertion in a steel tube, protective shield etc.). The ampoules should not exceed 250 ml in volume and should never be filled more than half full. In the filling process it is necessary to ensure that none of the contents remains adhering to the upper part of the vessel, otherwise they may decompose during sealing off with resulting serious disturbance of the reaction. This can easily be prevented with the devices shown in Figure 2.6. Arrangement 2.6a is suitable for work with solids: the weighed reactants are introduced through the long funnel; then the ampoule is evacuated and filled with nitrogen. For liquids, the apparatus 2.6b is to be preferred since it avoids having to evacuate the filled ampoule (which can cause wetting of the walls and loss of volatile substances). The ampoule,

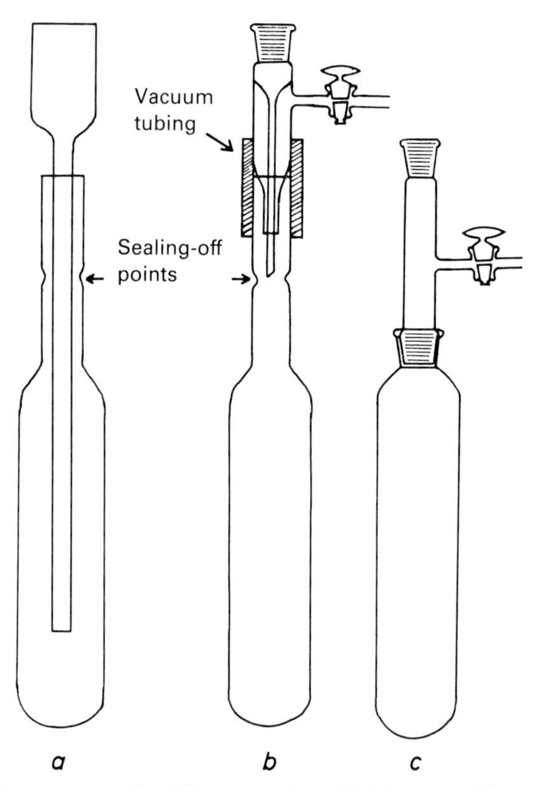

FIGURE 2.6 Arrangements for filling ampoules with (a) solids, (b) and (c) liquids.

into which the initiator has previously been weighed, is attached to the outer tube of the head by means of a piece of rubber tubing. It is then evacuated and filled with nitrogen through the tap on the side-arm. A measured amount of monomer can now be run in from an attached reservoir (Figures 2.3 and 2.4), if necessary, applying slight suction through the side-arm. Finally, the ampoule is pulled downwards so that the dropping tube is clear of the constriction which is then sealed off. The apparatus 2.6c is suitable for filling tubes and flasks fitted with ground joints.

For reactions in which only a slight pressure rise is expected (e.g. emulsion polymerization of gaseous monomers), pressure bottles (mineral water bottles) can be used; for higher pressures, autoclaves are to be preferred.

Polymerizations in aqueous medium and some ionic polymerizations can be conducted in normal multi-necked flasks[1] (with or without ground joints). In heterogeneous polymerizations, where the number and size of particles (precipitation polymerization, suspension and emulsion polymerization), or diffusion processes (e.g. polymerization of gaseous olefins with organometallic mixed catalysts) play a decisive part, the use of a stirrer motor with revolution counter is to be recommended. For polymerizations initiated by catalyst suspensions (e.g. alkali metals), a suitable high-speed stirrer[2] is used. It may also be mentioned that high speed mixers can be used as reaction vessels for interfacial polycondensations.

Flasks with self-sealing closures, used in combination with hypodermic syringes, are very suitable as reaction vessels, especially for ionic polymerizations. They have the advantage that catalyst can be injected, or samples removed, with practically complete exclusion of air and moisture (conversion-time curve, see Example 3–13). If suitable flasks are not available, ordinary flasks with ground joints can be adapted as shown in Figure 2.7. Bottles with screw caps can also easily be fitted with a self-sealing closure. First a hole is bored in the cap which is then fitted with a 2 mm strong rubber disc and, on top, a thin disc of soft PVC; the cap is then screwed back on the bottle. After the sealing disc had been pierced, the hole in the screw cap can be carefully covered with a piece of adhesive film. Syringes with glass barrels are to be preferred (corrosion, cleaning).

---

[1] Suppliers: e.g. Schott und Gen., Mainz; Quickfit (England).

[2] For example Ultra-Turrax from the firm Janke und Kunkel, or Stir-O-Vac from Labline Inc., Chicago, Ill.

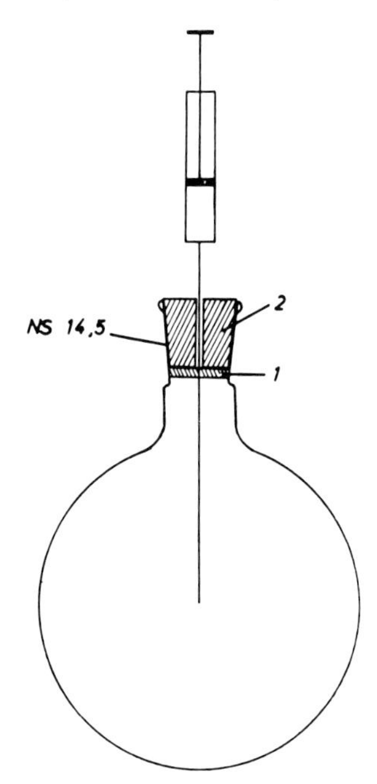

FIGURE 2.7 Self-sealing closure for flask with ground joint; 1: self-sealing disc (septum as used in gas chromatography); 2: bored rubber stopper.

If the polymerization is to be initiated by radiation, one generally uses ampoules or cells with well defined dimensions. A proper geometrical relationship between the vessel and the radiation source is important. Details of experimental arrangements can be found in the literature.[1,2] For kinetic investigations of homogeneous polymerizations, a variety of methods and apparatus have been developed. The dilatometric method[3,4,5] is especially to be mentioned on account of its simplicity and general applicability (see Figure 2.8). This procedure depends on the measurement of the contraction of volume that results from the different densities of the monomer and polymer. The conversion of the volume contraction to the

[1] *A. Chapiro*, "Radiation Chemistry of Polymers", Interscience Publishers, New York, London 1962; *A. Henglein, W. Schnabel and J. Wendenburg*, "Einführung in die Strahlenchemie", Verlag Chemie GmbH, Weinheim 1959; *M. Dole*, "The Radiation Chemistry of Macromolecules", Academic Press, London, 1972.

[2] See *Houben-Weyl 14/1* (1961) 52.

[3] *G.V. Schulz* and *G. Harborth*, Angew. Chem. *59* (1947) 90.

[4] *G.V. Schulz, G. Henrici* and *S. Olivé*, Z. Elektrochem. *60* (1950) 303.

[5] *H. Schuller*, Angew. Makromol. Chem. *2* (1968) 64.

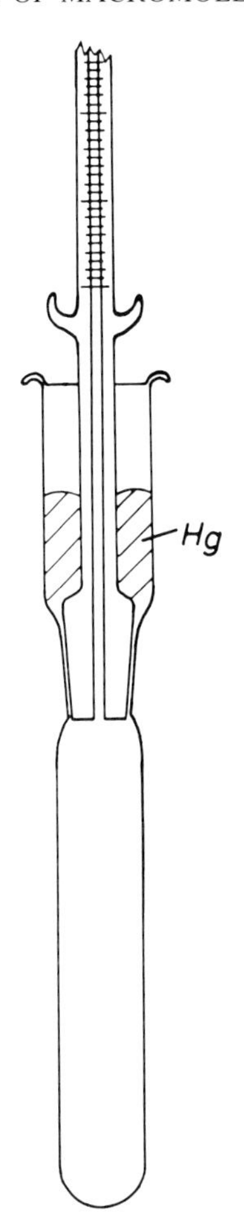

FIGURE 2.8 Dilatometer.

yield of polymer can be made by means of a gravimetrically determined calibration curve or by calculation from the specific volumes (see Example 3–12).

With appropriate precautions condensation and addition polymerization reactions can be carried out in the same apparatus as customarily used for organic preparative work (see Section 4.1 and 4.2). In order to obtain high molecular weights by polycondensation in solution, a special recycling apparatus can be used with advantage[1] (Figure 2.9).

---

[1] *Houben-Weyl 14/2* (1963) 18.

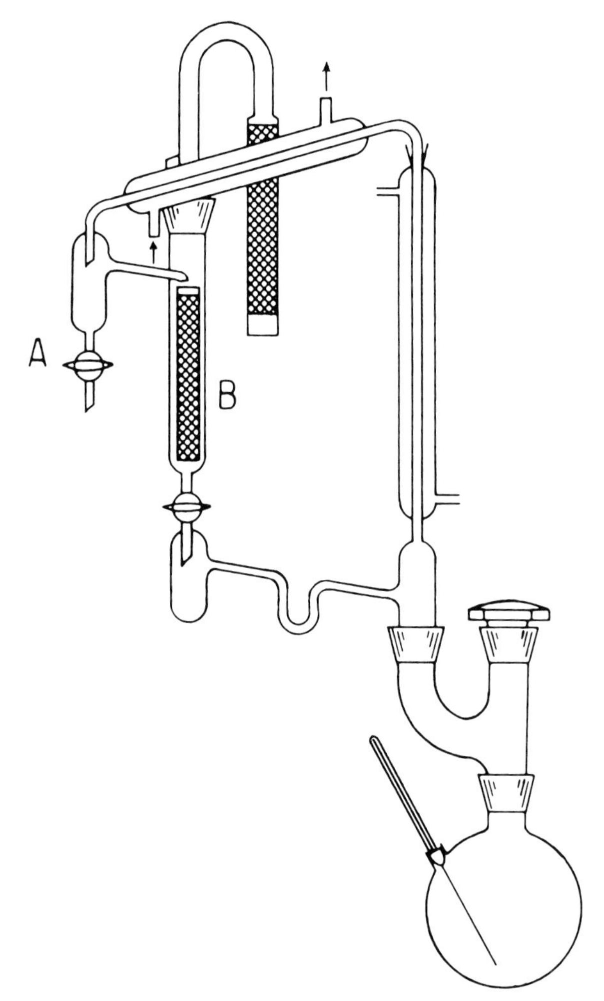

FIGURE 2.9 Recycling apparatus for the preparation of polyesters; see p. 52.

2.1.4. Temperature control in polymerization reactions

Exact temperature control is very important in polymerization reactions, since, amongst other things, the rate and degree of polymerization are strongly dependent on temperature. For accurate work a water or paraffin oil bath may be used, thermostatted in the normal way with the aid of a contact thermometer and immersion heater. For reactions carried out in ampoules or autoclave tubes, vapour baths prove very convenient. For this purpose, a tube sealed at one end (diameter 3–8 cm) is filled quarter-full with a suitable liquid and heated to boiling. The reaction vessel to be thermostatted is then suspended a few centimetres above the boiling liquid so that the vapour condenses as completely as possible. A list of compounds that are suitable for use as liquids in vapour baths over extended periods of time may be found in the literature.[1] For investigations on polymers at higher temperatures (e.g. for degradation experiments), it is more practical to use air thermostats[2] which can be regulated continuously up to 400°C with a precision of ±1°C; this is adequate for most purposes. For thermostatting below room temperature, the usual freezing mixtures may be used; alternatively cryostats (down to about −80°C) are available for use at the bench.[3]

2.1.5. Execution of polymerization reactions

Polymerizations may be classified according to the chemistry involved, as discussed in Chapter 1. They may also be divided according to the way in which they are carried out, which is especially convenient for a discussion of experimental methods. Accordingly they are grouped here as polymerizations in bulk, in solution and in dispersion.

2.1.5.1. Polymerizations in bulk

Against the advantages of polymerization in bulk (high molecular weight, high rates of polymerization, mostly very pure polymers), must often be set the following disadvantages: the difficulty of removal of the heat of polymerization; in some cases the insolubility of the polymer in the monomer; and, in highly viscous systems, side reactions such as chain transfer with the polymer (cf. Chapter 3), which not only make for technical difficulties, but also often affect the properties of the resulting polymer. As examples of bulk polymerizations, sometimes used in commercial processes, we may quote the radical polymerizations of styrene (Example 3–02), vinyl acetate (Example 3–06) and methyl methacrylate (Example 3–05), also the high pressure polymerization of ethylene, the polymerization of ϵ-caprolactam (Example 3–42), and the polymerization of trioxane (Example 3–40). For kinetic investigations, where polymerization is usually only carried to low conversions, the method offers many advantages.

---

[1] *Sorensen-Campbell*, p. 5.
[2] For example, from Heraeus GmbH, Hanau.
[3] For example, from Messgeräte-Werk Lauda, Lauda/Tauber.

The products of most stepwise condensation and addition polymerizations are obtained, both in the laboratory and in large scale production, by reactions in bulk above the melting points of both the starting compound and the resulting polymer. As the reaction advances the molecular weight and therefore the melt viscosity increases. The removal of the readily volatile reaction products (water, alcohol) from the reaction mixture, even under reduced pressure, becomes ever more difficult, so that it is frequently necessary to raise the reaction temperature steadily. In order to attain a high molecular weight, one is forced in some cases to work at temperatures of 250°C or higher towards the end of the reaction. Melt condensation is, therefore, only applicable when both the reactants and the resulting condensation polymer are thermally stable; otherwise noticeable side reactions may occur, leading to coloration, crosslinking, or lowering of molecular weight through chain scission. For this reason, it is difficult to prepare a series of high-melting polyamides in this way; they can, however, be made by solution or interfacial polycondensation of diamines and dicarboxylic acid chlorides. Molecular weights attainable by melt polycondensation do not usually exceed 50 000.

Some polymerizations can be carried out not only in the liquid phase but also in the solid state[1], for example the polymerization of acrylamide or of trioxane (Example 3–40b). The so-called post-condensation, for example of polyesters (see Example 4–03), also proceeds in the solid phase. Finally, condensation polymers containing reactive heterocycles in the main chain may likewise undergo ring closure reactions in the solid state, for example to polyimides (Example 4–20).

2.1.5.2. Polymerizations in solution

For the reasons given in Section 2.1.5.1, it is often expedient, especially in addition polymerizations, to carry out the reaction in a solvent. When both the monomer and the resulting polymer are soluble in the solvent, one speaks of a homogeneous solution polymerization; on the other hand, if the polymer precipitates during the course of the reaction, it is called a precipitation polymerization. For homogeneous polymerizations in inert solvents, at constant initiator concentration, both the reaction rate and degree of polymerization fall off with decreasing monomer concentration[2]; on the other hand, precipitation polymerizations frequently show abnormal kinetics.

---

[1] Summary: *A. Chapiro*, J. Polym. Sci., Part C*4* (1964) 1551.
[2] See Section 3.1.

In many cases the solvent takes part in the reaction, so that additional deviations from the normal course of polymerization are observed. For example, in radical polymerizations the solvent molecules can undergo transfer reactions with the growing macro-radicals, whereby the mean degree of polymerization is reduced, but the rate of polymerization is unchanged.[1] For solvents with high transfer constants, this can go so far as to yield exclusively products of low molecular weight, containing fragments of the chain transfer agent as end groups (telomerization).[1]

In cationic polymerizations, the influence of the solvent is still more pronounced[2]; besides transfer reactions, reactions with the initiator may also occur (e.g. alkyl halides with Lewis acids). Furthermore, the relative permittivity of the solvent is an important factor. In certain anionic polymerizations, the solvent also influences the configurational sequence of the CRU's (see Example 3–29). Thus, the solvent for polymerization must be chosen very carefully and certain solvents avoided.

Condensation polymerizations can also be conducted in solution, although this is not often done. Condensation of diols with dicarboxylic acids in solution is advantageous for the preparation of polyesters which do not withstand the high temperatures necessary for melt polycondensations, or when molecular weights above 30 000 are required (see Example 4–02). For this purpose, the polycondensation is carried out with an approximately 20% solution of reactants in an inert solvent. Especially preferred are hydrophobic solvents, such as benzene, toluene, xylene, or chlorobenzene, which not only form an azeotrope with the liberated water, but also prevent the back reaction by providing a protective solvation shell for the ester linkages already formed. The low viscosity of the 20% solution compared with that resulting from melt condensation allows the resulting water to be removed much more easily; hence, solution condensation can be carried out at relatively low temperature, controlled by the boiling point of the solvent. However, in order to obtain a sufficiently high esterification rate, a catalyst, usually an acidic compound such as *p*-toluenesulfonic acid, is necessary. When one of the starting components (diol or dicarboxylic acid) is insoluble in the desired solvent, it is possible first to carry out a pre-condensation in the melt at about 120–150°C and then to subject the resulting low-molecular-weight polyester to condensation in solution. Solution condensations are conveniently

---

[1] See Section 3.1.

[2] *Houben-Weyl 14/1* (1961) 783; *J.P. Kennedy and A.W. Ranger*, Adv. Polym. Sci., *3* (1964) 508; *P.H. Plesch* (Ed.). "The Chemistry of Cationic Polymerization", Pergamon Press, Oxford 1963.

performed in a recycling apparatus[1] (Figure 2.9). The water distilling off as an azeotrope is drawn off from the separator (A) and the solvent returned to the reaction vessel through the drying tube (B).

Condensation polymers, especially polyamides, can also be prepared in solution by the Schotten-Baumann reaction at low temperature[2] (Example 4–10). For this purpose, two rapidly reacting monomers, for example diamine and dicarboxylic acid dichloride, are mixed together with stirring in an inert solvent; the eliminated hydrogen chloride is trapped with an acid acceptor. One usually works at room temperature in approximately 10% solution in chloroform, dichloroethane, benzene or ethyl methyl ketone, with a tertiary amine (pyridine) as acid acceptor. This procedure has the following advantages: the polycondensation is carried out at low temperature (0–40°C); it is nevertheless very fast, the reaction usually being over in a few minutes. At low temperatures practically no side reactions occur. Furthermore, the equivalence of the two reactants need not be as exact as in other polycondensation procedures. Disadvantages are the following: relatively large amounts of solvent must be purified and handled; and large amounts of salts are formed as by-products. Condensation in solution at low temperature is, therefore, above all a laboratory method, in which these disadvantages are not so important; it is to be particularly recommended in that it yields a high-molecular-weight condensation polymer in a short time with little expenditure on apparatus.

The Schotten-Baumann reaction between dicarboxylic acid dichlorides and diamines can be performed not only in organic solvents, but also, at room temperature, by means of a special experimental technique known as interfacial polycondensation (Example 4–10). The two components are separately dissolved in two immiscible or only partially miscible solvents, the solutions then being brought together carefully to form an interface; the polycondensation can now take place only at the interface of the two liquid phases, whereby the practically instantaneously formed thin polyamide film impedes further diffusion together of the two reactants. Only when this film is pulled carefully upwards from the boundary layer, can the polycondensation carry on; the process can thus be run continuously in a simple way. Interfacial polymerization can also be performed in dispersion (Example 4–11); for this purpose the solution of acid dichloride is dispersed in the aqueous solution of diamine by vigorous stirring (if necessary in the presence of a water-soluble dispersion stabilizer). The

---

[1] Also see *Houben-Weyl 14/2* (1963) 18.

[2] *P.W. Morgan* and *S.L. Kwolek*, J. Polym. Sci. Part A2 (1964) 181, 209, 2693. *P.W. Morgan*, "Condensation Polymers by Interfacial and Solution Methods", Polym. Rev. *10* (1965).

polycondensation then takes place at the surface of the fine droplets. Water is especially suitable as solvent for the diamine component, while aliphatic chlorinated hydrocarbons are best for the dicarboxylic acid dichlorides.

Interfacial polycondensation can be carried out not only with aliphatic but also with aromatic dicarboxylic acid dichlorides. With disulfonic acid dichlorides the corresponding polymeric sulfonamides are formed. Of the diamines, however, only primary and secondary aliphatic amines are suitable; aromatic diamines react too slowly. An exact equivalence of the two reactants is not absolutely necessary. If, however, very high molecular weights are the goal, the optimum conditions must be sought by varying the relative concentrations and, if necessary, special solvents must be chosen. A big advantage of interfacial polycondensation is the high reaction rate; in many cases the reaction is finished in a few minutes. In addition, the apparatus required is very unpretentious and one can work at low temperature. Thereby side reactions, such as transamidation, as well as oxidative and thermal decomposition are avoided; these almost always occur during melt condensation. In this way one can prepare successfully even high-molecular-weight, very high-melting polyamides, that are obtainable by the usual methods only in low molecular weights, if at all. Furthermore, reactants can be used that still carry reactive groups, for example hydroxyl groups, C=C bonds or C≡C bonds. An advantage over solution condensation at low temperature is that the eliminated hydrogen chloride is not thrown out of solution as a salt, later to be separated, but remains in solution as the amine hydrochloride. The molecular weights attainable by interfacial polycondensation are at least as high as those obtained by melt condensation (10 000–30 000) and are often very much higher.

2.1.5.3. *Polymerizations in dispersion*

Polymerizations, especially addition polymerizations, can also be carried out under heterogeneous conditions. To this end, the liquid monomer is dispersed in a liquid in which it is insoluble or only slightly soluble. During reaction there is a change in aggregation state of the disperse phase, since the macromolecules formed are solids; thus the original emulsion becomes a suspension. If the polymer is insoluble in the monomer, this transition occurs early in the reaction; if, on the other hand, it is soluble in or swollen by the monomer, the status of the emulsion changes only at high conversion.

Suspension polymerization and emulsion polymerization thus occur in colloidally similar systems; both start from a dispersion of liquid monomer in a liquid (emulsion) and finish as a dispersion of solid polymer in a liquid (suspension). The term "suspension polymerization" was coined with respect to the final state, while the term "emulsion polymerization" refers to the initial state of the system. Despite this formal similarity, the two processes differ in some essential respects, for example in the size of the resulting polymer particles (0.1–0.5 μm in emulsion polymerization, 0.5 μm–2 mm in suspension polymerization) and in the kinetics of reaction (see Sections 2.1.5.3.1 and 2.1.5.3.2).

These methods offer the following advantages: the heat of polymerization is readily conducted away; and the reaction mixture remains very mobile even at high conversion, which considerably facilitates the work-up of the polymer. Both condensation and addition polymerizations can be carried out in dispersion (interfacial polycondensation, see Sections 2.1.5.2 and 4.1.2.3).

2.1.5.3.1. Polymerizations in suspension

In suspension polymerization[1] the liquid monomer, usually containing a dissolved water-insoluble initiator (e.g. dibenzoyl peroxide, 2,2′-azoisobutyronitrile), is finely dispersed by vigorous stirring in a suitable medium in which it is insoluble or only sparingly soluble. Polymerization takes place in the monomer droplets, and hence the rate of polymerization, average molecular weight, and properties of the product are comparable with those obtained by bulk polymerization under analogous conditions. The dispersion of the monomer in water can be assisted by the addition of small quantities (ca.0.1%) of a protective colloid (see Section 2.1.5.3.2) or finely divided inorganic substance (e.g. barium sulfate, calcium phosphate, or magnesium phosphate). These prevent both the coalescence of the monomer droplets and, in the later stages of polymerization, coagulation of the polymer particles swollen by monomer.

Water is almost exclusively used as dispersion medium for radical polymerization (two to ten times the amount of monomer). When the monomer is partially soluble in water or the polymer is insoluble in the monomer, the polymer precipitates in the form of discrete but irregularly shaped particles. On the other hand, if the monomer and initiator are insoluble in water but the polymer is soluble in the monomer, the polymer is produced in the form of regular beads whose diameter can be anywhere between 0.5 μm and several millimetres according to the experimental

[1] *Houben-Weyl 14/1* (1961) 133 and the literature there cited.

conditions. This particular case of suspension polymerization is referred to as bead polymerization.

2.1.5.3.2. Polymerizations in emulsion

Working in emulsion is essentially limited to addition polymerization reactions. As in suspension polymerization, the basic principle is finely to disperse a water-insoluble or sparingly soluble monomer in water and bring about polymerization in this state. There are, however, some essential differences between the two procedures. The dispersion of the monomer takes place in the presence of substances (emulsifiers) that can form micelles, into which the monomer is attracted. The particle size of the resulting polymer is much smaller than in suspension polymerization (diameter about 0.1 μm). Water-soluble compounds (potassium peroxodisulfate; redox systems) are always used as initiators, apart from a few special cases.[1] Polymerization takes place not in the monomer droplets but in the micelles which thereby slowly swell to latex particles. This means that the rate of polymerization (at constant initiator concentration) depends on the number of micelles and therefore on the emulsifier concentration.[2] Emulsion polymerization makes possible the preparation of very high-molecular-weight polymers at high rates of polymerization. The required reaction temperatures are low and can even be below 20°C when redox systems are used for initiation (see Example 3-22).

The ingredients for an emulsion polymerization consist essentially of four components:

— water;
— a monomer, insoluble or sparingly soluble in water;
— a water-soluble, radical-generating initiator;
— an emulsifier.

Emulsifier molecules consist of a hydrophilic part and a hydrophobic part. According to the electrical charge on the hydrophilic portion, one distinguishes between anionic, cationic, and non-ionic emulsifiers.[3,4]

---

[1] *J.W. Breitenbach* and *H. Edelhauser*, Makromol. Chem. *44/46* (1961) 196.

[2] Details of the kinetics of emulsion polymerization can be found in: *H. Gerrens*, Adv. Polym. Sci., *1* (1959) 234; *E. Bartholomé*, *H. Gerrens*, *R. Herbeck* and *H.M. Weitz*, Z. Elektrochem. *60* (1956) 334; *J.C.H. Hwa* and *J.W. Vanderhoff* (Eds.) "New Concepts in Emulsion Polymerization", J. Wiley and Sons, New York (1969); *D.C. Blackley*, "Emulsion Polymerization: Theory and Practice", Applied Science Publishers, London, 1975.

[3] Summary in *Houben-Weyl 14/1* (1961) 190 *et seq.*

[4] On the use of emulsifiers for polymerization in "inverted emulsions" (water-in-oil) see *H. Bartl* and *W. von Bonin*, Makromol. Chem. *57* (1962) 74.

Examples of anionic emulsifiers are the potassium, sodium or ammonium salts of fatty acids, sodium dodecyl sulfate, and salts of alkyl-substituted benzene- or naphthalene-sulfonic acids. Examples of cationic emulsifiers are quaternary ammonium salts that possess at least one hydrophobic substituent. Typical non-ionic emulsifiers are phenolic ethers, alcohols, carboxylic acids, and block copolymers of ethylene oxide and propylene oxide.

In very dilute aqueous solution, emulsifiers behave as isolated molecules or as electrolytes. With increasing emulsifier concentration, however, there occurs an abrupt change in some physical properties of the solution, e.g. the surface tension, viscosity, electrical conductance, osmotic pressure etc. The concentration, at which these abrupt changes are observed, is called the critical micelle concentration (CMC).[1] It has a characteristic value for each emulsifier. Below the CMC the emulsifier is dissolved in molecular form, but above the CMC the emulsifier molecules cluster together to form molecular aggregates, the so-called micelles, in which the hydrophobic residues are turned inwards and the hydrophilic residues are turned outwards towards the aqueous phase. These micelles have a diameter of about 35 Å. In emulsion polymerization one generally uses 0.5–5 wt. % of emulsifier relative to monomer. With the usual oil-in-water emulsions the water content varies from half to four times the amount of monomer.

The monomer is present partly as emulsifier-stabilized monomer droplets and partly within the afore-mentioned micelles. The size of the micelles is significantly increased by the addition of monomer (diameter (45–50 Å). However, the diameter of the droplets is about 10 000 Å, very much larger than that of the micelles (cf. Figure 2.10). The micelles are present at a concentration of about 10^{18} per cm^3 of liquor and each micelle contains around 100 monomer molecules. In contrast, the number of monomer droplets is only about 10^{10} per cm^3. Thus despite the larger volume of monomer droplets, the micelles offer a very much larger surface area. A radical formed in the aqueous phase will thus encounter a monomer-filled micelle much more often than a monomer droplet.
The polymerization may therefore be expected, on a statistical basis, to take place practically only in the micelles and not in the monomer droplets. This has been confirmed experimentally. The monomer consumed in the micelles is replaced by diffusion from the monomer droplets through the aqueous phase.

[1] *R. Wintgen*, Kolloid-Z. Z. Polym. *124* (1951) 141.

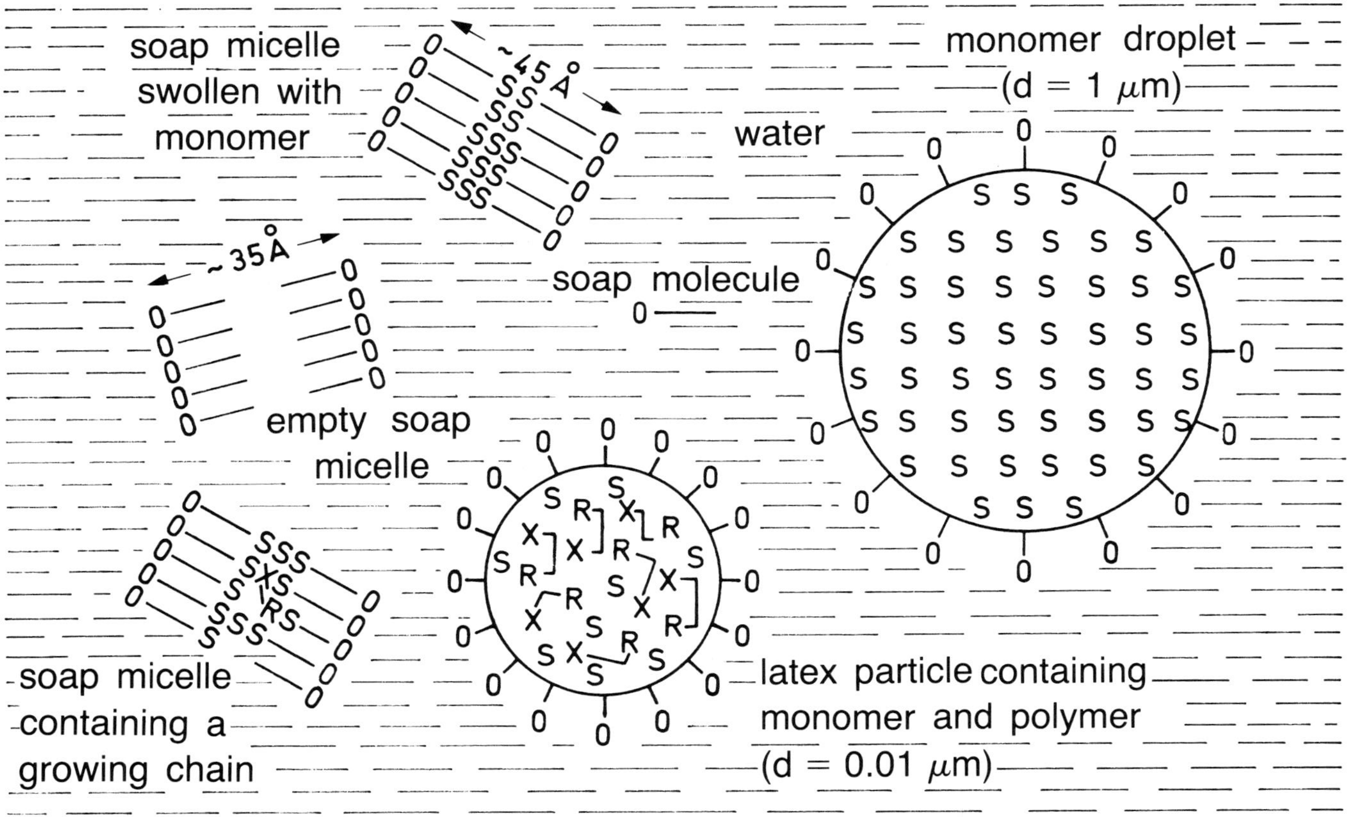

FIGURE 2.10 Schematic distribution of components in an emulsion polymerization.

According to the theories of Harkins and of Smith and Ewart, the kinetic course of an emulsion polymerization is divided into three periods. At first the micelles increase rapidly in size as the polymerization advances and are transformed into so-called latex particles, containing both monomer and polymer. These are still very much smaller than the monomer droplets and have an initial diameter of about 200–400 Å, corresponding to about 10^{14} particles per cm^3 liquor. The monomer used is continuously replaced from the monomer droplets via the aqueous phase. More and more emulsifier molecules are adsorbed from the aqueous phase on to the surface of the latex particles, assisting their stabilization; the micelles that still contain no polymer thereby slowly disappear. The concentration of free emulsifier finally falls below the critical micelle concentration; at this moment the surface tension increases significantly. From then on, therefore, practically no new latex particles can be formed. The first phase of the emulsion polymerization, the so-called particle-formation period, is complete after about 10–20% conversion.

From this point onwards, the polymerization still occurs only in the latex particles, whose number, however, remains constant. During this second period, the reaction is thus of zero order. When the polymerization has proceeded so far that all the monomer droplets have vanished, which occurs after 60–80% conversion, the residual monomer is all located in the latex particles. The monomer concentration in the particles now declines as polymerization proceeds further, i.e. in this final period the reaction is first order. At the end of the polymerization the emulsion consists of polymer particles with a size distribution between 500 and 1500 Å, which is larger than the original micelles (≈ 50Å), but smaller than the original monomer droplets (≈10 000 Å).

The changes of surface tension γ and overall rate of polymerization R_p with conversion y are shown in Figure 2.11. In contrast to emulsifiers, protective colloids do not form micelles. Their function, as implied by their name, is essentially to prevent the monomer droplets as well as the latex particles from coming together and coagulating. Suitable as protective colloids are water-soluble polymers such as starch, pectin, alginates, and gelatine; modified natural products and synthetic polymers such as hydroxyethyl cellulose, methyl cellulose, carboxymethyl cellulose, poly(vinyl alcohol), poly(acrylic acid), and poly(vinylpyrrolidone) can also be used.

2.1.5.4. Control and termination of polymerization reactions

In polymerization reactions the average molecular weight of the polymer, and the properties dependent thereon, can be varied within certain limits by proper choice of reaction conditions. Thus in radical polymerization an

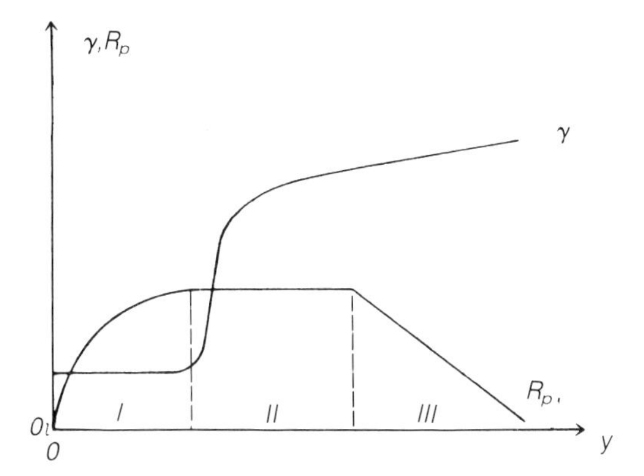

FIGURE 2.11 Overall rate R_p and surface tension γ as a function of conversion y in emulsion polymerization (schematic). The three stages I, II and III are indicated.

increase of reaction temperature or amount of initiator causes an increase in the number of growing radicals. Since the rate of the propagation reaction is first order with respect to the concentration of growing radicals, while that of the termination reaction is second order, the average molecular weight is reduced (with simultaneous increase in rate of polymerization). A decrease of monomer concentration also leads to lower molecular weights, but the rate of polymerization then falls as well. Since side reactions may intervene at high temperature or high initiator concentration, the molecular weight is often better controlled by the addition of regulators, i.e. substances with high transfer constants (see Section 3.1 and Example 3–14). Even at low concentrations such compounds decrease the average molecular weight markedly by terminating the growth of polymeric chains; at the same time, a new chain is started so that, as a rule, the rate of polymerization is unaffected. The fragments of the regulator are built into the macromolecule as end groups. Especially suitable as regulators are thiols (1-butanethiol, 1-dodecanethiol) and other organic sulfur compounds, e.g. bis(isopropoxycarbothioyl)disulfane, (diisopropyl xanthogen disulfide), aliphatic halogen compounds, aldehydes and acetals. Regulators play an important role in commercial processes, for example in emulsion polymerization and especially in the preparation of polybutadiene. In principle, regulation is also possible in ionic polymerization.[1]

---

[1] See e.g. *V. Jaacks*, *H. Baader* and *W. Kern*, Makromol. Chem. *83* (1965) 56, and *H.D. Hermann*, *E. Fischer* and *K. Weissermel*, ibid, *90* (1966) 1.

In many cases (e.g. in kinetic investigations, in determination of reactivity ratios or in the preparation of unbranched polymers) it is not appropriate to allow polymerization to proceed to complete conversion of monomer. Polymerization reactions can be stopped in different ways. Sometimes the reaction can be brought to a halt simply by cooling. In most cases the polymerization can be ended by pouring the reaction mixture, or a sample thereof, into a sufficient quantity of precipitant, whereby the polymer formed is precipitated and the residual monomer and initiator are highly diluted. Polymerizations are most effectively stopped by addition of an inhibitor (see p. 132) or some compound that destroys the initiator; the details of this method depend on the type of polymerization. For radical polymerizations one uses, for example, hydroquinone or *N*-phenyl-*β*-naphthylamine. Most ionic polymerizations can be stopped by addition of water, acids, or bases. When an ionic polymerization is carried out at low temperature, it is essential to terminate it by destruction of the initiator, since warming to room temperature is likely to make the reaction go more quickly. Organometallic initiators (also Ziegler-Natta catalysts) can be destroyed with water or alcohols, and Lewis acids (BF_3) with amines.

Control of molecular weight is much simpler for those polymerizations (both condensation and addition) where the macromolecules are built up stepwise. In such cases the reaction can be stopped at any stage, for example by cooling. In principle, the molecular weight of the polymer formed in these reactions can also be controlled by changing the mole ratio of the two bifunctional reactants or by adding an appropriate amount of monofunctional compound.

2.1.6. Reactions of polymers

Reactions of natural or synthetic macromolecules differ in some respects from those of low-molecular-weight compounds (see Chapter 5); this also applies to experimental procedures. In the following Sections, therefore, we deal with some features to which special attention must be paid in the reactions of polymers.

2.1.6.1. Peculiarities in reactions of polymers

When two low-molecular-weight compounds A and B react with one another, a by-product D is generally formed in addition to the desired product C:

$$A + B \longrightarrow C + D \tag{1}$$

The reaction mixture also usually contains unreacted A and/or B.

To apply this general reaction scheme to the reactions of polymers, A must be regarded as the reactive group on the macromolecule M. Consequently the reaction with reagent B yields group C (main product) and group D (by-product), that are likewise bound to the macromolecule by primary valencies; hence, in contrast to conversions of low-molecular-weight compounds, C and D cannot be separated:

$$\begin{array}{l} M(A)_n + B \longrightarrow M\ (A)_x(C)_y(D)_z \\ (x + y + z = n) \end{array} \qquad (2)$$

Depending on the structure and composition of the initial polymer (e.g. homopolymer or copolymer), n can have any value between unity and the degree of polymerization; in the latter case, each CRU has a reactive group.[1] The number of groups A, C and D in the reacted polymer are denoted by x, y, and z, respectively. It must be particularly emphasized that not all macromolecules in the product will contain the same number of these groups: x, y and z are thus average values. The situation $x = z = 0$ and $y = n$, means that the desired reaction has gone to completion, but this is seldom attained.

In addition to the reaction conditions (see Section 2.1.6.2), steric and statistical factors are important in the reactions of polymers. Since the reactive groups are situated on the macromolecules, they are, from the outset, not fully mobile. Reactive side groups are shielded on one side by the main chain of the macromolecule. Shielding, and therefore hindrance of the reaction, is also caused by the fact that the macromolecules are present in solution in a more or less strongly coiled form. This shape will generally change during the course of the reaction, so that the reaction may become either easier or more difficult as the reaction proceeds. In the reactions of double bonds in polymeric dienes (e.g. epoxidation), there are significant differences in the rates of reaction of 1,2- and 1,4-units which can be exploited analytically. Finally, in the reactions of stereoregular macromolecules (e.g. in the hydrolysis of polymethacrylates), an effect of the type and degree of tacticity is observed.

A statistical limit to the yield is always to be expected when two neighbouring groups of the macromolecules are involved simultaneously in the reaction, as, for example, in the formation of acetals from poly(vinyl alcohol) (see Example 5–02), where two hydroxyl groups of neighbouring CRU's react with an aldehyde molecule by ring closure. According to the calculations of Flory[2], a maximum conversion of only 86.5% of all the functional groups is possible in such reactions. This is understandable when

---

[1] Similar considerations apply to polymers with several reactive groups in the CRU.
[2] *P.J. Flory,* J. Am. Chem. Soc. *61* (1939) 1518; *64* (1942) 177.

one considers that, for statistical reasons, some of the functional groups become isolated and can no longer take part in the reaction when the necessary neighbouring group is missing:

$$
\begin{array}{c}
\ldots -CH_2-\underset{\displaystyle OH}{\underset{|}{CH}}-CH_2-\underset{\displaystyle OH}{\underset{|}{CH}}-CH_2-\underset{\displaystyle OH}{\underset{|}{CH}}-CH_2-\underset{\displaystyle OH}{\underset{|}{CH}}-CH_2-\underset{\displaystyle OH}{\underset{|}{CH}}- \ldots \\
\Big\downarrow + RCHO \\
\ldots -CH_2-\underset{\displaystyle O}{\underset{|}{CH}}-CH_2-\underset{\displaystyle O}{\underset{|}{CH}}-CH_2-\underset{\displaystyle OH}{\underset{|}{CH}}-CH_2-\underset{\displaystyle O}{\underset{|}{CH}}-CH_2-\underset{\displaystyle O}{\underset{|}{CH}}- \ldots \\
\quad \diagdown \underset{\displaystyle R}{\underset{|}{CH}} \diagup \qquad\qquad\qquad\qquad \diagdown \underset{\displaystyle R}{\underset{|}{CH}} \diagup
\end{array}
$$

Reactions of macromolecules with bifunctional reagents can proceed in both intra- and inter-molecular fashion. In the latter case crosslinking occurs and the product, therefore, becomes insoluble and usually infusible. Such reactions are frequently carried out intentionally (see Example 4–24); however, they can also appear as undesirable side reactions, markedly affecting the solubility of the product as well as making it more difficult to recover. For example, in the reactions of poly(acryloyl chloride) water must be carefully excluded, otherwise free carboxyl groups are formed that can react with the acid chloride groups of other macromolecules so leading to crosslinking.

2.1.6.2. Choice of reaction conditions

Before carrying out a reaction on a macromolecular compound, it is expedient in every case first to study the reaction of an appropriate low-molecular-weight model compound. The compound chosen as model should be similar to the CRU in structure as well as in its reactive group. In the case of addition polymers, the corresponding monomer is not generally suitable, since it contains a double bond that is not present in the polymer. For example, as a model for polystyrene one should take cumene, not styrene; for poly(vinyl ester)s, an isopropyl ester; and for polymethacrylates, the corresponding derivative of trimethylacetic acid. Furthermore, since there will be a mutual influence of neighbouring reactive groups in a macromolecule, model compounds should also be selected to correspond to dimers or trimers, e.g. 2,4-pentanediol as model for poly(vinyl alcohol), and derivatives of glutaric acid, namely α-methylglutaric acid or 1,3,5-pentanetricarboxylic acid, as models for derivatives of poly(acrylic acid). Experiments with such model compounds serve to determine optimum reaction conditions and the type of by-products to be expected.

At the same time, one obtains compounds that are models for the macromolecular reaction products on which, for example, solubility and analytical investigations may be performed (determination of functional groups, reference spectra in the u.v. or i.r. regions, pyrolytic gas chromatography). The information so obtained, however, is not necessarily valid for reactions of polymers, especially with regard to choice of solvent, reaction temperature, and work-up and purification procedures.

While reactions of low-molecular-weight compounds can sometimes be carried out in the gas phase, this technique is not applicable to macromolecular substances since they are involatile. A low-molecular-weight reagent can, however, be reacted in gaseous form with the solid or dissolved polymer[1], as, for example, in the commercial preparation of methylcellulose by reaction of alkali cellulose with gaseous methyl chloride.

If the polymer has to be used in solid form, e.g. if it is sparingly soluble, or is insoluble because of crosslinking (as in the case of ion-exchangers, see Examples 5–11 to 5–13), or if the reaction is to take place only at the surface, it is expedient to work with material that has been ground as finely as possible (see Section 2.4.1 concerning size reduction of polymers). The polymeric substance is then suspended in an inert medium, to which a swelling agent can often be added with advantage, thereby swelling the polymer either superficially or throughout, so favouring access by the reagent. Sometimes the reagent itself can act as swelling agent, as, for example, in the acetylation of the semi-acetal groups of polyformaldehyde (see Example 5–09).

If the reaction is carried out in homogeneous phase, one must consider the high viscosities, and hence the hindrance to material–and heat-exchange, in polymer solutions. In practice, this means that one must use strong and effective stirring to obtain good mixing and to avoid local overheating.

When the low-molecular-weight reagent is liquid at the desired reaction temperature, and dissolves both the reacting polymer and its product, one can sometimes work without additional solvent. The reagent is then used in large excess and the polymeric reaction product isolated, as necessary, from the resulting solution by conventional means (see Section 2.2). Examples of this type are the reactions of cellulose (see Examples 5–06 to 5–08) and of polyacrolein.[2] When this technique is not possible, one is obliged to use a solvent. First, it is necessary to determine which solvents

[1] *W. Kern* and *R.C. Schulz*, Angew. Chem. *69* (1957) 168.
[2] *R.C. Schulz*, Angew. Chem. *76* (1964) 357.

dissolve the polymer, the reagent, and the catalyst when such is needed. However, it must be remembered that the solubility of the polymer can change quite sharply during the course of the reaction, even after low conversion of the reactive groups; the polymer will then precipitate prematurely, markedly hindering further reaction. In this case, as the reaction progresses, one may try adding portions of a solvent for the product. Should it not be possible to find a common solvent or solvent mixture for both starting polymer and end product, it is usually preferable to run the reaction in a medium that is a solvent for the product. One then begins with a dispersion of very finely divided polymer; as the reaction advances the partially converted macromolecules dissolve and can then react to completion in a homogeneous system.

Special care must be given to the choice of reaction temperature, since this can significantly affect, amongst other things, the proportion of by-products. With macromolecular substances the different reaction products within the chain cannot be separated from one another, so that it is essential to choose a temperature that favours the main product, even when this must be bought at the expense of longer reaction times. Low temperatures are especially to be recommended if crosslinking, thermal degradation, or chain fission by autoxidation (e.g. with polymeric dienes) or hydrolysis (e.g. with cellulose) are to be feared, since these can cause a considerable change of molecular weight and physical properties. With autoxidizable polymers one should work under nitrogen or in the presence of an anti-oxidant.

In most cases, the recovery and purification procedures for the macromolecular product differ from those of the model product except for its separation from low-molecular-weight reagents and by-products; this is further discussed in Section 2.2.

2.1.6.3. Analytical problems in reactions of polymers

In order to assess the extent of reaction of a polymer, one cannot always apply the methods normally used to follow the reactions of low-molecular-weight compounds. Suitable methods for the characterization of polymers are, therefore, discussed in detail in Section 2.3.

One may first check qualitatively whether the desired reaction has taken place, especially by means of solubility tests and u.v. or i.r. spectroscopy; it is also necessary to examine whether unreacted groups, or groups other than those desired, are present. Quantitative analysis is aimed at evaluating the proportions of A, C, and D in equation (2), for which one may call not only on the usual methods of determination, but also on special procedures, such as spectroscopy and pyrolytic gas chromatography (see

Section 2.3). It is also sometimes expedient to choose the low-molecular-weight reactant so that an easily determinable element is introduced (e.g. by using chloroacetic acid or dichloroacetic acid for acylations).

The results of quantitative analysis can be presented in different ways. The experimental value may be expressed as a weight percentage conversion, i.e. as grammes of analytically determined CRU's per 100 g of polymer. However, this method of expressing the conversion is not very informative, since it does not indicate the extent to which the CRU's have been converted. It is, therefore, clearer to state the conversion in mole %, i.e. to indicate how many CRU's per hundred have reacted in the appropriate manner. It is then necessary to know the structure of any groups formed in side reactions, in order to calculate the molecular weight of all types of CRU present in the polymer. Making the simplifying assumption that no side reactions have occurred, the reaction product can be regarded as a two-component system, made up of unconverted groups A and product groups C. The following equation can then be used to convert wt. % into mole %.

$$a = \frac{100}{1 + \dfrac{100 - \alpha}{\alpha} \dfrac{M_A}{M_C}}$$

$$\mathrm{c} = 100 - a$$

M_A and M_C denote the molecular weights of the CRU's containing groups A and C respectively, α is the fraction of CRU's containing group A, expressed in wt. %, and a and c are the proportions of groups A and C, respectively, in mole %.

If many analyses of the same type have to be evaluated, it is convenient to prepare a diagram in which the analytically determined content of the appropriate group or element is plotted against the composition both in wt. % and in mole % (see Figure 2.12). It should again be pointed out that the observed values are generally less than the theoretical values for the reasons already mentioned (see Section 2.1.6.1).

Instead of giving the conversion in mole %, an average conversion factor $\bar{x}$ can be used:

$$\bar{x} = \text{amount (in mol) of groups A per CRU} \times \frac{\text{total conversion (in mole \%)}}{100}$$

This mode of expression is particularly recommended whenever the macromolecule contains more than one reactive group per CRU, e.g. three hydroxyl groups per CRU in cellulose (see Examples 5–06 and 5–07).

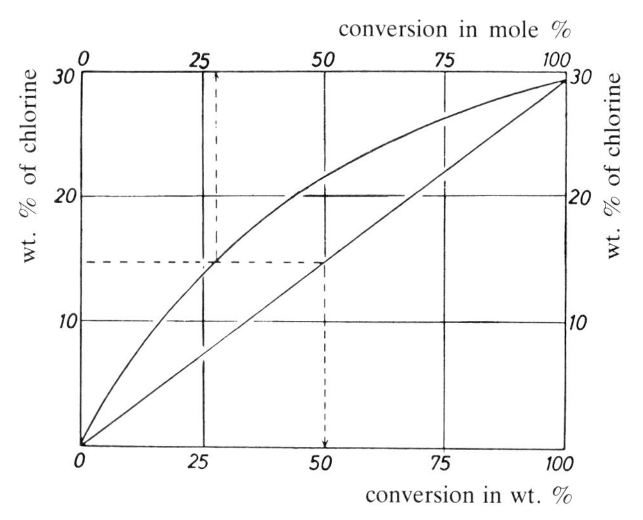

FIGURE 2.12 Example of graphical determination of wt. % conversion (lower line) and mole % conversion (upper line) from the chlorine content in the esterification of poly(vinyl alcohol) with chloroacetic acid to give $[C_4H_5ClO_2]_n$ at 100% conversion (29.5 wt. % of chlorine). Dashed lines indicate values corresponding to an observed composition of 14.75 wt. % of chlorine.

The calculations illustrated above can, of course, be extended to cover reactions involving the formation of three or more different groupings, for example, when performing successive reactions on a polymer. In such cases, the number of groups that has to be taken into account often becomes unmanageably large because of side reactions or because quantitative analysis becomes very difficult; calculation of the conversion is then possible only on the basis of simplifying assumptions that must be decided individually. In such cases, the analysis will be limited to the determination of the expected product groups and the yield expressed in wt. % appropriate to the particular reaction step.

2.2. ISOLATION AND WORK-UP OF POLYMERS

2.2.1. Isolation of polymers

Isolation is simplest if the polymer is insoluble in the reaction mixture and precipitates during formation (precipitation and suspension polymerization, interfacial polycondensation; see Section 4.1.3). In these cases the product can be separated by filtration or centrifugation. Aqueous solutions can be filtered through paper filters; solutions in organic solvents are better filtered through cloth or sintered glass discs. If the polymer remains dissolved in the reaction mixture, there are two possible ways of proceeding: either the excess monomer, solvent and other volatiles can be distilled off under vacuum or the polymer can be precipitated by addition of a precipitant. The first procedure is only applied in exceptional cases, since it generally leads to a resinous or horny polymer contaminated with initiator residues and especially with trapped monomer and solvent. The most usual procedure is, therefore, precipitation by means of a precipitant which should satisfy the following requirements. It must be miscible with the monomer and solvent, and dissolve all additives (e.g. initiator) as well as by-products (e.g. oligomers). Furthermore, the polymer should be insoluble and should separate in flocculant (not oily or resinous) form. Finally, it should be readily volatile and be absorbed or occluded by the polymer as little as possible.

The general procedure is to drop the reaction mixture or polymer solution into a 4- to 10-fold amount of precipitant under vigorous stirring. The concentration of polymer solution (generally not above 10%) and the amount of precipitant are chosen so that the polymer precipitates in flocculant, readily filtrable form. It often happens that the precipitated polymer remains in colloidal suspension; in this case, it may help either to lower the temperature (external cooling or by addition of dry ice) or to add electrolytes (solutions of sodium chloride or aluminium sulfate, dilute hydrochloric acid, acetic acid, or ammonia). Coagulation of the polymer can also sometimes be achieved by prolonged vigorous stirring or shaking. Polymers that tenaciously retain solvent and tend to resinify can be precipitated with advantage by the spray method.[1] For this purpose, the polymer solution is sprayed in the form of a mist into the precipitant, thereby producing a fine floccular precipitate, the large surface area of which favours the outward diffusion of unpolymerized monomer and other compounds present.

2.2.2. Purification and drying of polymers

Careful purification and drying of polymers is important not only for analytical characterization, but also because mechanical, electrical and optical properties are strongly influenced by impurities. Not the least

[1] *R.C. Schulz* and *A. Sabel*, Makromol. Chem. *14* (1954) 115.

important aspect of purification is the fact that even traces of impurities may cause or accelerate degradation or crosslinking reactions.

The conventional techniques for the purification of low-molecular-weight compounds, such as distillation, sublimation and crystallization, are not applicable to polymers. In some cases, it is possible to remove the impurities by cold or hot extraction of the finely divided polymer with suitable solvents or by steam distillation. Separation of low-molecular-weight components from water-soluble polymers (e.g. poly(acrylic acid), poly(vinyl alcohol), polyacrylamide) can be accomplished by dialysis[1] or electrodialysis.[1] However, the most widely used method of purification is by reprecipitation in which the solution of polymer (concentration less then 5%) is dropped into 4- to 10-fold excess of precipitant, with stirring. If necessary, this operation is repeated with other solvent/precipitant pairs until the impurities are no longer detectable.

The drying of polymers often presents great difficulty, since many polymers tenaciously retain or trap solvent or precipitant; this phenomenon is termed "occlusion". The magnitude of this effect can be judged by the following examples. Cyclohexane is occluded so strongly by cellulose that, after drying for two days at 100°C and 0.1 torr, one cyclohexane molecule is retained for every six cellulose CRU's. A 0.2 mm thick polystyrene film, prepared from a solution in benzene or tetrahydrofuran by drying to constant weight in a stream of nitrogen at 75° C, still retains 1.7% benzene or 12.7% tetrahydrofuran respectively. There is no general rule for the prevention or avoidance of occlusion. In some cases, a change of solvent/precipitant system may help to achieve this goal. Raising the drying temperature is also beneficial.

An important pre-requisite for successful drying is to subdivide the polymer as finely as possible (see Section 2.4); freeze-drying[2–4] is of particular significance in this respect. In some cases a combination of spray precipitation (see Section 2.2.1) and freeze drying is to be recommended. For example, one can spray the polymer solution into a mortar, the bottom of which is covered with pieces of solid carbon dioxide the size of a hazel nut. The pieces are then ground more finely, the mortar placed in a desiccator and evacuated with an oil pump. The polymer solution can also be sprayed into a liquid cooled to low temperature, the liquid being

---

[1] *Houben-Weyl 1/1* (1958) 653.
[2] *Houben-Weyl 1/1* (1958) 939.
[3] *R.C. Schulz*, Chem.-Ing-Techn. *28* (1956) 296.
[4] *C. Duclairoir* and *J.C. Brial*, J. Appl. Polym. Sci. *20* (1976) 1371; (gives instructions about apparatus and examples of freeze-drying of urea-formaldehye resins).

immiscible with the solvent for the polymer, e.g. an aqueous solution into low-temperature ether. The polymer then precipitates in the form of a light flocky snow; decantation of the ether is followed by evacuation as described above.

2.2.3. Stabilization of polymers[1]

Even in the purest state, many polymers undergo chemical changes that affect their mechanical and physical properties. For example, changes can occur through autoxidation or the action of light, by hydrolysis or acidolysis, or by elimination of low-molecular-weight compounds (e.g. hydrogen chloride from poly(vinyl chloride)). Stabilization of polymers by use of suitable additives is, therefore, unavoidable, especially in commercial applications. Even for laboratory work, stabilization is sometimes essential, for example against autoxidation when handling polymeric dienes or polymeric olefins (viscosity measurements at high temperature). For this purpose, the addition of 0.1–0.5% *N*-phenyl-β-naphthylamine proves very effective. A compilation of the various stabilizers, often developed for specific purposes, may be found in the literature.[2] The blending of a stabilizer with a polymer can be achieved by dropping the polymer solution into a precipitant containing the stabilizer or by dispersing the finely divided polymer in an ethereal solution of stabilizer and then slowly evaporating the ether. For large amounts of polymer, blending is often best carried out on heated rollers.

2.3. CHARACTERIZATION OF POLYMERS[3,4]

For the unequivocal characterization of a low-molecular-weight compound it is sufficient to specify a few physical or chemical properties, for example boiling point, melting point, refractive index, and elemental analysis. If two low-molecular-weight samples have the same characteristic properties they may be considered as identical.

The characterization of macromolecular substances is considerably more difficult. Owing to the high intermolecular forces macromolecules are not

---

[1] For literature see Section 5.3.

[2] e.g. *Houben-Weyl 14/1* (1961) 441 and other Chapters.

[3] Also see *M. Hoffmann* and *P. Schneider in Houben-Weyl 14/2* (1963) 917; *M. Hoffmann* in *Houben-Weyl 14/2* (1963) 960.

[4] For determination of technological data see "Kunststoff-Handbuch". Vol. I. Grundlagen (Eds. R. Vieweg and D. Braun), Carl Hanser Verlag, Munich, Vienna 1975.

volatile without decomposition so that no boiling point can be determined. The melting points of partially crystalline polymers are generally not sharp. Amorphous polymers frequently show only sintering or softening, often accompanied by decomposition. In addition to elemental analysis other data must, therefore, be determined, for example, solubility, viscosity of the solution, mean molecular weight, molecular weight distribution, and degree of crystallinity.

The fundamental difficulty remains that polymeric substances cannot be obtained in a structurally and molecularly uniform state, unlike low-molecular-weight compounds.[1] Thus macromolecular materials of the same analytical composition may differ not only in their structure and configuration (see Section 1.2) but also in molecular size and molecular weight distribution; they are polymolecular, i.e. they consist of mixtures of molecules of different size. Hence, it is understandable that the expression "identical" is not, in practice, applicable to macromolecules. Up to the present time there is no possibility of preparing macromolecules of absolutely uniform structure and size.[2] It follows, therefore, that physical measurements on polymers can only yield average values. The aforementioned peculiarities of macromolecular substances mean that the methods of characterization suitable for low-molecular-weight compounds are frequently not applicable or only applicable in very modified form; completely new methods of investigation must often be employed.

Since the properties of a polymer can be noticeably influenced by small variations in the molecular structure, and these in turn depend on the preparation conditions, it is necessary when reporting data to indicate not only the type of measurement (e.g. molecular weight by end group analysis; crystallinity by infrared measurement; etc.) but also the type of preparation (e.g. radical polymerization in the bulk at 80°C; polymerization with a particular organometallic mixed catalyst at 20°C).

2.3.1. Solvents and solubility[3]

When studying a polymer, one should first determine its solubility. This is very characteristic for polymeric compounds and can serve as a means of characterization, for example in the recognition of crosslinking (see

[1] *W. Kern*, Angew. Chew. *71* (1959) 585.

[2] Even polymers with narrow molecular weight distribution, prepared under special conditions (see Section 3.2.1), possess a certain non-uniformity. In contrast, nature has the capacity to generate macromolecules of uniform structure and molecular size.

[3] See Gnamm: Lösungsmittel und Weichmachungsmittel". 8th Edn., Wissenschaftliche Verlagsanstalt, Stuttgart, in preparation.

Example 3–50 and 4–06), for separating and distinguishing between tactic and atactic macromolecules (see Example 3–28), or for the characterization of copolymers (see Example 3–43). Moreover, solubility is a prerequisite for most physical measurements.

When determining the solubility it has to be remembered that macromolecular compounds show extremes of behaviour; they are frequently either infinitely soluble or practically insoluble or only swellable to a limited extent; saturated solutions in contact with unswollen solid phase, such as normally observed with low-molecular-weight compounds, are a very rare occurrence. The "goodness" of a solvent can, therefore, not be indicated in terms of an equilibrium constant for equilibrium between dissolved and solid forms. It is much better expressed in terms of the amount of a certain precipitant that must be added to a solution of polymer before precipitation commences. A more exact measure is provided by a comparison of the second virial coefficients of osmotic pressure of the corresponding solutions[1], or a comparison of viscosity numbers in different solvents (see Section 2.3.2.1).

Swelling in solvents is a typical feature of macromolecules that exceed a certain molecular weight. One aspect of this is that macromolecular compounds can take up large amounts of solvent, forming a gel, with marked increase of volume (see Section 1.4.1). If this process comes to a halt before a homogeneous solution has been formed, one speaks of limited swelling; unlimited swelling is synonymous with solution. The extent of swelling depends on the chemical nature of the polymer, the molecular weight, the swelling medium, and the temperature. For crosslinked polymers, which are of course insoluble, it is a measure of the degree of crosslinking.

Although there are many thermodynamic theories for the description of polymer solutions[2], one must still rely in many cases on empirical rules and arguments by analogy.[3] As a guide, some solvents and non-solvents are indicated in Table 2.2. However, not every combination of solvent and non-solvent yields a useable system[4] for the reprecipitation of a polymer; this must be ascertained by preliminary experiments. Suitable solvent/precipitant combinations for a number of polymers may be found as specific examples in Chapters 3, 4 and 5.

---

[1] *G.V. Schulz*, Angew. Chem. *64* (1952) 553.

[2] *A Münster*, "Löslichkeit und Quellung" in *H. Stuart* (Ed.), "Physik der Hochpolymeren" *2* (1953) 193; *P.J. Flory*, Disc. Faraday Soc. *49* (1970) 7; *J. Biros*, *L. Zeman* and *D. Patterson*, Macromolecules *4* (1971) 30.

[3] *O. Fuchs*, Kunststoffe *43* (1953) 409; Makromol. Chem. *18/19* (1956) 166.

[4] Solvents and non-solvents must be miscible with one another in the range used.

TABLE 2.2

Solubility of various polymers

Polymer	Solvents	Non-solvents
polyethylene, poly(1-butene), isotactic polypropene	*p*-xylene[a], trichlorobenzene[a] decane[a], decalin[a]	acetone, diethyl ether, lower alcohols
atactic polypropene	hydrocarbons, pentyl acetate	ethyl acetate, propanol
polyisobutene	hexane, benzene, carbon tetrachloride, tetrahydrofuran	acetone, methanol, methyl acetate
polybutadiene, polyisoprene	aliphatic and aromatic hydrocarbons	acetone, diethyl ether, lower alcohols
polystyrene	benzene, toluene, chloroform, cyclohexanone, butyl acetate, carbon disulfide	lower alcohols, diethyl ether, acetone
poly(vinyl chloride)	tetrahydrofuran, cyclohexanone, ethyl methyl ketone, dimethylformamide	methanol, acetone, heptane
poly(vinyl fluoride)	cyclohexanone, dimethylformamide	aliphatic hydrocarbons, methanol
polytetrafluoroethylene	insoluble	—
poly(vinyl acetate)	benzene, chloroform, methanol, acetone, butyl acetate	diethyl ether, petroleum ether, butanol
poly(isobutyl vinyl ether)	2-propanol, ethyl methyl ketone, chloroform, aromatic hydrocarbons	methanol, acetone
poly(methyl vinyl ketone)	acetone, 1,4-dioxane, chloroform	water, aliphatic hydrocarbons
polyacrylates and polymethacrylates	chloroform, acetone, ethyl acetate, tetrahydrofuran, toluene	methanol, diethyl ether, petroleum ether
polyacrylonitrile	dimethylformamide, dimethyl sulfoxide, conc. sulfuric acid	ethanol, diethyl ether, water, hydrocarbons
polyacrylamide	water	methanol, acetone
poly(acrylic acid)	water, dil. alkali, methanol, 1,4-dioxane, dimethylformamide	hydrocarbons, methyl acetate, acetone
poly(vinyl sulfonic acid)	water, methanol, dimethyl sulfoxide	hydrocarbons, acetone
poly(vinyl alcohol)	water, dimethylformamide[b], dimethyl sulfoxide[b]	hydrocarbons, methanol, acetone, diethyl ether
starch	water, chloral hydrate, copper (II) ethylene diamine	acetone, methanol
cellulose	aqueous tetraamminecopper (II) hydroxide, aqueous zinc chloride, aqueous calcium thiocyanate	methanol, acetone

TABLE 2.2 Continued

Polymer	Solvents	Non-solvents
cellulose triacetate	acetone, chloroform, 1,4-dioxane	methanol, diethyl ether
cellulose trimethyl ether	chloroform, benzene	ethanol, diethyl ether, petroleum ether
carboxymethyl cellulose	water	methanol
aliphatic polyesters	chloroform, formic acid, benzene	methanol, diethyl ether, aliphatic hydrocarbons
poly(ethylene terephthalate)	*m*-cresol, 2-chlorophenol, nitrobenzene, trichloroacetic acid	methanol, acetone, aliphatic hydrocarbons
polyamides	formic acid, conc. sulfuric acid, dimethylformamide, *m*-cresol	methanol, diethyl ether, hydrocarbons
polyurethanes	formic acid, γ-butyrolactone, dimethylformamide, *m*-cresol	methanol, diethyl ether, hydrocarbons
polyoxymethylene	γ-butyrolactone[b], dimethylformamide[b] benzyl alcohol[b]	methanol, diethyl ether, aliphatic hydrocarbons
poly(ethylene oxide)	water, benzene, dimethylformamide	aliphatic hydrocarbons, diethyl ether
poly(tetrahydrofuran)	benzene, methylene chloride, tetrahydrofuran	aliphatic hydrocarbons, diethyl ether
polydimethylsiloxane	chloroform, heptane, benzene, diethyl ether	methanol, ethanol

[a] Often only soluble at high temperature.
[b] Only on heating.

For the investigation of the solubility of a polymer one may proceed as follows. 30–50 mg samples of finely divided polymer are placed in small test tubes with 1 ml solvent and allowed to stand for several hours. From time to time the contents are stirred or shaken, and examined for the appearance of streaks. The solution process is significantly influenced by the state of sub-division of the polymer. If no solution occurs after several hours at room temperature one can slowly raise the temperature, if necessary to the boiling point of the solvent. Any coloration or gas formation is indicative of decomposition of the polymer (see Section 2.2.3 concerning stabilizers). If the polymer dissolves at higher temperature, which may require a long time, the solution should be allowed to cool slowly to check whether the polymer comes out of solution again and if so at what temperature (important for subsequent measurements). If the polymer simply swells, without going into solution, the procedure is repeated with other solvents or solvent mixtures. If it swells in all solvents, without dissolving, one may assume that it is crosslinked.

Considering the rather complicated processes that take place during dissolution it is not to be wondered that some systems show peculiar behaviour. For example while solubility generally increases with temperature there are also polymers which exhibit a negative temperature coefficient of solubility in certain solvents. Thus, poly(ethylene oxide) dissolves in water at room temperature and precipitates again on warming. Solutions of poly(methyl vinyl ether) in water show the same behaviour.[1]

Surprising effects are also often observed when using solvent mixtures.[2] There are cases where mixtures of non-solvent act as a solvent; conversely a mixture of two solvents may behave as a non-solvent. For example polyacrylonitrile is completely insoluble in both nitromethane and water but dissolves in a mixture of the two. Similar behaviour is shown by the systems polystyrene/acetone/heptane and poly(vinyl chloride)/acetone/carbon disulfide. Examples of systems in which the polymer dissolves in each solvent separately but not in the mixture are provided by polyacrylonitrile/malononitrile/dimethylformamide and poly(vinyl acetate)/formamide/acetophenone. These peculiarities are especially to be taken into account if one is adjusting certain solution properties (e.g. for fractionations) by adding one solvent to another.

Finally should be mentioned the phenomenon of incompatibility of mixtures of polymers in solution[3], exhibited by practically all polymers. When two polymers are separately dissolved in the same solvent and the two homogeneous solutions mixed, phase separation occurs. For example, if 10% solutions of polystyrene in benzene and of poly(vinyl acetate) in benzene are mixed a phase separation occurs in spite of the use of the same solvent. This at first manifests itself as an appreciable turbidity, leading finally to the formation of two separate phases. One phase contains mainly the first polymer, the other phase mainly the second polymer, but in both phases there is still present a certain proportion of the other polymer.

This limited compatibility of polymer mixtures can be explained thermodynamically and depends on various factors[3,4], such as the structure of the macromolecule, the molecular weight, the mixing ratio, the overall polymer concentration, and the temperature. Quite a small structural difference often suffices to bring about incompatibility of macromolecules. For

---

[1] For further examples see *O. Fuchs*. Farbe Lack *71* (1965) 104.
[2] *B.A. Wolf*, Adv. Polym. Sci. *10* (1972) 109; Makromol. Chem. *177* (1976) 1073.
[3] *O. Fuchs*, Makromol. Chem. *90* (1966) 293; Angew. Makromol. Chem. *1* (1967) 29; ibid. *6* (1969) 79.
[4] *R. Casper* and *L. Morbitzer*, Angew. Makromol. Chem. *58/59* (1977) 1.

example polystyrene is incompatible with poly (α-methylstyrene), and poly(methyl acrylate) with poly(ethyl acrylate). Copolymers made from the same monomers, and differing in composition by quite small amounts, are often no longer compatible.[1] The phenomenon of incompatibility of polymers is even more pronounced in the solid state.

2.3.2. Determination of molecular weight of polymers

Knowledge of the molecular weight of a polymer is indispensable for a number of investigations (e.g. for kinetic, physical and technological measurements). Conventional physical methods such as boiling-point elevation and freezing-point depression fail above a molecular weight of about 20 000 even if the precision is substantially improved.[2] Therefore, a number of new physical methods has been developed which can be divided into absolute and relative methods. A necessary pre-requisite for their application is that the polymer is soluble in a suitable solvent.

Absolute methods yield directly the molecular weight and degree of polymerization in which the calculation from the experimental data requires only universal constants such as the gas constant and the Avogadro constant, apart from readily determinable physical properties such as density, refractive index, etc. The most important methods[3] in use today are:

— osmotic pressure method,
— ultracentrifuge method (determination of sedimentation and diffusion constants),
— light scattering method (determination of the intensity of the light scattered from the Tyndall beam as function of angle of observation, at a particular wavelength).

These methods require considerable outlay in apparatus and experimental effort so that they are not available in every laboratory. Therefore, chemical methods (determination of end groups), despite certain limitations, are important because of their relative simplicity (see Section 2.3.2.2).

Relative methods measure some property that depends clearly on molecular weight, for example the precipitability or the viscosity of the polymer solution. However, such measurements can only be evaluated if

---

[1] *R. Casper* and *L. Morbitzer*, Angew. Makromol. Chem. *58/59* (1977) 1.
[2] *G.V. Schulz* and *K.G. Schön*, Z. Phys. Chem. (Frankfurt/Main) *2* (1954) 197.
[3] See *Houben-Weyl 3/1* (1955) 371; *H.A. Stuart* (Ed.). "Physik der Hochpolymeren" *2* (1953) 373.

an experimental calibration curve has first been established by comparison with an absolute method. Amongst the relative methods, viscosity measurements, first introduced by Staudinger[1], deserve special mention on account of their general applicability and simple execution. They will, therefore, be discussed in detail separately in Section 2.3.2.1.

Synthetic macromolecular substances consist practically always of mixtures of macromolecules of different molecular weights and are said to be polydisperse, so that only an average value of the molecular weight can be determined. Because of the polydispersity it is necessary to distinguish between those methods which give the size of the macromolecules and so yield a weight-average molecular weight, and those which depend on the number of macromolecules and so yield a number-average molecular weight. The weight-average molecular weight, $\bar{M}_w$, is given by light scattering, ultracentrifuge and viscosity measurements[2], the number-average molecular weight, $\bar{M}_n$, by osmotic measurements or determination of end groups. They are defined by the following equations (N_i = amount in mol corresponding to molecules with molecular weight M_i):

$$\bar{M}_n = \frac{\Sigma N_i M_i}{\Sigma N_i}$$

$$\bar{M}_w = \frac{\Sigma N_i M_i^2}{\Sigma N_i M_i}$$

For molecularly uniform (monodisperse) substances $\bar{M}_w = \bar{M}_n$; for polydisperse substances $\bar{M}_w$ is always larger than $\bar{M}_n$. By comparison of the weight-average with the number-average molecular weight one therefore obtains a measure of the polydispersity[3] of a particular polymer:

$$U = \frac{\bar{M}_w}{\bar{M}_n} \quad (U \text{ is from the German Uneinheitlichkeit}) \qquad (5)$$

The quotient U is generally larger than 2. With special initiators and under special experimental conditions, nearly monodisperse polymers (U = 1.1 to 1.3) can sometimes be obtained (see Section 3.2.1).

[1] *H. Staudinger*, "Die hochmolekularen organischen Verbindungen", Springer-Verlag, Berlin 1932, p. 52.

[2] For a finer distinction between weight average, sedimentation average, and viscosity average of the degree of polymerization see *H.A. Stuart* (Ed.) "Physik der Hochpolymeren" *2* (1953) 280; *G. Meyerhoff*, Adv. Polym. Sci. *3* (1961) 59, *Houben-Weyl 3/1* (1955) 445; *H.G. Elias*, *R. Bareiss* and *J.G. Watterson*, Adv. Polym. Sci. *11* (1973) 111.

[3] Occasionally the expression $U' = (\bar{M}_w/\bar{M}_n) - 1$ is used to characterize the polydispersity ($U' = U - 1$).

When conducting molecular weight determinations in solution one must ensure that the dissolved particles are present as separate macromolecules and not in the form of associates of macromolecules. Proof of this can be obtained by carrying out reactions on functional groups of the polymer that do not lead to cleavage of the polymer chains. If the degree of polymerization of the original polymer agrees with that of the converted polymer, then association can be excluded. Values of molecular weight determined in different solvents should also be in agreement if association is absent. For example micellar solutions of soaps in water can be clearly distinguished from molecular solutions of the same soaps in other solvents by this method.

2.3.2.1. *Determination of solution viscosity of polymers*[1]

The viscosimetric method of molecular weight determination, introduced by Staudinger, is based on the observation that thread-like molecules, even in relatively low concentration, considerably raise the viscosity of the solvent in which they are dissolved, to an extent which is greater, the higher the molecular weight. This method is applicable only to linear and slightly branched molecules; it fails with sphere-like or strongly branched molecules (globular proteins, glycogens).

Since the molecular weight determination depends not on the absolute viscosity but on the relative increase of viscosity, the measurements are expressed in the form of the specific viscosity (η_{sp}):

$$\eta_{sp} = \frac{\eta - \eta_0}{\eta_0} \tag{6}$$

where η is the viscosity of the solution, and η_0 that of the pure solvent.

If the measurements are made in a capillary viscometer of specified dimensions (see Figure 2.15) and at low concentrations (so that the density of the solution is approximately the same as that of the solvent), then the viscosities η and η_0 can be replaced to a good approximation by the flow times t (solution) and t_0 (solvent). Equation (6) then becomes:

$$\eta_{sp} = \frac{t - t_0}{t_0} \tag{7}$$

[1] For detailed treatments see *Houben-Weyl 3/1* (1955) 431: *H.A. Stuart* (Ed.) "Physik der Hochpolymeren" *2* (1953) 280.

If this value is divided by the concentration c of polymer in solution, one obtains the reduced specific viscosity η_{sp}/c. However, this quantity is concentration-dependent and it is more exact to use the limiting value (intrinsic viscosity or limiting viscosity number)[1,2], $[\eta]$, as a measure of the viscosimetric behaviour of the thread-like molecule, as defined by the following expression:

$$[\eta] = \lim_{c \to o} (\eta_{sp}/c) \tag{8}$$

Since η_{sp} is dimensionless, $[\eta]$ has units of reciprocal concentration (e.g. $l\,g^{-1}$ or $dl\,g^{-1}$); more recently the unit $cm^3\,g^{-1}$ has also come into use. Hence in viscosity measurements the concentration units must always be stated.

Measurements cannot be made at infinite dilution so it is necessary to work down to as low a concentration as possible and extrapolate to zero concentration. For this purpose the η_{sp}/c values are determined at various concentrations (10, 5, 2.5, 1.25 $g\,l^{-1}$) and plotted against concentration. Extrapolation to zero concentration yields $[\eta]$ as the intercept on the ordinate. For polymers with low or moderate molecular weight one generally obtains straight lines, while for very high molecular weight samples the curves are frequently bent upwards (see Figure 2.13). In the latter case additional measurements should be carried out at lower concentrations (e.g. 0.6 $g\,l^{-1}$) to facilitate extrapolation from the more linear part of the curve.

Quite different viscosity behaviour is shown by solutions of polyelectrolytes[3] in polar solvents (e.g. polymeric acids in water, see Examples 3–10 and 5–03). The η_{sp}/c-values at first fall off with decreasing concentration as for uncharged polymers but then climb steeply again (see Figure 2.13, curve a). Addition of neutral salt to the solution of polyelectrolyte (e.g. 5% sodium chloride in aqueous solutions) restores the normal behaviour (Figure 2.13, curve b).

This is connected with the fact that in polyelectrolytes the shape and density of the macromolecular coils is affected by the degree of ionization. In the ionized state the like charges distributed along the length of a

---

[1] In the English literature the following are also frequently used: relative viscosity $\eta_{rel} = t/t_0$; inherent viscosity $\eta_{inh} = (\ln \eta_{rel})/c$.

[2] In industry the *Fikentscher* K-value is still widely used for characterizing polymers: this value is empirically related to the molecular weight; see *Houben-Weyl 14/1* (1961) 83. For the conversion of K-values into relative viscosities see Tables in *P.E. Hinkamp*, Polymer *8* (1967) 381.

[3] *W. Kern,* Z. Phys. Chem., A *181* (1938) 249, 283; *184* (1939) 197, 302; also see *H.A. Stuart* (Ed.), "Physik der Hochpolymeren" *2* (1953) 321, 680, 695.

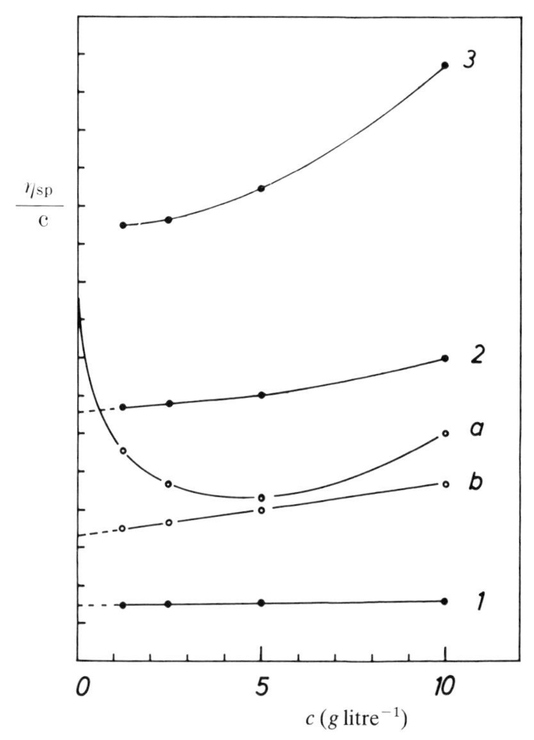

FIGURE 2.13 Graphical evaluation of the limiting viscosity number (intrinsic viscosity) from viscosity measurements at different concentrations. Curves 1, 2 and 3: same polymer and solvent, but increasing molecular weight. Curves a and b:polyelectrolyte in water (a) before and (b) after addition of a low-molecular-weight electrolyte.

macromolecule repel each other, leading to a marked coil expansion and hence a considerable increase in viscosity. Every factor that causes an increase in degree of dissociation, therefore, leads to a rise in the solution viscosity, and vice versa. The viscosity behaviour of aqueous solutions of polymeric acids (e.g. poly(methacrylic acid), see Example 3–10) of various concentrations can then be explained as follows. On dilution of the aqueous solution the normal effect is first observed, i.e. the viscosity decreases. With further dilution the increasing degree of dissociation of the carboxylic groups begins to make itself felt. Finally the effect caused by the higher degree of dissociation outstrips that resulting from dilution and η_{sp}/c rises again. On addition of sodium chloride the degree of ionization is essentially held steady, and therefore also the coil expansion; the rise in viscosity with decreasing concentration of polymer is thus suppressed.

There are a number of empirical equations that permit the limiting viscosity number $[\eta]$ to be calculated from a single viscosity measurement, for example the equation of Schulz and Blaschke[1]:

$$[\eta] = \frac{\eta_{sp}/c}{1 + K_\eta \eta_{sp}} \tag{9}$$

For low concentrations of polymers K_η frequently has the value of 0.28, independent of the solvent and the dissolved polymer. This equation is not always applicable so that for a new polymer it is always advisable to check the agreement between the $[\eta]$-value calculated in this way and that found graphically.

So far it has not been possible to give a satisfactory theoretical interpretation of the concentration dependence of the solution viscosity of macromolecules. Since the limiting viscosity number depends not only on the size of the macromolecule but also on its shape and on the solvent, there is no simple relationship for the direct calculation of molecular weight from viscosity measurements. One is, therefore, always obliged to establish for each polymer a calibration curve or calibration function[2] by comparison with an absolute method. This, however, is only valid for a given solvent and temperature. The original Staudinger equation

$$[\eta] = K_m M \tag{10}$$

is only valid in exceptional cases. More generally applicable is the equation first proposed by W. Kuhn:

$$[\eta] = K_m M^\alpha \tag{11}$$

Mathematical evaluation of M by this equation is somewhat inconvenient and a graphical method is preferable. Equation (11) can be expressed in logarithmic form

$$\log [\eta] = \log K_m + \alpha \log M \tag{12}$$

so that a double logarithmic plot of $[\eta]$ against M gives a straight line whose slope corresponds to the exponent α (see Figure 2.14). The constant K_m can then be calculated from equation (12). The exponent α in equation (11) depends on the form of the macromolecule in solution. For rigid spheres $\alpha = 0$; however most macromolecules are present in solution as more or less strongly expanded coils to which solvent molecules are

[1] *G.V. Schulz* and *F. Blaschke*, J. Prakt. Chem. *158* (1941) 130.

[2] One must take into account the type of molecular weight average used for the calibration. See *G. Meyerhoff*, Adv. Polym. Sci. *3* (1961) 59.

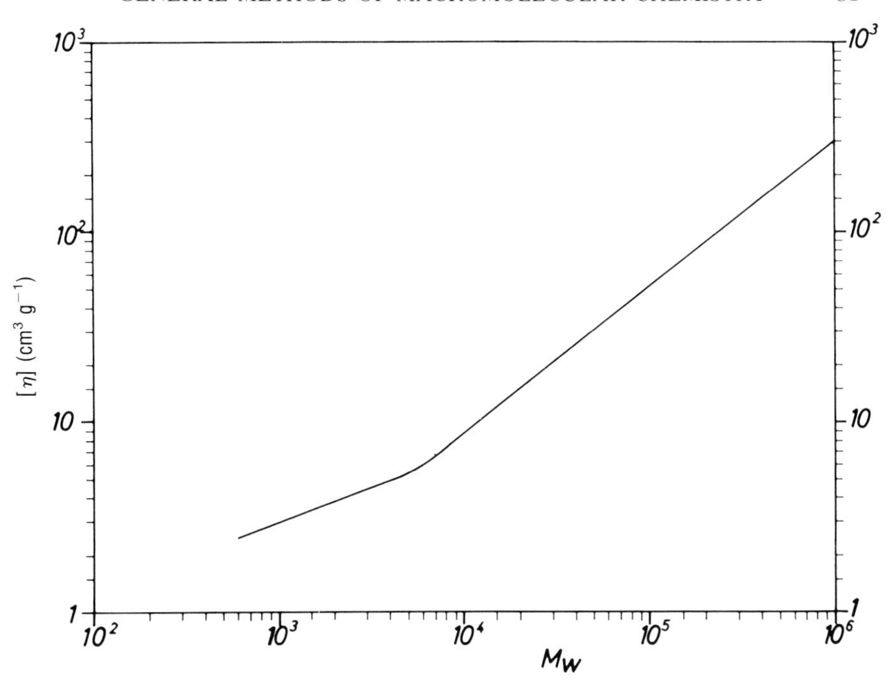

FIGURE 2.14 Relation between viscosity and molecular weight for polystyrene in benzene at 20°C, according to Meyerhoff.[1]

bound by solvation forces. Accordingly for most polymers[2] α-values lie between 0.5 and 1.0, 0.5 being the extreme value for non-draining coils (for definition see Section 1.4.4), and 1.0 for fully draining coils. Cases are also known where α is greater than 1. This occurs with particularly stiff and elongated macromolecules, which approximate to the model of a rigid rod in solution, for which $\alpha = 2$.

Since the degree of expansion of the polymer coils is directly dependent on the solvating power of the solvent, under otherwise comparable conditions, both α and $[\eta]$ provide a measure of the "goodness" of a solvent (see Section 2.3.1): high values of α and $[\eta]$ (at constant molecular weight and temperature) indicate marked coil expansion and therefore a good solvent; low values of α and $[\eta]$ indicate a bad solvent. For example the values of α for poly(vinyl acetate) in methanol and acetone are 0.60 and 0.72, respectively.

---

[1] *G. Meyerhoff*, Z. Phys. Chem. (Frankfurt/Main) *4* (1955) 335.

[2] An exception is the very highly branched and therefore near-spherical glycogen, for which $\alpha = 0$ in aqueous solution.

The interactions between solvent and polymer depend not only on the nature of the polymer and type of solvent but also on the temperature. Increasing temperature usually favours solvation of the macromolecule by the solvent (the coil expands further and α becomes larger), while with decreasing temperature the association of like species, i.e. between segments of the polymer chains and between solvent molecules, is preferred. In principle, for a given polymer there is for every solvent a temperature at which the two sets of forces (solvation and association) are equally strong; this is designated the theta-temperature. At this temperature the polymer exists in solution in the form of a non-draining coil, i.e. the exponent α in equation (11) has the value 0.5. This situation is found for numerous polymers; e.g. the theta-temperature is 34°C for polystyrene in cyclohexane, and 24°C for polyisobutene in benzene.

K_m- and α-values for some polymer/solvent systems are indicated in Table 2.3. More comprehensive collections can be found in the literature.[1,2,3]

The following apparatus is needed for carrying out viscosity measurements: a capillary viscometer with suitable mounting, a thermostatted bath, a stopwatch (0.1 s), several graduated 10 ml flasks, and graduated 5 ml and 3 ml pipettes. For the reasons already given (Section 1.4.1) the measurements are performed only on dilute solutions.

The most commonly used capillary viscometer is the Ostwald viscometer[4] (Figure 2.15). Care must be taken in its construction to ensure that the connections to capillary 5 are funnel-shaped with smooth walls (see Figure 2.15). The mark M2 should be just above the entrance to the capillary so that the liquid meniscus can be very clearly seen. The diameter of the capillary (in general between 0.3 and 0.4 mm) is chosen so that the flow time for the solvent is about 60–150 s. The flow times for a number of solvents at 20°C are indicated in Table 2.4; they must of course be determined experimentally for each viscometer. Special versions of the Ostwald viscometer have been developed for measurement of solution viscosity at higher temperatures.[5]

Since the viscosity of a solution depends very much on temperature, good thermostatting is necessary (to within 0.05 to 0.1°C). Suitable thermostats (for temperatures up to 250°C) are commercially available[6];

---

[1] *H.A. Stuart* (Ed.), "Physik der Hochpolymeren" *2* (1953) 304.
[2] *G. Meyerhoff*, Adv. Polym. Sci. *3* (1961) 59.
[3] "Polymer Handbook", Section IV-I.
[4] The *Ubbelohde* viscometer is a capillary viscometer with suspended level (see *Houben-Weyl 3/1* (1955) 435). Supplier: Schott and Gen., Mainz, W. Germany.
[5] *L.L. Böhm*, G-i-T-Fachzeitschrift für das Laboratorium *20* (1976) 879.
[6] E.g. from Firma Messgeräte-Werk Lauda, Lauda/Tauber, W. Germany.

TABLE 2.3

K_m- and α-values for the calculation of molecular weights from viscosity measurements according to equation (12). $[\eta]$ in $cm^3 g^{-1}$.

Polymer	Solvent	Temp. (°C)	$10^3 K_m$	α	Lit.
polystyrene	benzene	20	12.3	0.72	1
poly(α-methylstyrene)	toluene	25	7.81	0.73	2
polyisobutene	cyclohexane	24	107	0.50	3
polybutadiene	cyclohexane	20	36.0	0.70	4
polyisoprene	toluene	25	50.2	0.67	5
poly(vinyl acetate)	acetone	30	10.2	0.72	6
poly(vinyl alcohol)	water	25	300	0.50	7
poly(methyl methacrylate)	acetone	25	9.6	0.69	8
polyacrylonitrile	dimethylformamide	25	23.3	0.75	9
polyacrylamide	water	25	6.31	0.80	10
poly(ethylene terephthalate)	phenol/tetra-chloroethane (1/1)	25	21.0	0.82	11
polycarbonate from bisphenol A	tetrahydrofuran	20	39.9	0.70	12
Nylon-6,6	2M KCl in 90% HCOOH	25	142	0.56	13

[1] *G. Meyerhoff*, Z. Phys. Chem. (Frankfurt/M.) *4* (1955) 335.
[2] *H.W. McCormick*, J. Polym. Sci. *41* (1959) 327.
[3] *W.R. Krigbaum* and P.J. Flory, J. Polym. Sci. *11* (1953) 37.
[4] *P. Ribeyrolles, A. Guyot* and *H. Benoit*, J. Chim. Phys. *56* (1959) 377.
[5] *W.C. Carter, R.L. Scott* and *M. Magat*, J. Am. Chem. Soc. *68* (1946) 1480.
[6] *M. Matsumoto* and *Y. Ohyanagi*, J. Polym. Sci. *46* (1960) 441.
[7] *K. Dialer, K. Vogler* and *F. Patat*, Helv. Chim. Acta *35* (1952) 869.
[8] *S.N. Chinai, J.D. Matlack, A.L. Resnick* and *R.J. Samuels*, J. Polym. Sci. *17* (1955) 391.
[9] *R.L. Cleland* and *W.H. Stockmayer*, J. Polym. Sci. *17* (1955) 473.
[10] *W. Scholtan*, Makromol. Chem. *14* (1954) 169.
[11] *A. Conix*, Makromol. Chem. *26* (1958) 226.
[12] *G.V. Schulz* and *A. Horbach*, Makromol. Chem. *29* (1959) 93.
[13] *H.-G. Elias* and *R. Schumacher*, Makromol. Chem. *76* (1964) 23.

however, they are also readily constructed from ordinary laboratory equipment, especially for temperatures from 20 to 50°C. The arrangement sketched in Fig. 2.16 (a 10 litre beaker can be used as container) can give a control of at least ±0.1°C at room temperature by correct choice of distance of contact thermometer from the immersion heater and cooling coil and by proper control of the cooling rate. Correct mounting of the viscometer is also essential for accurate and reproducible measurements. The viscometer must be mounted vertically (control by means of a plumb-line) and free from vibration (stirrer motor should be mounted independently of the support for the viscometer; stirring should not be too vigorous).

The viscosity measurements are conducted as follows. 100 mg of well-dried polymer are weighed into a 10 ml graduated flask on an analytical

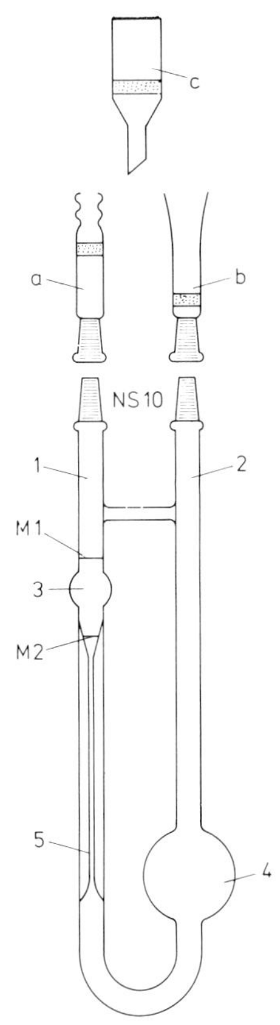

FIGURE 2.15 Ostwald viscometer.[1] Total length: 25 cm; capillary length: 10 cm; bulb 3: diameter 1.3 cm; bulb 4: diameter 2.2 cm; filling level: 3 ml; flow volume: 0.5 ml; a and b: mountings with sintered discs, porosity G1 or G2; c: sintered glass filter for filtration of solvent and polymer solution.

balance and dissolved in somewhat less than 10 ml solvent. After the solution has been brought to the temperature of measurement (suspend the flask in the viscometer bath), the solution is made up to the mark (polymer concentration $10\,g\,l^{-1}$). The polymer solution is now filtered through a sintered glass filter (Figure 2.15c) into a clean dry tube (no subsequent washing!) in order to remove dust particles, paper fibres etc.

[1] Standardized design according to *G.V. Schulz and H.J. Cantow*, Makromol. Chem. *13* (1954) 71, or *Houben-Weyl 3/1* (1955) 435. For automatic viscometers (supplier: Schott and Gen, Mainz), see *P. Höllbacher*, G-i-T-Fachzeitschrift für das Laboratorium *19* (1975) 302.

TABLE 2.4

Approximate flow times of some solvents in an Ostwald viscometer at 20°C (flow volume 0.5 ml).

Solvent	Capillary diameter (mm)	Flow time t_o (s)
carbon disulfide	0.3	58
chloroform	0.3	88
acetone	0.3	95
hexane	0.3	114
ethyl acetate	0.3	116
carbon tetrachloride	0.3	142
toluene	0.3	160
methanol	0.3	171
benzene	0.3	172
butyl acetate	0.4	62
pyridine	0.4	72
water	0.4	74
glacial acetic acid	0.4	86
1,4-dioxane	0.4	90
cyclohexane	0.4	92
formic acid	0.4	108
formamide	0.5	100
sulfuric acid	0.7	118
p-cresol	0.7	153

which might be present and which would otherwise seriously disturb the measurements. 3 ml of this solution (solution 1) are pipetted into arm 2 of the Ostwald viscometer which is suspended vertically in the bath as shown in Figure 2.16. After temperature equilibration (about 5 min at 20°C), the polymer solution is forced by means of pressure on the rubber bulb (Figure 2.16) from arm 2 into arm 1 until it is a little above mark M1. The pressure is now released by operation of the three-way tap (Figure 2.16), and the time required for the solution to flow from mark M1 to mark M2 is measured. The average of five measurements is taken as the flow time t. Depending on the total flow time, they should not deviate from one another by more than 0.2–0.4 s. The flow time of the solvent t_o is likewise determined with a filtered 3 ml sample; this determination should be carried out each time before beginning the measurements on the solutions since it provides a simple and accurate check of the entire set-up (temperature control, cleanliness of the viscometer etc.) The viscometer is now removed and the polymer solution poured out as completely as possible through arm 2. After attaching the head-pieces a and b, the viscometer is rinsed several times with pure solvent (application of slight vacuum at head-piece a) and then with purified acetone; it is finally dried

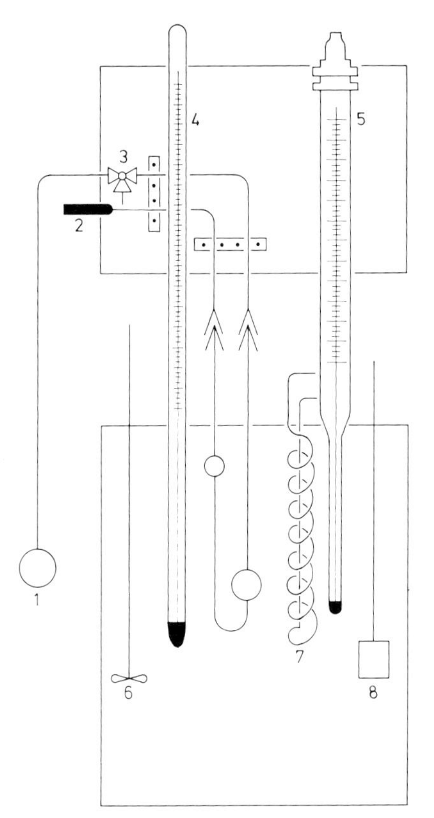

FIGURE 2.16 Thermostatted bath and mounting for Ostwald viscometer. 1: rubber bulb; 2: drying tube; 3: three-way tap; 4: thermometer (0.1°C); 5: contact thermometer; 6: stirrer; 7: cooling coil; 8: immersion heater.

by drawing air through the viscometer (the sintered glass filter b should be covered with a piece of filter paper). The viscometer is then ready for the next measurement. The sintered glass filter c is cleaned and dried similarly.

For the graphical determination of the limiting viscosity number $[\eta]$, the necessary measurements at lower concentrations are made as follows. 5 ml of solution 1 are pipetted into a graduated 10 ml flask and made up to 10 ml as described above (polymer concentration, $5\,g\,l^{-1}$) and its flow time determined after filtration (solution 2). After the viscometer and sintered glass filter c have been cleaned and dried the measurements are repeated at polymer concentrations of 2.5 and $1.25\,g\,l^{-1}$ obtained by successive dilution of solution 2. The series of concentrations can, of course, also be obtained by making separate weighings for each solution.

2.3.2.2 Determination of the end groups of polymers

If a macromolecular substance contains molecules with analytically identifiable end groups, its average molecular weight can be determined by chemical as well as by physical methods. The end groups of macromolecules formed by radical-initiated addition polymerization can be labelled by choosing a suitable initiator, such that the derived radical fragments become built into the polymer (see Chapter 3). In this case it is important to know the type of chain termination since this determines the number of labelled end groups per macromolecule (two for termination by combination, one for termination by disproportionation). Errors can occur if, for example, there is chain transfer to the monomer; this will reduce the number of labelled end groups per molecule, so that the method of end-group determination will lead to too high a molecular weight. Specific and very exact analytical methods must be applied to end-group determination since the groups to be estimated constitute only a small fraction of the macromolecule (less than 0.5%, depending on the molecular weight). Some of the suitable analytical methods are halogen analysis (e.g. by use of *p*-dibromobenzoyl peroxide[1]), ^{14}C analysis[2] (using appropriately labelled per-compounds[3] or azo-compounds[4]), or spectroscopic analysis (using azo-compounds with characteristic absorption bands[5]). The molecular weights of macromolecules made by polycondensation, or by addition polymerization involving two compounds, can also be obtained by end-group determination. In particular the hydroxyl and carboxyl end groups in polyesters can be estimated very precisely both by titration and by colorimetry.[6] Hydroxyl end groups (e.g. in polyoxymethylenes) can also be determined by acetylation or methylation.[7] The number-average molecular weight is calculated from the analytically determined end-group content according to the following relationship:

$$\bar{M}_n = 100\, zE/e \tag{13}$$

where E denotes the molecular weight of the end groups, z is their number per macromolecule, and e is the experimentally determined content of end groups expressed as a weight percentage.

---

[1] *W. Kern* and *H. Kämmerer*, Makromol. Chem. *2* (1948) 127.
[2] *J.C. Bevington*, Adv. Polym. Sci. *2* (1960) 1.
[3] *J.C. Bevington* et al., J. Polym. Sci. *20* (1956) 133; *22* (1956) 257.
[4] *J.C. Bevington*, Trans. Faraday Soc. *51* (1955) 1392.
[5] *H. Kämmerer, F. Rocaboy, K.-G. Steinfort* and *W. Kern*, Makromol. Chem. *53* (1962) 80.
[6] *W. Kern, R. Munk, S. Sabel* and *K.H. Schmidt* Makromol. Chem. *17* (1955/56) 201; *W. Kern, R. Munk* and *K.H. Schmidt*, Makromol Chem. *17* (1955/56) 219.
[7] *T.A Koch* and *P.E. Lindvig*, J. Appl. Polym. Sci. *1* (1959) 164.

2.3.3 Fractionation of polymers[1,2]

In the preparation of macromolecular substances one nearly always obtains, with few exceptions (see Example 3–27), mixtures that consist of macromolecules of various molecular weights. Many physical properties depend not only on the average molecular weight but also on the molecular weight distribution, while the latter gives valuable information concerning the reaction mechanism (e.g. type of chain termination, chain transfer, regulation). It is therefore important from many points of view to be able to determine the molecular weight distribution by fractionation. No separation procedure is yet known by which truly monodisperse fractions can be obtained from a polydisperse sample; but there are methods for separating fractions whose distribution about the molecular weight average is considerably sharper than in the original sample. These methods are generally based on the greater solubility of the low-molecular-weight components compared with the high-molecular-weight components. There are two types of method, namely fractional precipitation and fractional extraction.

To fractionate a sample by precipitation a precipitant is slowly added to the polymer solution (concentration of polymer 0.1–1 wt. %) at constant temperature until a persistent cloudiness appears. After some time the droplets separate out in the form of a second liquid (or swollen) phase. This fraction contains the highest molecular weight components and is separated by decantation or centrifugation. Further precipitant is added to the upper layer until further separation is observed, and so forth. Fractional precipitation can be conducted in a conical flask in a conventional thermostat; but a fractionating vessel, as shown in Figure 2.17, can be used with advantage. Here the precipitant is dropped into the stirred thermostatted solution from a burette until turbidity appears. The temperature is then raised until the solution clarifies and finally reduced again to the original temperature without stirring. The fraction separates out as an oily phase that can be run off through the tap. A disadvantage of fractional precipitation is that the residual solution becomes ever more dilute so that separation of the later fractions becomes progressively more difficult. Furthermore the method is rather time consuming since the formation of the gel phase occurs very slowly.

These disadvantages can be circumvented to some extent by means of the triangular fractionation technique of Meyerhoff.[3] In this, enough

[1] *M.J.R. Cantow* (Ed.), "Polymer Fractionation", Academic Press, New York, 1967.
[2] *L.H. Tung*, "Fractionation of Synthetic Polymers", Marcel Dekker, New York, 1977.
[3] *G. Meyerhoff*, Z. Elektrochem. *61* (1957) 325.

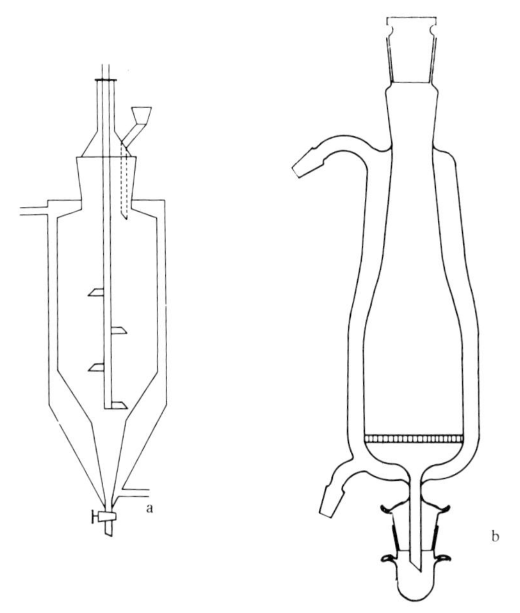

FIGURE 2.17 Thermostattable separating funnels for (a) fractional precipitation and (b) fractional extraction.

precipitant is added to the dilute polymer solution to bring about separation of approximately half the polymer. After separating the gel it is dissolved again so that one now has two solutions with which one proceeds as for the original solution, i.e. enough precipitant is added to bring about separation of approximately half the polymer in each case, and so on. In this way large volumes of solution can be avoided and there is a considerable saving in time, since several fractions can be worked up simultaneously.

In some cases the molecular weight distribution can be determined by turbidimetric titration[1,2], based on the principle of fractional precipitation. A precipitant is added to a very dilute solution of the polymer and the resulting turbidity is measured as a function of the amount of added precipitant; the preparative separation of the fractions is thereby avoided. If the polymer is chemically homogeneous, the mass distribution function can then be calculated. Turbidimetric titration is also suitable as a

[1] *J. Hengstenberg*, Z. Elektrochem. *60* (1956) 236.
[2] *J. Klein*, Angew. Makromol. Chem. *10* (1970) 21; 169.

means for establishing the best fractionation conditions (e.g. choice of solvent/precipitant combination, size of fractions etc.), before carrying out a full-scale fractionation by precipitation.

Fractional extraction is free from the disadvantages encountered in fractional precipitation. Here, the technique consists in extracting the polymer with a series of solvent/precipitant mixtures, the proportion of solvent being increased stepwise. Since one begins with the poorest solvent mixture—in contrast to fractional precipitation—the first fraction contains the low-molecular-weight components, and the final fraction the high-molecular-weight components. The physical state of the polymer is very important for the efficiency of fractional extraction. The polymer to be investigated can be prepared as a thin layer on aluminium foil (about 20–50 μm thick), following the method of Fuchs.[1] The clean foil (600–1000 cm^2 total area) is dipped into an approximately 10% solution of the polymer (0.5–1 g) in a readily volatile solvent, and is then removed and the solvent allowed to evaporate slowly. The amount of substance deposited should be about 100 mg per 100 cm^2 foil. This value can be lower but should not be greater by more than 50%. After the film has been completely dried (vacuum), the foil is weighed and cut up into strips. These are placed in a conical flask (250 ml), or better in a special fractionating vessel (Figure 2.17b) and treated successively with solvent/precipitant mixtures of different composition. Because of the low thickness of the polymer film (5–10 μm) the required equilibration time is only 5–10 min. After this the solution is decanted or drawn off from the foil and the dissolved polymer recovered by evaporation. In this way 10–20 fractions can be isolated from about 1 g of polymer in a relatively short time (1–2 days); see Example 3–16.

Especially sharp fractions can be obtained by column fractionation, which, in addition, permits continuous operation. In this procedure the polymer is deposited on an inert carrier (e.g. glass beads, sand or kieselguhr) which is then used to pack a vertical column. The column is now eluted with a suitable solvent/precipitant combination whose composition is changed continously, the eluted solution being collected in portions, preferably with a fraction collector. Column fractionation can be conducted either at constant temperature[2] or with simultaneous solvent and temperature gradients.[3] Fractionation at constant temperature is ex-

[1] *O. Fuchs*, Makromol. Chem. *5* (1951) 245; *7* (1951) 259.
[2] *V. Desreux*, Rec. Trav. Chim. Pays-Bas *68* (1949) 789.
[3] *C.A. Baker* and *R.J.P. Williams*, J. Chem. Soc. (1956) 2352; *G. Meyerhoff* and *J. Romatowski*, Makromol. Chem. *74* (1964) 222.

perimentally simpler since one can use a double-walled glass tube (Liebig condenser) as column and control the temperature by circulation of water from a thermostat.

Gel chromatography[1,2] is another method of fractionation, much used in recent years. The packing materials are usually crosslinked organic compounds which possess a pore structure that depends on the crosslink density and the preparation conditions. Inorganic packings, e.g. silica gel, can also sometimes be used. For the fractionation of water-soluble polymers, crosslinked dextran gels are available under the trade name Sephadex.[3] These swell in water and can be obtained with various crosslink densities. For polymers soluble in organic solvents, crosslinked polystyrenes or copolymers of methyl methacrylate and ethylene dimethacrylate can be used. The dry resin is first swollen to equilibrium in the solvent which is to be used for dissolving the polymer to be fractionated; the gel so obtained is then packed as uniformly as possible into the column. The polydisperse polymer is eluted in the same manner as for column chromatography but only solvent molecules and those macromolecules whose size is less than the prevailing pore size can diffuse into the pores of the swollen gel. Larger molecules cannot penetrate the pores of the gel completely and are therefore the first to leave the column on elution with solvent; accordingly the elution time increases with decreasing molecular size. The so-called exclusion limit gives an approximate indication of the limiting molecular weight up to which the macromolecules of the polymer to be fractionated can penetrate the network and therefore be separated; larger molecules cannot be separated from one another, since they cannot diffuse into the gel and are, therefore, not held back. Network structure and exclusion limit are closely related: the tighter the network, the smaller the exclusion limit.

The efficiency of fractionation by gel chromatography depends not only on the type of gel but also on the dimensions of the column. The internal volume V_i of the gel pores is determined by the amount of dry resin used

[1] Also called gel-permeation chromatography (GPC). See *H. Determann*, "Gelchromatographie", Springer-Verlag, Berlin, Heidelberg, New York 1967; *J. Seidl, J. Malinsky, K. Dusek* and *W. Heitz*, Adv. Polym. Sci. *5* (1967) 113; *M.J.R. Cantow* (Ed.), "Polymer Fractionation", Academic Press, New York 1967; *W. Heitz* and *W. Kern*, Angew. Makromol. Chem. *1* (1967) 150; *W. Heitz*, Angew. Chem. *82* (1970) 675; *L. Mandik*, Prog. in Org. Coat. *5* (1977) 131.

[2] Suitable automation for GPC analysis can be obtained from Waters Messtechnik GmbH, Herzog-Adolf-Strasse 4, 6240 Königstein, W. Germany, and from Du Pont Instrumente, Dieselstrasse 18, 6350 Bad Nauheim, W. Germany.

[3] Obtainable from Deutsche Pharmacia GmbH, Postfach 5480, D 7800 Freiburg, W. Germany.

and by its swellability, which in turn depends upon the eluting agent. The total volume of the gel bed V_t is thus made up of the volume of the gel framework, the internal volume V_i of the gel, and the external volume V_o between the gel particles. The external volume V_o is identical with the elution volume V_e of a substance with a molecular weight above the exclusion limit; macromolecules of this size cannot penetrate the network but pass through the column unimpeded. V_o can thus be readily determined.

Molecules, that are so small that not only the external volume V_o but also the total internal volume V_i is available to them, leave the column with an elution volume

$$V_e = V_o + V_i \tag{14}$$

For intermediate-sized macromolecules only a fraction K_d of the internal volume is accessible ($0 < K_d < 1$) and the value of V_e is then given by

$$V_e = V_o + K_d \; V_i \tag{15}$$

The constant K_d is the apparent distribution coefficient for the distribution of a substance between the swelling medium inside and outside the gel particles. K_d depends mainly on the molecular size and to a lesser extent on the shape of the molecule in solution.

The concentration of polymer in the eluate can be determined by precipitation, or spectroscopically, or by measurement of the refractive index or refractive index increment. The relationship between elution volume and molecular weight, which depends on the nature of the polymer, on the gel, and on other factors, can be determined by calibration with substances of known molecular weight. When the values of log M are plotted against the corresponding elution volumes, obtained under identical experimental conditions, a graph is obtained that is linear over a wide range and provides a calibration for the determination of molecular weight distribution (see Example 3–19).

Once the amounts and molecular weights of the fractions have been determined, the molecular weight distribution of a polydisperse material can be expressed graphically in the form of a distribution curve (see Example 3–16). The mass distribution function[1] is written as

$$m_p = f(P) \tag{16}$$

[1] For detailed theoretical derivations see *Houben-Weyl 3/1* (1955) 445; *H.A. Stuart* (Ed.), "Physik der Hochpolymeren" *2* (1953) 355, 726.

where m_p is the mass fraction of macromolecules with degree of polymerization P. Such an evaluation is illustrated below with results on the fractionation of a sample of polystyrene of average degree of polymerization 800 (Table 2.5).

TABLE 2.5

Fractionation of a polystyrene of average degree of polymerization 800.

Fraction No.	Amount in %	$I(P)$ in %	$10^4\,dm_p/dP$	P
8	3.4	1.7	1.10	169
7	3.7	5.25	3.45	363
6	7.3	10.75	4.35	433
5	16.8	22.8	6.95	680
4	24.9	43.7	7.55	900
3	9.9	61.6	5.30	1300
2	26.5	79.25	4.25	1470
1	7.5	96.25	1.40	2240

At first sight it might appear reasonable to present the fractionation results in the form of a histogram in which the amount in % (second column of Table 2.5) is plotted directly against the degree of polymerization P. However, this portrayal only gives a true picture of the molecular weight distribution if each individual fraction deviates from its molecular weight average by the same amount, i.e. if the breadths of the steps are all equal. Unfortunately this condition is rarely fulfilled in practice, as can be seen from the second column in Table 2.5. Thus fraction 2 is the largest in amount while fraction 3 is much smaller than fraction 4. The reason for this difference is not that the mass fraction of macromolecules with degree of polymerization 1470 is larger than that of macromolecules with degree of polymerization 1300, rather it arises from the fact that fraction 2 embraces a much wider range of degrees of polymerization than fraction 3. Therefore, it follows that the amount of the fraction should not be represented by a line (height of step), but by an area, i.e. that the mass distribution function (16) should be applied in integral form:

$$I(P) = \int_1^P f(P)\,dP \tag{17}$$

The integral distribution function is obtained from the experimental data (amount and molecular weight of individual fractions) in the following way. The mass distribution within a given fraction is approximately symmetrical. Designating the degree of polymerization of the mth fraction as P_m one can then assume that half the fraction has a smaller degree of polymerization than P_m, and the other half a larger value. One further assumes that fractions 1 to $(m-1)$ all have a degree of polymerization smaller then P_m. Thus, by summing the amounts of all fractions from 1 to $(m-1)$ and adding half the amount of the mth fraction one obtains the mass fraction of all degrees of polymerization from zero to P_m, and hence a pair of values of the integral function (17). For example, to obtain the value of $I(P)$ for fraction 5 in Table 2.5 one adds the percentage amounts of fractions 8, 7 and 6 $(3.4 + 3.7 + 7.3 = 14.4)$ and adds half the amount of fraction 5 $(14.4 + 8.4 = 22.8)$. The integral distribution curve (see Figure 2.18) is obtained by plotting the $I(P)$ values obtained in this way, against the corresponding degree of polymerization. This curve tells us, for example, that 10.75% of the macromolecules have degrees of polymerization up to 433, 43.7% up to degree of polymerization 900, and so on. Therefore, a polymer is more uniform with respect to molecular weight, the steeper the integral distribution curve.

The differential mass distribution function

$$dm_p = f(P)dP \tag{18}$$

in which dm_p is the mass fraction with degree of polymerization between P and $P + dP$, can be most simply obtained by graphical differentiation of the integral curve described by equation (17). For this purpose, the slopes of lines tangential to the integral curve are determined with the aid of a ruler at as many points of the curve as is feasible. This gives the values of

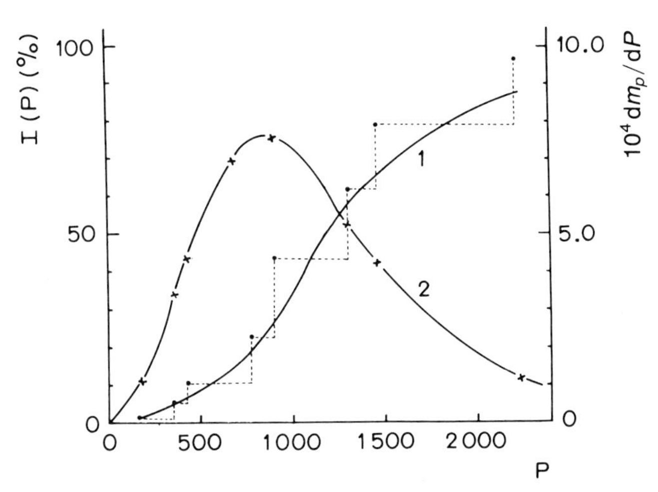

FIGURE 2.18 Integral (1) and differential (2) mass distribution function of the polystyrene of Table 2.5.

dm_p/dP listed in Table 2.5 and plotted in Figure 2.18. An inflection point in the integral curve corresponds to a maximum in the differential curve. This procedure gives information about the distribution that is more detailed, the sharper the fractions and the greater their number. This kind of differential distribution curve tells us how many macromolecules there are with a given degree of polymerization P in the polymer sample. It normally has a single maximum and resembles a Gaussian bell-shaped curve. Distribution curves with two or more maxima are an indication of side reactions during the preparation, or degradation reactions during fractionation.

With the exception of gel-permeation chromatography, the above-mentioned fractionation methods depend on differences in solubility. Fractionation according to molecular size alone can, therefore, only be expected if the macromolecules are chemically uniform. For polymers that are non-uniformly branched, or in which partial chemical conversions have taken place, or which are random, graft or block copolymers, the solubility depends on factors other than molecular weight. Fractionation can then lead to separation according to chemical composition[1,2] of the polymer. The results from a fractionation can, therefore, only lead to a reliable molecular weight distribution if there is convincing evidence for the chemical uniformity of the fractions (elementary analysis, i.r. analysis, pyrolytic gas chromatography). By changing the solvent or precipitant it is sometimes possible to choose between fractionation by molecular size and fractionation by chemical composition.

2.3.4. Determination of glass transition temperature, softening point, melting range and crystalline melting point

The thermal behaviour of polymeric substances differs markedly from that of low-molecular-weight compounds.[3] Highly crystalline polymers have moderately sharp melting points; poorly crystalline and amorphous polymers, on the other hand, melt over a wider temperature range. An important and, for polymeric substances, characteristic thermal quantity is the glass transition temperature.

---

[1] *O. Fuchs*, Makromol. Chem. *58* (1962) 65.
[2] *H.J. Cantow* and *O. Fuchs*, Makromol. Chem. *83* (1965) 244.
[3] Also see Section 1.4.4.

2.3.4.1. Determination of the glass transition temperature

Below a certain temperature any amorphous polymer behaves as a hard glass. When heated above this temperature individual segments of the macromolecules achieve greater mobility (micro-Brownian motion); as a result the polymer becomes soft and elastomeric.[1] The temperature at which this change sets in is called the glass transition temperature T_g. This temperature is very important in technological applications and depends, amongst other things, on the chemical nature of the polymer, on the configuration (e.g. *cis*- or *trans*-polymeric dienes), on the degree of crystallinity, on the length of the side chains and on the degree of branching. The glass transition temperature is generally determined with specialized equipment[2] and depends to some extent on the method used. One measures the temperature dependence of a particular physical quantity such as refractive index, elastic modulus (or torsion modulus), dielectric constant, heat capacity, expansion coefficient, or specific volume; all these change abruptly at the glass transition temperature. In amorphous polymers the glass transition temperature T_g frequently coincides with the softening point; in crystalline polymers, on the other hand, the crystalline melting point lies considerably above T_g. For crystalline polymers the glass transition temperature can be estimated rather well by rule of thumb (Boyer-Beaman rule): it is about two-thirds of the crystalline melting point, expressed in degrees Kelvin.[3]

In many cases both the glass transition temperature and some thermodynamic quantities can be determined by inverse gas chromatography. In this procedure the polymer is used as stationary phase and the interaction with given volatile probe molecules is studied as a function of temperature.[4]

2.3.4.2. Determination of softening point

The softening point is generally determined by slowly heating a test-piece under constant load until it experiences a certain deformation. The temperature at which this occurs is known as the softening point. Since the methods used involve empirical and arbitrarily chosen test parameters,

[1] See *H.A. Stuart* (Ed.), "Physik der Hochpolymeren" *3* (1955) 637.
[2] Also see Section 1.4.4.
[3] For compilation of glass transition temperatures, see *J. Brandrup* and *E.H. Immergut*, "Polymer Handbook", Section III-139.
[4] For summary see *J.M. Braun* and *J.E. Guillet*, Adv. Polym. Sci. *21* (1976) 107.

the softening point is physically less well defined than the glass transition temperature. In practice the softening point is generally determined by one of three relatively simple methods.

The most widely used method is that of Vicat in which a blunt steel needle (area of point 1 mm^2) is applied vertically to the surface of a test-piece under a load of 49 N. The oven temperature is then raised at 50°C h^{-1} and the temperature determined at which the needle has sunk 1mm into the test-piece; this is taken as the softening point (Vicat temperature).

In the method of Martens a test rod is mounted upright in a support and the upper free end is put under a bending stress via a small weighted lever. The rod is slowly heated in an oven until a specified deflection is attained. The softening point determined in this way is called the Martens temperature.

Finally, in the English-speaking world, there is another extensively used method in which a rod supported at its two ends is loaded at the centre and slowly heated in a liquid until a certain distortion is attained; the temperature at which this occurs is called the heat distortion temperature (HDT).

These methods are applicable both to thermoplastics and to crosslinked polymers; interconversion of the softening temperatures determined by the different techniques is not possible.

2.3.4.3. *Determination of melting range and crystalline melting point*

The melting point of a low-molecular-weight compound is an important quantity, which, through determination of the mixed melting point with a reference sample, can serve for its identification. On the other hand macromolecular substances seldom have a precise melting point but melt instead over a certain temperature range which is influenced by various factors. The melting point or the melting range of a polymer can thus assist its characterization, but not its identification in the sense used for low-molecular-weight compounds.

Recognition of the melting range is relatively simple if the polymer is partly crystalline, because the change of birefringence can then be observed using a hot-stage polarizing microscope.[1] First the approximate melting temperature of the polymer is determined on a Kofler hot-block.[2] A small amount of polymeric substance is then melted on a microscope

[1] For example the melting point microscope with polarizer made by Leitz.
[2] See *Houben-Weyl 2* (1953) 788.

slide, using the hot block at a temperature somewhat above the melting range. A normal cover slip, heated to about the same temperature, is finally pressed on to the molten polymer with a cork, so that a thin homogeneous film is produced. The slide is now placed on the hot stage of a melting-point microscope at a temperature about 20 °C below the melting point. The sample is observed under polarized light and slowly heated. The temperature range from the first noticeable change to the final disappearance of birefringence is noted. The mean value of the first and last readings may be taken as the crystalline melting point. Many polymers recrystallize on cooling (however, see isotactic polystyrene, Example 3–28) so that duplicate measurements are possible. The temperature at which crystallization commences on cooling from the melt is called the crystallization temperature. It is not identical with the melting point of the crystallites but lies somewhat below, on account of the hindered motion of the macromolecules in the molten state.

The determination of the melting range according to the method described can be upset by the fact that a non-crystalline polymer can also show birefringence if its macromolecules have been oriented by the action of external forces. For example orientation can occur when the cover slip is pressed on to the molten polymer, or, if a film is used, during the preparation or cutting of the test-piece. The disappearance of birefringence due to orientation does not then occur at the melting point of the sample. Such errors can be avoided by duplicate determinations on the same sample, since once the polymer has been melted it is unlikely to undergo any reorientation in the absence of external forces. A more exact determination of the crystalline melting point is possible by differential thermal analysis (DTA).[1]

For amorphous polymers with the same CRU the melting range depends, amongst other things, on the molecular weight distribution and the degree of branching. An effect of molecular weight is only observed in the oligomeric region (e.g. for polyoxymethylene dimethyl ethers[2] or for oligoamides[3,4] and oligoesters[3,4]); this arises from the usual increase in melting point with molecular size observed in a homologous series of low-molecular-weight compounds.

---

[1] Polymer Reviews *6* (1964) 347; "High Polymers" *12/2* (1962) 159; *C.B. Murphy*, Anal. Chem. *34* (1962) 298R; *D. Schultze*, "Differentialthermoanalyse", Verlag Chemie, Weinheim 1969.

[2] *W. Kern* in *H. Staudinger*, "Die hochmolekularen organischen Verbindungen" (1932) 224.

[3] *M. Rothe*, "Physical Data of Oligomers" in "Polymer Handbook", Section VII–1.

[4] *H. Zahn* and *G.B. Gleitsmann*, Angew. Chem. *75* (1963) 772.

Crosslinked polymers do not melt, but possess a softening range which generally lies in the vicinity of the decomposition temperature. There are also some uncrosslinked polymers that likewise have no melting point or range (e.g. polyacrylonitrile); they begin to decompose above a certain temperature.

2.3.5. Determination of melt viscosity (melt index) of polymers[1]

Two main types of viscometer are suitable for the determination of the viscosity of a polymer melt[2]: a rotation viscometer (Couette viscometer; Platte-Kegel viscometer) and a capillary viscometer or capillary extrusiometer. The latter are especially suitable for laboratory use since they are relatively easy to handle and are also applicable in the case of high shear rates. With the capillary extrusiometer the measure of fluidity is not expressed in terms of the melt viscosity η but as the amount of material extruded in a given time (10 min). The amount of extrudate per unit of time is called the melt index or melt flow index i. It is obviously necessary to specify the temperature and the shearing stress or load. Thus i_2 (190°C) = 9.2 g/10 min means that at 190°C and 2 kg load, 9.2 g of polymer melt are extruded through a standard nozzle in 10 min.

The capillary extrusiometer for the measurement of melt index consists of a heatable cylinder fitted with a standard nozzle at the lower end, the upper opening being closed by a piston that can be loaded with different weights.[3] The cylinder is heated to the required temperature and filled with powdered or granulated polymer; the piston is loaded with the desired weight. After a specified initial melting period (about 5 min) the piston catch is released so that the polymer melt is forced through the nozzle. At intervals of 1 min (longer for highly viscous melts) the extruded polymer is cut off at the nozzle and weighed; the time is measured by stopwatch. The melt index is calculated from the mean of at least five measurements.

2.3.6. Determination of crystallinity of polymers

The simplest way of establishing qualitatively the crystallinity of a polymer is by observation of birefringence under a suitable microscope, taking care

---

[1] *W. Philippoff*, "Viscosität der Kolloide", Steinkopff-Verlag, 1942; *E.T. Severs*, "Rheology of Polymers", Reinhold Publ., New York 1962; *V. Semjonow*, Kunststoffe *56* (1966) 7.
[2] Also see Section 1.4.2.
[3] Appropriate apparatus is available commercially.

to exclude the possibility of orientation birefringence (see Section 2.3.4.3).

X-ray diffraction allows a quantitative determination of the degree of crystallinity as well as the usual crystallographic data.

In some cases crystalline polymers show additional absorption bands in the infrared, as in polyethylene[1] ("crystalline" band at 730 cm^{-1}, "amorphous" band at 1 300 cm^{-1}) and polystyrene[2] (bands at 982, 1 318 and 1 368 cm^{-1}). By determining the intensity of these bands it is possible to follow in a simple way the changes of degree of crystallinity caused, for example, by heating or by changes in the conditions of preparation.

The degree of crystallinity is also reflected in the density of the polymer so that the determination of density provides at least a relative measure for an increase of crystallinity.

Differential scanning calorimetry (DSC) is frequently applied to the determination of the crystallinity from the heat of crystallization or melting (also see Section 2.3.4.3).

2.3.7. Determination of density of polymers

The densities of polymers[3] can be determined by the pyknometer technique or by the flotation method.

In the pyknometer technique the liquid volume displaced by the polymer sample is determined by weighing. Most polymers have a density greater than that of water which can, therefore, be used as the liquid. Polymers in the form of powders or pressed discs tend to adsorb or occlude air bubbles which can lead to serious errors. This can be largely prevented by careful degassing of the pyknometer and polymer sample under vacuum before filling with liquid, or by addition of a small amount (0.1%) of commercial detergent to lower the surface tension of the water.

In the flotation method, which is especially suitable for powdered polymers, a liquid of higher density is added to a liquid of lower density until the test particles neither sink nor rise to the surface. The densities of the solid and liquid mixture are then equal and it remains only to determine the density of the latter. The experiment is conducted as follows. The powdered polymer is placed in a small beaker with a certain amount of less dense liquid. The heavier liquid is then run in from a burette

---

[1] *H.G. Zachmann* and *H.A. Stuart*, Makromol. Chem. *44/46* (1961) 622; *H.A. Stuart* (Ed.), "Physik der Hochpolymeren" *3* (1955) 244.

[2] *D. Braun*, *W. Betz* and *W. Kern*, Naturwiss. *46* (1959) 444.

[3] Also see *Houben-Weyl 3/1* (1955) 188.

with gentle stirring until the state of suspension is attained. The density of the liquid mixture can then be determined by pyknometer or can be derived from a previously determined calibration curve. For the determination of densities less than unity (e.g. polyethylene), ethanol/water mixtures are suitable; for densities greater than unity one may use mixtures of water with aqueous salt solutions (40% $CaCl_2$ solution: $d_4^{20} = 1.40\,g\,ml^{-1}$; 72% $ZnCl_2$ solution: $d_4^{20} = 1.95\,g\,ml^{-1}$).

The density gradient method, which is an elegant variation of the flotation method, should also be mentioned.[1]

2.3.8. Degradation of polymers

Valuable information about stability, composition and structure of polymers can frequently be obtained from degradation experiments.[2] Degradation may be brought about by chemical or thermal methods (see Section 5.3) as well as by high energy radiation[3], ultrasonics or mechanical means.

Such degradation experiments can be evaluated in different ways according to the type of method. In the first procedure only the weight loss or decrease of molecular weight is measured, the volatile products being ignored. This provides a measure of the stability of the polymer under the given conditions. If on the other hand the nature and composition of the low-molecular-weight fragments are determined one can make deductions about the structure of the corresponding polymer and the proportions of individual CRU's of a copolymer (see Section 2.3.11.) Thermal or chemical methods of degradation are generally used for such investigations, which do not require elaborate equipment.

2.3.8.1. Thermal degradation of polymers

The thermal degradation of a polymer must be carried out under purified nitrogen in order to exclude the additional effect of oxygen. Depending on the structure of the polymer and the decomposition temperature, various kinds of low-molecular-weight fragments can be formed (see Section 5.3).

A suitable apparatus for carrying out degradation experiments is shown in Figure 2.19. It allows several samples to be decomposed simultaneously in a nitrogen stream or in vacuum. The temperature of the vessel may be

[1] *A. Weissberger*, "Technique of Organic Chemistry", 3rd Edn., *1* Part 1 (1959) 182.
[2] Also see *Houben-Weyl 14/2* (1963) 917; *N. Grassie*, "Developments in Polymer Degradation", Applied Science Publishers, London 1977.
[3] *A. Chapiro*, "Radiation Chemistry of Polymeric Systems", High Polymers *15*, Interscience Publishers, New York, London 1962.

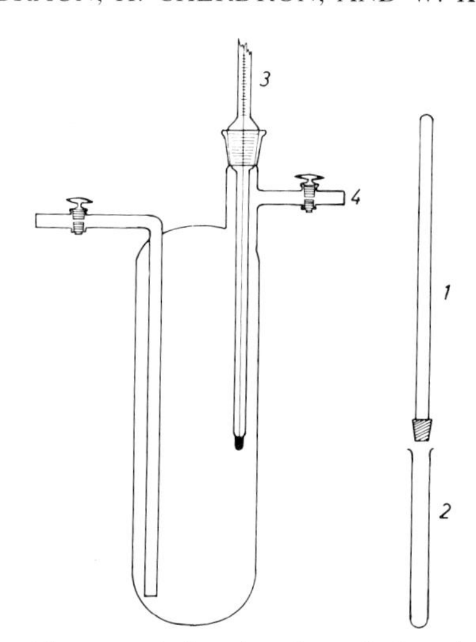

FIGURE 2.19 Decomposition vessel for the thermal degradation of polymers under nitrogen or vacuum. 1, glass rod with rubber stopper, used to remove the test tubes from the decomposition chamber; 2, test tube (0.7 × 5 cm); 3, thermometer with ground-in joint; 4, outlet for the volatile decomposition products.

controlled by a liquid thermostat (up to about 250°C) or by an air thermostat (up to about 400°C). For higher temperatures one can use baths of liquid metal or electrically heated muffle furnaces. To eliminate variations due to diffusion of the gaseous degradation products through the polymer melt, the test tubes (2) should all have the same internal diameter (7 mm tubes are suitable for 50–100 mg samples) and the same amount of polymer should be used in each experiment.

If only the stability of the polymer is required, then the percentage weight loss is determined by weighing the test tubes after various times. The results are plotted for various temperatures. If the plot of log (amount of undecomposed polymer in %) against time gives a straight line, the reaction is first order and the slope gives the rate constant at the corresponding temperature. The residual polymer can be removed from the test tubes to conduct supplementary experiments such as solution tests, viscosity measurements or optical and analytical studies, as required. The thermal stability of various polymers can be compared by measuring the decomposition rate at 350°C (in % per min); alternatively the "half-value temperature" T_h, the temperature at which the polymer loses half its weight on heating under vacuum[1] for 30 min, can be used for comparison.

[1] Compilation in *B.G. Achhammer*, *M. Tryon* and *G.M. Kline*, Kunststoffe *49* (1959) 600.

If the qualitative and quantitative composition of the degradation products is also of interest, they must be trapped and analyzed in a suitable way. This is most easily done by attaching a cold trap to the outlet (4) in Figure 2.19 and analyzing the condensate. For this purpose rather more polymer is used (1–5 g) than for simple stability tests.

Gas chromatography (pyrolytic gas chromatography[1]) is very appropriate for qualitative and quantitative degradation experiments on polymers. The decomposition can be carried out in one of three ways: either directly in the gas chromatograph[2] by means of an electrically heated coil; or in a suitable external vessel[3], if necessary in the presence of a catalyst[4]; or in a sealed glass tube in an appropriate heating bath (ampoule technique[5]). Reproducible pyrolysis temperatures between 300 and 900°C can be achieved with a Curie-point high-frequency pyrolyzer.[6] The amounts of material required for pyrolytic gas chromatography are small (a few μg to mg); a mass spectrometer can be used as detector.[7] We may note that thermodynamic data of high polymers can be determined by the technique of inverse gas chromatography.[8]

2.3.8.2. *Chemical degradation of polymers*

Acidolysis and hydrolysis are the chief methods used for analytical studies under this heading. With few exceptions (e.g. polyacetals) their application is therefore limited to polysaccharides, polypeptides, proteins, products of condensation polymerization and the products of addition polymerization of two monomers. Oxidative degradation, such as that caused by atmospheric oxygen at elevated temperatures, often leads to a multitude of degradation products so that analytical evaluation, and deductions about the structure and composition of the polymers, are rendered difficult; the main reason for carrying out degradation experiments of this type is to test the stability of the polymer and the effectiveness of added stabilizers under oxidative conditions.

Chemical degradation reactions of polymers are generally found to be random processes. For example, the hydrolysis of a polyester does not

---

[1] *M. Hoffmann*, *H. Krömer* and *R. Kuhn*, "Polymeranalytik" I, Thieme Verlag, Stuttgart 1977; *M.P. Stevens*, "Characterization and Analysis of Polymers by Gas Chromatography", Marcel Dekker, New York and London 1969.

[2] For example, *J. Voigt*, Kunststoffe *51* (1961) 18, 314.

[3] *D. Braun*, Farbe Lack *69* (1963) 820.

[4] *K.H. Burg*, *E. Fischer* and *K. Weissermel*, Makromol. Chem. *103* (1967) 268.

[5] *H. Cherdron*, *L. Höhr* and *W. Kern*. Angew. Chem. *73* (1961) 215.

[6] *J.Q. Walker*, Chromatographia *5* (1972) 547.

[7] *D.O. Hummel*, "Polymer Spectroscopy", Verlag Chemie, Weinheim 1974.

[8] *J.M. Braun* and *J.E. Guillet*, Adv. Polym. Sci. *21* (1976) 107.

begin at the chain ends followed by shedding of monomeric units, but begins at any ester group within the chain. The resulting fragments are, therefore, still of high molecular weight; only by the continuous repetition of the hydrolytic scission are they finally broken down completely to entities containing a single CRU. Since even the low-molecular-weight fragments are generally involatile under the prevailing hydrolysis or acidolysis conditions (in contrast to thermal degradation), these degradation reactions cannot be followed gravimetrically. When working in solution it is often possible to follow the progress of reaction by titration of the unused reagent as a function of time. A more sensitive indication of the degradation is provided by the determination of the viscosity of the polymer solution during the degradation process (see Examples 5–18 and 5–19).

2.3.9. Optical investigations on polymers[1]

Ultraviolet and infrared spectroscopy are the main optical methods for investigating polymers. U.v. spectroscopy[1] is limited to soluble polymers with chromophores, for example with aromatic groups, polyene sequences or carbonyl groups; it can be used with advantage for the quantitative determination of the composition of copolymers (e.g. copolymers of styrene[2]) or for the determination of residual monomer in polymers.[3]

Infrared spectroscopy[4] is especially important since it can also be applied to insoluble (crosslinked) polymers. It can be used not only for purely analytical purposes[5] (determination of functional groups and end groups[6]) but also for structural investigations[7] (e.g. for polymeric dienes, see Example 3–30; and branching in polyethylene[8]). In some cases it is still the most reliable method for the determination of the composition of copoly-

[1] *D.O. Hummel*, "Polymer Spectroscopy", Verlag Chemie, Weinheim 1974.
[2] *W. Kern* and *D. Braun*, Makromol. Chem. *27* (1958) 23; *R. Brüssau* and *D. Stein*, Angew. Makromol. Chem. *12* (1970) 59.
[3] *I.E. Newell*, Anal. Chem. *23* (1951) 445.
[4] Detailed compilation: *S. Krimm*, Adv. Polym. Sci. *2* (1960) 51; Fourier-transform infrared analysis (FTIR) of polymers: J. Macromol. Sci. (Macromol. Rev.) C*16* (1978) 197.
[5] *Hummel-Scholl*, "Atlas der Kunststoff-Analyse", Verlag Chemie, Weinheim, and Hanser Verlag, München 1978.
[6] *H. Kämmerer*, *F. Rocaboy*, *K.-G. Steinfort* and W. Kern, Makromol. Chem. *53* (1962) 80; *H. Kämmerer* and *G. Sextro*, Makromol. Chem. *137* (1970) 183.
[7] *G. Schnell*, Ber. Bunsenges. Phys. Chem. *70* (1966) 297; *M. Hoffmann*, *H. Krömer* and *R. Kuhn*, "Polymeranalytik", Thieme Verlag, Stuttgart 1977.
[8] For summary see *H.A. Stuart* (Ed.), "Physik der Hochpolymeren" *3* (1955); *H.G. Zachmann*, Adv. Polym. Sci. *3* (1964) 581; *Houben-Weyl 14/1* (1961) 610.

mers (e.g. ethylene/propylene copolymers[1]). The determination of crystallinity by infrared spectroscopy is dealt with in Section 2.3.6.

The techniques normally used in infrared spectroscopy can be readily applied to polymers. If it is not possible to work in solution the measurements can be made with films of thickness 30 to 300μm. These can be obtained by melting the sample directly onto a rock salt plate (for low-melting polymers), or by the method described in Section 2.4. When using the potassium bromide disc technique or the suspension technique (mull in paraffin oil or perfluorinated hydrocarbons) it is especially important to ensure that the polymer is as finely divided as possible (see Section 2.4.1). If necessary the potassium bromide disc must be reground in a mortar and the disc remade.

The use of the polarizing microscope and of X-rays have already been mentioned in connection with measurements of crystallinity (see Section 2.3.6); electron microscopic investigations are also important (especially for the observation of crystalline structure[2]). Nuclear magnetic resonance spectroscopy of polymers[3] is especially valuable for the determination of tacticity and sequence distributions.

2.3.10. Determination of important groups and elements

Qualitative and quantitative elemental analysis of polymers[4] can be carried out by the conventional methods[5] used for low-molecular-weight compounds so that a detailed description here is unnecessary. The same is true for readily identifiable groups, such as acetyl, methoxyl or amide groups (also see the references on chemical analysis of polymers[6]). Difficulties are sometimes encountered with polymers, requiring the experimental conditions to be modified (e.g. by extending the reaction time). Special methods for the determination of end groups have already been mentioned in Section 2.3.2.2.

---

[1] *T. Gössl*, Makromol. Chem. *42* (1960) 1.

[2] For summary see: *H.A. Stuart* (Ed.), "Physik der Hochpolymeren" *3* (1955); *H.G. Zachmann*, Adv. Polym. Sci. *3* (1964) 581.

[3] For summary see *F.A. Bovey* and *G.V.D. Tiers*, Adv. Polym. Sci. *3* (1963) 139; *F.A. Bovey*, "High Resolution NMR of Macromolecules", Academic Press, New York 1972.

[4] *Houben-Weyl 14/2* (1963) 918.

[5] *Staudinger-Kern-Kämmerer*, "Anleitung zur organischen qualitativen Analyse", Springer-Verlag, Berlin-Heidelberg-New York 1968.

[6] *J. Harwood*, Angew. Makromol. Chem. *4/5* (1968) 279; *D. Braun*, J. Polym. Sci., Polym. Symp. *50* (1975) 149 and Angew. Makromol. Chem. *76/77* (1979) 351; *R.C. Schulz* and *O. Aydin*, J. Polym. Sci., Polym. Symp. *50* (1975) 497.

Elemental analysis or determination of functional groups is especially valuable for copolymers or chemically modified polymers. For homopolymers, where the elemental analysis should agree with that of the monomer, deviations from the theoretical values are an indication of side seactions during polymerization; however they can also sometimes be traced to inclusion or adsorption of solvent or precipitant, or, in commercial polymers, to the presence of added stabilizers. The preparation of the sample for analysis must, therefore, be very carefully handled (several reprecipitations, if necessary using various solvent/precipitant combinations; thorough drying).

2.3.11. Characterization of copolymers[1]

In the characterization of copolymers one may distinguish between qualitative analysis, designed to test whether the material is a genuine copolymer or only a physical mixture of homopolymers, and quantitative analysis of the weight fraction of the incorporated comonomers. These investigations start from a knowledge of some physical and chemical properties of the homopolymers prepared under similar conditions, and their physical mixtures.

Qualitative analysis is relatively simple if the homopolymers differ in their solubility behaviour, for example when one homopolymer is soluble in benzene and the other not. In this case a sample of the supposed copolymer is extracted with benzene, and a second sample extracted with a solvent for the other homopolymer. The reprecipitated extracts and residues are examined for composition. The extraction must, however, be very carefully carried out and repeated several times since polymer mixtures are frequently quite difficult to separate by extraction.[2] If no pure homopolymer is isolated in this way one can be sure that the sample is a genuine copolymer. If the solubility properties of the original homopolymers are insufficiently different it is sometimes possible to induce such differences through chemical transformation, for example by hydrolysis (for copolymers of vinyl acetate, and of acrylic or methacrylic esters), or by epoxidation and hydroxylation (for copolymers of dienes). The qualitative investigation of copolymers is considerably more difficult when the homopolymers cannot be distinguished by their solubility. In this case other physical data of the supposed copolymer can be compared with the

---

[1] *D. Heinze*, Makromol. Chem. *101* (1967) 166; *M. Hoffmann*, *H. Krömer* and *R. Kuhn*, "Polymeranalytik", Thieme Verlag, Stuttgart 1977.

[2] *H. Dexheimer* and *O. Fuchs*, Makromol. Chem. *96* (1966) 172.

corresponding data for various physical mixtures of homopolymers, for example softening point and melting range, density and crystallinity. Copolymers can frequently be distinguished from physical mixtures of homopolymers by the qualitative and quantitative composition of the pyrolysis products (see Section 2.3.8).

Quantitative analysis of copolymers is relatively simple if one of the comonomers contains a readily determinable element or functional group (see Examples 3–44, 3–45 and 3–46); however C,H-analyses are only of value when the difference between the carbon or hydrogen content of the two comonomers is sufficiently great. If the composition cannot be determined by elemental analysis or chemical means the problem can be solved either by spectroscopic methods[1], for example by u.v. measurements (e.g. styrene copolymers), by i.r. measurements (e.g. olefin copolymers), and n.m.r. measurements, or by gas chromatographic methods[2] after thermal or chemical decomposition of the samples.

In principle the composition of a copolymer may also be determined by analyzing the composition of the residual monomer by a suitable method after polymerization. It will usually be necessary first to separate the copolymer by precipitation, followed by careful recovery of the filtrate containing the residual monomer; but the direct method of analysis of the copolymer will generally be preferred. Block and graft copolymers can be characterized in the same manner; however, consideration must be taken of the fact that they usually contain large amounts of the homopolymers which must first be removed.

The more refined characterization of a random copolymer involves the determination of the reactivity ratios r_1 and r_2 (copolymerization parameters), also the calculation of Q and e (or q and ε) values (see Section 3.3 and Examples 3–44 to 3–46).

2.3.12. Mechanical measurements on polymers

The determination of the mechanical characteristics of a polymer serves ultimately to establish its usefulness and applicability as an industrial material. Although at first sight such measurements are of a purely applied character, some methods of investigation yield data that are not only useful for engineering practice but also allow deductions about composition, structure and state of aggregation of the polymeric material. Thus, they

---

[1] See Section 2.3.9.
[2] See Section 2.3.8.

TABLE 2.6

List of the most important mechanical properties for the initial characterization of a polymer

Property	Symbol	Units
tensile strength (yield strength)	σ_B	$N\,mm^{-2}$
tensile strength at break	σ_R	$N\,mm^{-2}$
elongation at yield	ε_B	%
elongation at break	ε_R	%
modulus of elasticity or Young's modulus	E	$(J\,m^{-1})mm^{-2}$ $N\,mm^{-2}$
impact strength	a_n	$N\,mm^{-1}$
notched impact strength	a_k	$N\,mm^{-1}$
ball hardness	H	$N\,mm^{-2}$

supplement the methods of characterization of polymers discussed in Sections 2.3.1 to 2.3.11. The following mechanical properties can, in this sense, serve for the physical characterization of a polymer in the solid state: strength and elongation, stiffness (modulus of elasticity), brittleness and toughness, and hardness. They are collected in Table 2.6 and discussed separately in Sections 2.3.12.1 to 2.3.12.4.

The numerous other tests employed in engineering practice to determine mechanical (and other) properties[1], as well as the special methods for testing rubbers, foams, coatings, films, and fibres, will not be dealt with here.[2]

The results of mechanical tests on polymers can obviously only be compared with one another when they are obtained at the same temperature, since the physical properties of polymers change markedly with temperature (see Section 1.4). Furthermore, the manner of preparation and pretreatment (conditioning) of the test specimen is decisive for the reliability and reproducibility of mechanical tests on polymers.[3,4] For the

[1] In Table 2.6 only the so-called short-term tests are listed, in which the specimen is stressed only once, since they suffice for initial characterization. On the other hand, when the polymer is to be used as an engineering material, it is important to carry out long-term tests in which the specimen is subjected over a long period (months or years) to a constant or alternating stress.

[2] See *Nitsche/Wolf*, "Praktische Kunststoffprüfung" 1961; *B. Carlowitz*, "Tabellarische Übersicht über die Prüfung von Kunststoffen", Umschau-Verlag, Frankfurt (Main) 1966: *Houwink/Stavermann*, "Chemie und Technologie der Kunststoffe", Vol. 3 (1963): "Typisierung und Prüfung der Kunststoffe".

[3] See *Nitsche/Wolf*, "Praktische Kunststoffprüfung" (1961), p. 24.

[4] *J. Dasch*, Kunststoffe *57* (1967) 117; *S. Wintergerst*, Kunststoffe *57* (1967) 188.

determination of constants that are characteristic of the material, it should be as isotropic as possible, that is, exhibit the same properties in all directions, and therefore be free of internal stresses; in addition the temperature and humidity should be held constant for all measurements. While the latter conditions are relatively easy to fulfil, the preparation of completely stress-free and therefore isotropic test specimens is quite difficult to achieve for the reasons already given (see Section 1.4). Flow orientation in test specimens is especially liable to occur in thermoplastic materials (less so in thermosetting polymers), and in extreme cases can lead to values of tensile strength or impact strength that are two or three times as high when measured in the flow direction compared with those at right angles to this direction. Possible ways of suppressing the anisotropy have been suggested in the literature.[1,2] Test specimens can be fabricated either directly by injection or compression moulding or casting (thermosetting plastics) in suitable forms, or by milling, sawing or punching of sheets of the polymer in question.[1]

2.3.12.1. *Stress-strain measurements*

In a stretching experiment a test specimen is placed under tension, causing the length to increase and the cross-section to decrease, until finally it breaks. For these stress-strain measurements the test specimen[3] has shoulders at both ends, such that the break occurs in the desired place, namely at the position of lowest cross-section. The specimen is held at its broader parts in the clamps of the testing machine. The machine then pulls the clamps apart at constant speed, whereby a force is transmitted to the test specimen; the latter is plotted continuously against the change of length by means of a coupled recorder. The maximum tension P_{max} during the experiment is not always the same as the tension at break.

The prevailing tension divided by the smallest cross-section F_o of the test specimen at the beginning of the experiment gives the corresponding stress σ, which is thus the tension per unit cross-section ($1\,mm^2$). The ultimate tensile strength σ_B is obtained by dividing the maximum load P_{max} by the initial cross-section F_o:

$$\sigma_B = \frac{P_{max}}{F_o} \; (\mathrm{N\,mm^{-2}})$$

[1] See *Nitsche/Wolf*, "Praktische Kunststoffprufung" (1961), p. 24.
[2] *J. Dasch*, Kunststoffe 57 (1967) 117; *S. Wintergerst*, Kunststoffe 57 (1967) 188.
[3] The dimensions of test specimens are standardized, but vary from one country to another.

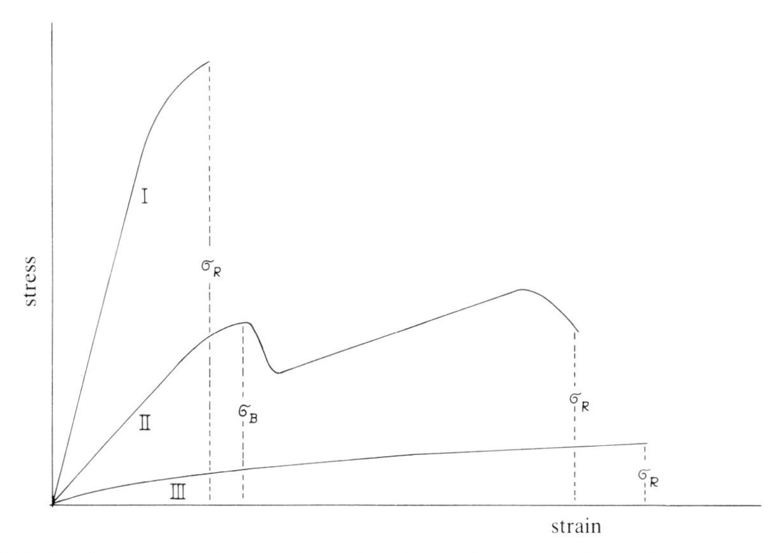

FIGURE 2.20 Stress-strain diagrams for various types of polymers (for explanation, see text). σ_B = yield strength; σ_R = tensile strength at break.

The elongation ε is generally understood to be the extension with respect to the original length. The elongation at yield is accordingly the extension, $\Delta l = l - l_o$, at maximum load P_{max}, divided by the initial length l_o:

$$\varepsilon_B = \frac{100\,\Delta l}{l_o}\ (\%)$$

From the values of the yield strength and the corresponding elongation one thus obtains a measure of the ultimate load that can be carried by the material. However, it is very much more informative for the characterization of a polymer to observe not just the values of yield strength σ_B and elongation at yield ε_B, but the whole stress-strain experiment shown graphically as a plot of stress σ against ε (stress-strain diagram). Plastics may be divided into three main categories according to the shape of such a stress-strain diagram[1] (see Figure 2.20).

The first group comprises materials whose stress-strain curve is very steep and almost linear[2], and flattens only slightly near the break point

[1] There are also intermediate cases.

[2] The point in the stress-strain curve where deviation from linearity sets in is called the proportionality limit. This means that Hooke's law is no longer obeyed above this point. The position of the proportionality limit naturally depends very much on the temperature, since the mechanical properties of polymers change markedly with temperature (see Section 1.4).

(curve I in Figure 2.20). Like metals, these materials deform only to a small extent at relatively high loads. Amongst these may be numbered all thermosetting plastics as well as some thermoplastics such as polystyrene and poly(methyl methacrylate), i.e. substances that are only slightly elastic and rather brittle.

The second group exhibit the phenomenon of drawability. This manifests itself in the stress-strain behaviour (curve II in Figure 2.20) as follows. At first these materials behave in a similar way to those of curve I; the proportionality limit[1] lies at low values, and the deformation with increasing load is also quite small. Then, suddenly, a large extension occurs, even though the load remains constant or becomes smaller; the material begins to flow[2] and the stress-strain curve sometimes runs nearly parallel to the abscissa. The point at which the flow begins is called the upper yield point. The stress at this point is called the yield strength σ_B although the specimen has not broken. When all the macromolecules have been brought into a new position (orientated) by the flow process, the flow ceases and the stress increases again[3] until finally the sample breaks. The stress σ at this point is called the tensile strength at break σ_R, associated with the elongation at break ε_R. Many thermoplastics belong to this group, such as polyolefins, Nylon-6 and Nylon-6,6, and unplasticized poly(vinyl chloride) (rigid PVC).

The third group comprises materials that show relatively large deformations even at low loads. In the stress-strain diagram (curve III in Figure 2.20) there is no sudden drop in the stress, i.e. no flow limit. Furthermore, in the middle range the curve is not quite so flat as with drawable materials (group II); i.e. the increase in strength resulting from reorientation of the macromolecules is a gradual one. A further increase in load leads eventually to failure; the stress at this point is called the tensile strength at break σ_R, and the corresponding elongation at break is ε_R. To this group belong all plasticized thermoplastics (e.g. soft PVC) as well as rubbers (elastomers).

Finally, the modulus of elasticity E (Young's modulus), which is a measure of the stiffness of the polymer, can be calculated from the stress-strain diagram. According to Hooke's law there is a linear relation between the stress σ and the strain ε:

$$\sigma = E\varepsilon$$

---

[1] See footnote[2] on p. 110.
[2] Concerning the mechanism of flow in polymers, see Sections 1.4.1 to 1.4.3.
[3] Use is made of the increased strength in the drawing of fibres and films.

so that the elastic modulus is given simply by the slope of the stress/strain curve in the linear region, i.e. below the proportionality limit. Thus,

$$E = \frac{\text{stress}}{\text{relative change of length}} = \frac{\text{N mm}^{-2}}{\text{mm mm}^{-1}} = \text{N mm}^{-2}$$

Hence, the elastic modulus corresponds in principle to the force per square millimetre that is necessary to extend a rod by its own length. Materials with low elastic modulus experience a large extension at quite low stress (e.g. rubber, $E \approx 1\,\text{N mm}^{-2}$); on the other hand materials with high elastic modulus (e.g. polyoxymethylene, $E \approx 3\,500\,\text{N mm}^{-2}$) are only slightly deformed under stress. Different kinds of elastic modulus are distinguished according to the nature of the stress applied.[1] For tension, compression and bending, one speaks of the intrinsic elastic modulus (*E*-modulus).[2] For shear stress (torsion), a torsion modulus (*G*-modulus)[3], can be similarly defined, whose relationship to the *E*-modulus is described in the literature.

2.3.12.2. *Dynamic-mechanical measurements*

In dynamic-mechanical measurements the test specimen is not destroyed. The measurements are called dynamic because the mechanical properties are determined under oscillatory conditions. Of the numerous methods of measurement the oscillatory torsion experiment, using the so-called torsion pendulum[4,5], is one of the most widely used. In this test the specimen, in the form of a strip, is clamped firmly at the upper end while an oscillating disc is fastened to the lower end, the turning motion of which can be followed optically and continuously recorded. The test specimen is surrounded by a thermostatted chamber which can be heated or cooled. After imparting an initial impulse to the torsion pendulum the decay of oscillation with time can be analyzed to yield two pieces of information. The vibration period allows the calculation of the torsion modulus (*G*-modulus)[6], which is a measure of the rigidity. From the decrease of

---

[1] In contrast to metals the time factor plays a much greater role in physical measurements on polymers. If such materials are subjected to a constant stress, they experience a deformation ε which increases with time; this process is known as creep. Conversely if a sample is extended by a constant amount, the initial stress σ dies away with time. This decay is called stress relaxation. Since σ and ε of polymers are thus functions of time, it is important to state the conditions under which the elastic modulus was determined.

[2] The expressions tensile-*E*-modulus, compression-*E*-modulus, and bending-*E*-modulus are often used.

[3] See Section 2.3.12.2.

[4] See *Nitsche/Wolf*, "Praktische Kunststoffprüfung" (1961) p. 148.

[5] *K. Schmieder* and *K. Wolf*, Kolloid-Z. *127* (1952) 65; *134* (1953) 149.

[6] Definition, see Section 2.3.11.1.

amplitude with time one obtains a measure of the "internal mechanical absorption" called the mechanical loss factor *d* (or logarithmic decrement). From such measurements it is possible to make valuable deductions about the molecular motions; not only about the motions of chain segments within a macromolecule (micro-Brownian motion) but also concerning the motions of the entire macromolecule relative to others (macro-Brownian motion).[1]

Since these processes are strongly dependent on temperature[2], it is appropriate to carry out torsion-oscillation experiments over a wide range of temperature, and to plot the values for the modulus and *d* against temperature (see Figure 2.21); an absorption maximum in the *d*-curve corresponds to an inflection in the modulus curve. Such diagrams are especially useful in determining the positions of the softening point (glass transition temperature), melting point, melting range and transition points, as well as indicating the influence of temperature on stiffness (*G*-modulus or *E*-modulus).

As for stress-strain measurements, the value of the modulus and the shape of the modulus curve allow deductions concerning not only the state of aggregation but also the structure of polymers. Thus, by means of torsion-oscillation measurements, one can determine the proportions of amorphous and crystalline regions, crosslinking and chemical non-uniformity[3], and can distinguish random copolymers from block copolymers. This procedure is also very suitable for the investigation of plasticized or filled polymers, as well as for the characterization of mixtures of different polymers (polyblends).[4] This is illustrated in Figures 2.21 and 2.22.

Curve 1 in Figure 2.21 corresponds to an elastomer, characterized by a low value of the elastic modulus over a wide temperature range and a sudden increase at low temperature corresponding to the transition from the elastic to the brittle (glassy) state at the glass transition temperature (−50°C). Curves 3 and 4 are characteristic of partially crystalline polymers. Here the modulus has values covering three powers of ten, but only at the crystalline melting point does it drop to the much lower level

[1] Dynamic-mechanical measurements are sometimes referred to as "mechanical spectra" by analogy with infrared spectra which likewise result from the absorption of vibrational energy by molecular segments.

[2] The measuring frequency, as well as temperature, is important; in general this lies between 1 Hz and several hundred Hz, and must be kept as constant as possible in a given series of experiments.

[3] On the other hand these measurements give no information on molecular non-uniformity (molecular weight distribution).

[4] *L. Bohn*, Kolloid-Z. *213* (1966) 55.

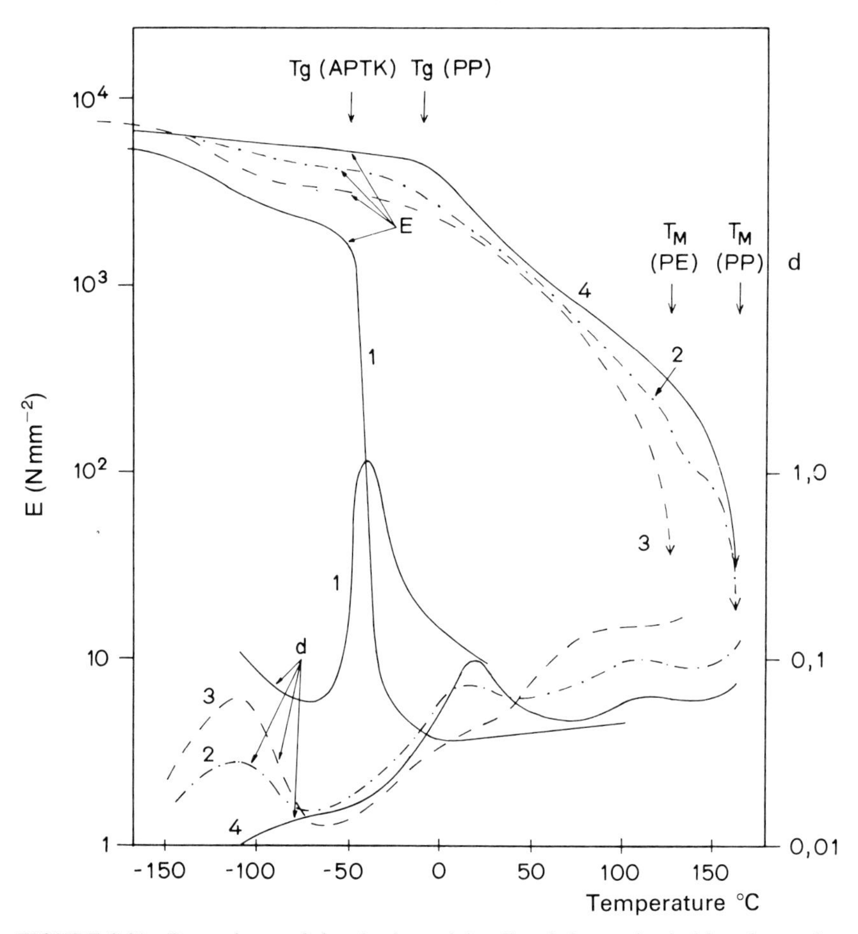

FIGURE 2.21 Dependence of the elastic modulus E and the mechanical loss factor d on temperature for various polymers. Curves 1: elastomer (statistical copolymer of ethylene and propene, 70:30); curves 2: block copolymer of ethylene and propene (50:50); curves 3: partially crystalline polymer with very low glass transition temperature (polyethylene); curves 4: partially crystalline polymer with moderately low glass transition temperature (isotactic polypropene).

characteristic of the elastic state. The corresponding curves for the mechanical loss factor d show the following characteristics (see Figure 2.21). For elastomers the transition to the glassy state is accompanied by a pronounced "mechanical absorption" (curve 1); on the other hand, with crystalline polymers (curves 3 and 4), two absorption maxima are evident: the maxima at −100°C (polyethylene) and 0°C (isotactic polypropene), respectively, correspond to the glass transition temperature, and the maxima at 120°C and 145°C, respectively, to the crystalline melting point. The extent to which the behaviour of a random copolymer differs from that of a block copolymer can be seen by comparison of curves 1 and 2. Random copolymers of ethylene and propene are fully amorphous and rubber-elastic and show a dynamic-mechanical behaviour that is typical of an elastomer (curve 1). On the other hand block copolymers, consisting of long sequences of ethylene and long sequences of propene units, combine additively the properties of the two homopolymers: they possess two softening ranges above room temperature that are to be assigned to the crystalline melting points and two maxima below room temperature at the two glass transition temperatures.

A further example of the application of torsion-oscillation measurements is the investigation of the influence of plasticizers on the properties of poly(vinyl chloride) (PVC). Poly(vinyl chloride) is a polymer with the typical properties of an amorphous thermoplastic, that is, the stiffness is maintained up to the softening point (which is practically identical with the glass transition temperature in all amorphous polymers) and then falls very steeply (Figure 2.22, curve 1). By addition of plasticizers the glass transition temperature is shifted to lower temperatures. Furthermore the decrease of modulus occurs over a wider temperature range, though the extent of the fall remains the same (curve 2). When the plasticizer content is further increased one finally obtains a curve (curve 3) the shape of which is very like that of an elastomer (PVC synthetic leather). These curves have a parallel in the three characteristic stress-strain curves in Section 2.3.12.1. Curve 1 represents a polymer with high elastic modulus at room temperature and with low extensibility; curve 2 represents a plasticized polymer with yield point; and curve 3 represents a polymer/plasticizer system with low elastic modulus and high extensibility.

2.3.12.3. *Determination of impact strength and notched impact strength*

Besides the test methods in which the load is applied over a relatively long period and the deformation rate is small (as in stress-strain and hardness measurements), there are other methods of interest in which the material is placed suddenly under high stress. Such measurements include that of impact strength, which gives an idea of the brittleness and toughness of the material. For this purpose, the test specimen is broken by means of a weighted pendulum, the energy of fracture (N mm = mJ) being measured. The impact strength is given by the fracture energy divided by the area of cross-section and has the units mJ mm^{-2}. The test specimens may be either smooth or notched rectangular bars; accordingly one speaks of (normal) impact strength (a_n) and of notched impact strength (a_k). The notched

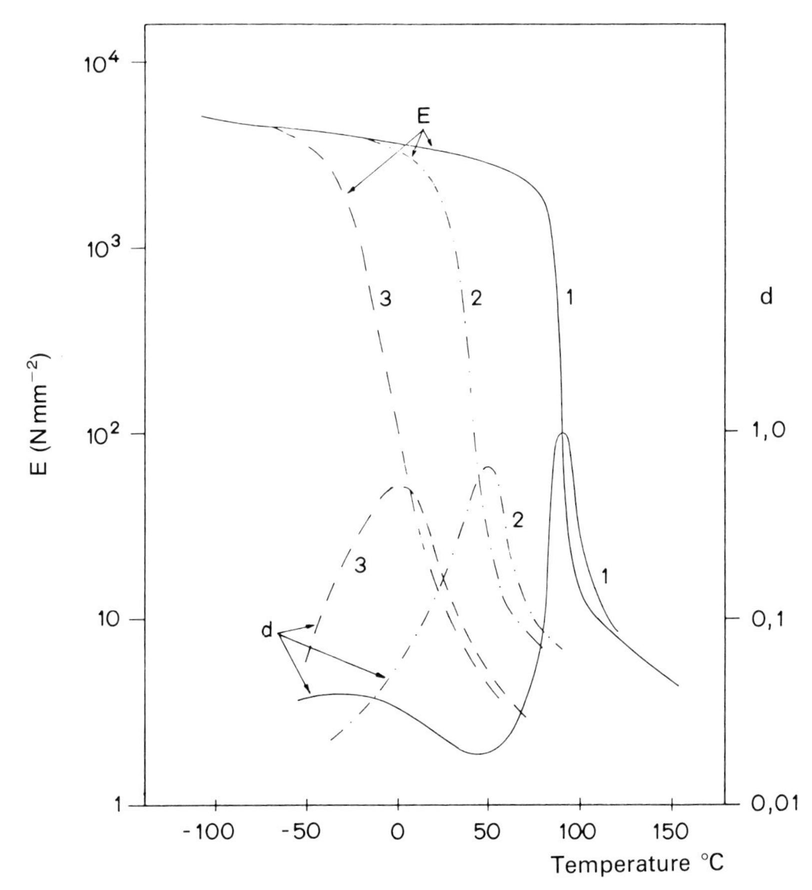

FIGURE 2.22 Dependence of the elastic modulus E and the mechanical loss factor d on temperature for poly(vinyl chloride) with various contents of plasticizer. Curves 1: unplasticized; curves 2: with 20 wt. % plasticizer (dioctyl phthalate); curves 3: with 40 wt. % plasticizer (dioctyl phthalate).

impact method is simply a variant on the normal impact test, the specimen being notched in a V-shape before the test. On impact the shear is thereby concentrated at a particular point on the test bar.

There are two principle methods for measuring impact and notched impact strength, which in practice differ only in the way in which the test bar is held. In the Charpy method the test piece is suspended at both ends and is struck in the centre by a weighted pendulum; in the Izod method the test piece is rigidly supported at one end only and is struck at the free end by the pendulum.

2.3.12.4. Determination of hardness

By hardness is understood the resistance that one body offers against penetration by another. Hence, to judge the hardness of a material one measures the force that is required to obtain a certain depth of penetration. This force or depth of penetration is dependent not only on temperature but also on many factors not characteristic of the material, such as the form of the penetrating body (ball or needle) and the time factor. Unfortunately there is no universal hardness test and there are numerous methods in use[1], many of which do not cover the entire range of possible hardness and can, therefore, not be used on all polymers.[2] These methods can be divided into two groups: in one group the depth of penetration is measured after removal of the load (e.g. Brinell test); in the other the depth of penetration is measured under full load.[3] The latter methods are the most suitable for thermoplastics and thermosetting plastics. A very popular method is the ball hardness test: a steel ball (diameter 5 mm) is pressed into the sample (4 mm thick plate) with constant force and the depth of penetration measured after 10 s and after 60 s loading. Since the area of deformation must be taken into account, ball hardness has the dimensions $N\ mm^{-2}$.

It must be emphasized that the values of hardness measured by the different methods cannot be interconverted.

2.4. PROCESSING OF POLYMERS

Processing of a polymer can be performed with the polymer in various states of aggregation: in solution, in dispersion, and in the melt. The method chosen will depend on whether the polymer melts without decomposition or is soluble. However, the nature of the application is also decisive. In practice a moulded object can be prepared from a thermoplastic only via the melt, while for textile coatings the only feasible method is to process from solution or dispersion. But processing of polymers is not only of industrial interest. It is also indispensable for the pre-fabrication of

---

[1] See *Nitsche/Wolf*, "Praktische Kunststoffprüfung" (1961).

[2] An indication of the hardness of a polymer can also be obtained from the elastic modulus (higher E-modulus corresponds to greater hardness), with the advantage that here the whole range of elasticity, and hence of hardness, can be embraced.

[3] Evaluation of hardness may differ enormously, depending on the principle of measurement. Thus, in testing the hardness of an elastic material, such as a rubber eraser, the Brinell test would indicate that it is very hard (since it measures only the residual deformation), while the ball hardness method would indicate that it is very soft since it determines both permanent and elastic deformation.

defined test specimens required for a number of physical and chemical investigations. Although there is suitable equipment for carrying out the usual industrial processing procedures, such as injection moulding and extrusion, the following simpler methods suffice for preliminary investigations on a laboratory scale (50–1000 g).

2.4.1. Size reduction of polymers

In so far as a polymer may not be formed in a suitable finely divided form during preparation or recovery (spray precipitation or freeze drying), it may be necessary to reduce the size of the particles before carrying out the main test or investigation. Conventional grinding in a mortar is usually ineffective because of the toughness of many polymers, unless the sample is first made brittle by cooling to low temperature with liquid nitrogen. During grinding, the polymer frequently acquires an electrical charge; this can be prevented by moistening the sample with a little ether. Grinding can also be done in suitable mills. However the generation of heat may cause low melting polymers to begin to flow, leading to formation of lumps; this can be avoided by the addition of small pieces of dry ice. Especially suitable are cooled analytical mills[1], which have the additional advantage of small milling volume.

2.4.2. Melt processing of polymers

The processing of polymers in the melt is the method most extensively used in technology. It is an essential requirement that the product does not undergo any significant degradation at the high processing temperature, which must generally be considerably higher than the softening or melting temperature in order to reduce the high viscosity of the polymer melt. In this processing technique the polymer is heated above its melting or softening point, the viscous melt is brought into the desired form by mechanical forces and the formed object is finally cooled. This procedure is very widely applicable and allows the preparation of objects in practically any size or shape, for example of moulded bodies (by pressing, injection moulding or extrusion), of films and sheets (by extrusion and callandering), and of fibres (by extrusion).

Polymers that are difficult to process in the melt, either because the melt viscosity is too high or because they decompose, can often be processed at

[1] For example, as supplied by Firma Janke und Kunkel, Staufen/Brsg., W. Germany.

low temperatures through the addition of a plasticizer, leading to improved flow properties. Suitable plasticizers are high-boiling liquids, such as full esters of phosphoric acid, phthalic acid (e.g. dioctyl phthalate), and various aliphatic dicarboxylic acids; they must of course be compatible with the polymer. The effect of the added plasticizers can be explained in terms of a reduction in the intermolecular forces between the macromolecules as a result of the presence of plasticizer molecules between the polymer chains. The mobility of the polymer chains is thereby enhanced and the glass transition temperature lowered (for internal plasticizer action see Section 3.3.1). Polymers that contain plasticizers are more flexible and of lower hardness than the untreated products (see Example 3–49). Plasticizers play an important technological role in many polymers.

2.4.2.1. *Preparation of pressed films*

Thin polymer films are very suitable for a number of physical investigations (microscopy, infrared spectroscopy, mechanical measurements). They can be prepared in the laboratory as follows. A certain amount of finely powdered polymer is spread on a thin (0.1 mm) aluminium foil (15 cm × 15 cm), which is then covered with a second foil and the whole placed between the plates of a hydraulic press heated to the melting point of the polymer. After pressing for 30–60s the foil sandwich is removed from the press, and cooled with water or between two metal plates. Finally the two aluminium foils are pulled carefully away from the thin polymer film. If a certain film thickness is required a suitable template can be laid between the aluminium foils. The optimum conditions, such as amount of polymer, press-temperature, -pressure and -time, must be determined empirically in every case. If the film is opaque, the temperature was probably too low; if it is too thin or contains gas bubbles (decomposition) the temperature was too high. Finally, the rate of cooling can also affect the properties of the film. Sometimes the aluminium foils can only be removed from the polymer film with difficulty; rapid cooling (with water) or prior coating of the aluminium foils with silicone oil or with an aqueous dispersion of polytetrafluoroethylene usually prevents sticking.

If a suitable press is not available, one may improvise as follows.[1] The heating plates of two electric irons are first bored to take a thermocouple and then connected in parallel to the power supply through a variable transformer. A calibration curve is determined for the temperature

[1] *W.R. Sorenson* and *T.W. Campbell*, "Preparative Methods of Polymer Chemistry" 2nd Edn., Interscience Publishers, New York 1968.

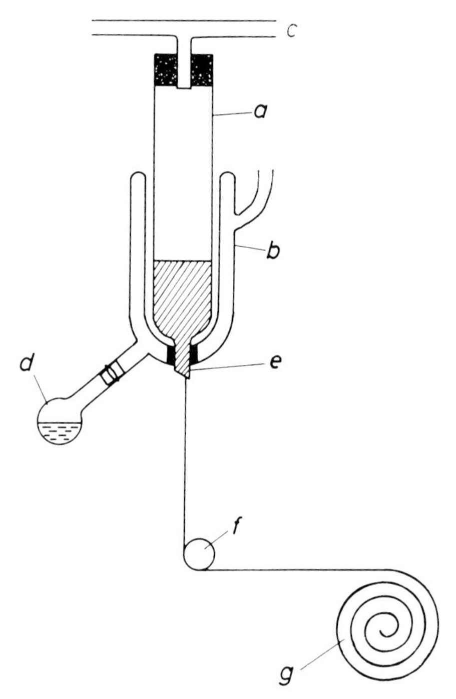

FIGURE 2.23 Apparatus for melt-spinning of polymers

attained at different voltages. For the preparation of a film the finely powdered polymer is placed, as described above, between the two hot-plates, which are then pressed together horizontally in a suitable vice. In this arrangement the pressure cannot be measured; but with a little skill the optimum conditions can be found.

2.4.2.2. *Melt-spinning*

The simplest method of melt-spinning is to melt the polymer in a test tube, dip a glass rod in the melt and pull it out slowly. The threads obtained in this manner are short and irregular so that they are not well suited for subsequent investigations. A simple apparatus[1] for continuous melt-spinning is shown in Figure 2.23. A thick-walled glass tube (a) (diameter 3–4 cm), drawn out to a short, sealed-off capillary at one end, is placed in

[1] *W.R. Sorenson* und *T.W. Campbell*, "Preparative Methods of Polymer Chemistry" 2nd Edn., Interscience Publishers, New York 1968; for other devices see *Houben-Weyl*, *14/2* (1963) 138.

a double-walled vessel (b) and half-filled with polymer; the T-piece (c) is then evacuated and filled with nitrogen. Finally the polymer is melted by heating with the vapours of the boiling liquid contained in flask (d) (see Section 2.1.4). After the tip of the capillary (e) has been cut off, the polymer melt can be forced out in the form of a fine thread by application of a moderate pressure of nitrogen. The fibre is passed over the guide-roll (f) and wound up on the spool (g). The temperature in the capillary can be regulated by adjusting the extent to which it projects from the double-walled vessel (b).

2.4.3. Processing of polymers from solution

If a suitable arrangement for processing the polymer melt is not available, or if the polymer will not melt, or melts only with decomposition, the processing may be carried out from solution. This technique, however, is limited to the fabrication of films and fibres.

2.4.3.1. Preparation of films

The simplest way of making films in the laboratory is to pour a highly viscous solution of the polymer onto a glass or metal plate and to allow the solvent slowly to evaporate. The polymer film is then carefully peeled off. The following points have to be watched. The solvent should not evaporate too quickly, otherwise the resulting film can wrinkle and tear; if necessary the experiment is carried out in an atmosphere of solvent (use desiccator). On the other hand, the boiling point should not be too high, otherwise the last residues of solvent are difficult to remove. The concentration of the polymer should be such that the solution can still be poured, but does not run off the glass plate. An approximately 20% solution is usually about right but the best conditions must be established by experiment. In order to obtain a film of uniform thickness the glass plate must be as level as possible and the polymer solution evenly distributed. This may be achieved most easily with the aid of a glass rod wrapped at both ends with some tape or twine. For more precise work metal devices with adjustable layer thickness are preferable. When ready the film is lifted at one edge with the aid of a razor blade or small knife and then slowly peeled away from the glass plate. Instead of using a glass plate it is also possible to cast the film on the surface of mercury contained in a flat dish.

Films can also be made from aqueous polymer dispersions at room temperature if the glass transition temperature of the polymer is not too high. For example an aqueous poly(vinyl acetate) dispersion (see Example 3–07) is painted on to a flat support and allowed to dry. The polymer particles, originally separated from one another by the action of protective colloids and emulsifiers, agglomerate irreversibly under these conditions and form a coherent film that cannot be redispersed in water.

2.4.3.2. Solution-spinning

There are two methods of spinning fibres from solution: dry spinning and wet spinning. In dry spinning the viscous polymer solution is forced through a jet (spinneret) into a chamber filled with hot air (or nitrogen); the solvent is thereby evaporated leaving behind a ready-made thread. This spinning process cannot be conducted in a simple manner and is not readily carried out in the laboratory. On the other hand wet spinning, in which the polymer solution is injected into a suitable precipitant, so coagulating the polymer in the form of a thread, is readily conducted on a small scale. The quality of the thread depends very much on the precipitation conditions (precipitant, bath temperature) which must be determined experimentally in each case. The precipitant must be chosen so that complete coagulation does not occur too quickly; if necessary a solvent-precipitant mixture can be used (e.g. polyacrylonitrile dissolved in dimethylformamide, precipitated in dimethylformamide/water mixture). For injecting the polymer solution into the precipitant bath, hypodermic syringes are very useful; they allow control of both the thread diameter and rate of spinning, in a simple manner (see Example 3–20).

2.4.4. Preparation of foamed polymers (foam plastics)

Foam plastics are materials with a porous, cell-like structure. They are of two types, according to whether the pores are open or closed. Amongst the first group are materials used as sponges; examples of the latter are heat-insulating and sound-proofing materials.

There are two main methods of making foam plastics. The first method starts from a ready-made polymer which is then foamed in a separate process. All foamed materials prepared from thermoplastics, such as polystyrene (see Example 3–04) and the so-called foam rubbers, are fabricated in this way. With thermoplastics the porous structure is achieved by incorporating a "blowing agent" into the polymer which is then heated above the softening point so that it is expanded by the gas or vapour from the blowing agent. Suitable expanding agents are low-boiling inert solvents (e.g. pentane or halogenated aliphatic hydrocarbons), as well as compounds that decompose with gas evolution on heating (hydrogencarbonates, azo-compounds). In the preparation of foam rubber from an aqueous emulsion of polymer, use is made of the fact that the detergent in the emulsion causes it to foam very strongly on stirring, especially when air is blown in at the same time. By subsequent crosslinking (curing) of the emulsified polymer the fine-pored cell structure is fixed so that it does not disappear on further treatment (e.g. on washing with water).

In the second method of preparation of foamed polymers the foaming process proceeds simultaneously with the formation of polymer. All thermosetting foams, for example foams of urea/formaldehyde resins (see Example 4–15), or foams of epoxy-resins, are made in this way. For this purpose, the requisite components for making the thermosetting resin are mixed together with an expanding agent (usually a low-boiling liquid), the mixture poured into a mould, and allowed to react at the required temperature. The liberated heat of reaction evaporates the expanding agent, causing the reaction mixture to foam. A special case of this technique is to be found in the preparation of polyurethane foams, since here no expanding agent need be added; under the special conditions of the addition polymerization of diisocyanates to diols, the inclusion of a little water generates carbon dioxide from part of the isocyanate, causing the polymer to foam (see Section 4.2.1.2).

3. Synthesis of Macromolecular Substances by Addition Polymerization of Single Compounds

As already explained in Section 1.1 polymerization reactions can proceed by various mechanisms and be catalyzed by initiators of diverse kinds. For addition polymerization of single compounds, initiation of chains may occur via radical, cationic, anionic or so-called coordinative-acting initiators, but some monomers will not polymerize by more than one mechanism. Both thermodynamic and kinetic factors can be important, depending on the structure of the monomer and its electronic and steric situation[1,2] (cf. Table 3.1).

3.1. RADICAL HOMOPOLYMERIZATION

Radical polymerization is induced by an initiation step in which radicals[3] are formed. Radicals can be generated thermally from the monomer, although such a mechanism of initiation has only been completely verified in the case of styrene (see Example 3–01). Radicals are usually generated by decomposition of an initiator (frequently called the catalyst). The radicals so formed then react with the monomer; successive additions of further monomer molecules produce growing radicals (macroradicals) which are finally removed by a termination reaction. Termination usually occurs by combination or disproportionation of two macroradicals; the radicals generated by the decomposition of the initiator are incorporated into the macromolecules as end groups, giving two end groups per macromolecule for termination by combination, but only one for termination by disproportionation.

---

[1] *A.D. Jenkins* and *A. Ledwith*, "Reactivity, Mechanism and Structure in Polymer Chemistry", Wiley-Interscience, London 1974.
[2] *K.J. Ivin*, Angew. Chem. *85* (1973) 533.
[3] For the general chemistry of free radicals see for example *W.A. Pryor*, "Free Radicals", McGraw-Hill, New York 1966; *I. Ernest*, "Bildung, Struktur und Reaktionsmechanismen in der organischen Chemie", Springer Verlag, Wien, New York 1972, p. 277 et seq.

TABLE 3.1

Initiators for addition polymerizations

Radical	Cationic	Anionic	Coordinative
Inorganic and organic percompounds, e.g. peroxodisulfates, peroxides, hydroperoxides, peresters Azo-compounds Substituted ethanes e.g. benzpinacol Redox systems with inorganic and organic components Heat U.v. radiation High-energy radiation	Protonic acids Lewis acids with or without co-initiators Carbonium ions Iodonium ions Ionizing radiation	Proton acceptors Lewis bases Organometallic compounds Electron-transfer agents, e.g. alkali metals, alkali-aromatic complexes, alkali metal ketyls	Organometallic compounds Mixed catalysts (Ziegler-Natta catalysts) π-complexes with transition metals Activated transition metal oxides

In transfer reactions the growth of a chain is ended, for example, by transfer of a hydrogen atom from the molecule ZH, but at the same time a new polymer chain is started by the radical $Z\cdot$ that is formed at the same time. Thus several macromolecules result from one primary radical; therefore, the kinetic chain length, namely the total number of monomer molecules induced to polymerize by one primary radical, is much larger than the degree of polymerization of the macromolecules formed. A general scheme is as follows.

Primary act:

$$\text{Initiator} \longrightarrow 2R\bullet$$

$$R\bullet + CH_2{=}\underset{X}{\underset{|}{C}}H \longrightarrow R{-}CH_2{-}\underset{X}{\underset{|}{\dot{C}}}H$$

Propagation:

$$R{-}CH_2{-}\underset{X}{\underset{|}{\dot{C}}}H + n\,CH_2{=}\underset{X}{\underset{|}{C}}H \longrightarrow R{-}[CH_2{-}\underset{X}{\underset{|}{C}}H]_n{-}CH_2{-}\underset{X}{\underset{|}{\dot{C}}}H$$

Chain transfer:

$$R{-}[CH_2{-}\underset{X}{\underset{|}{C}}H]_n{-}CH_2{-}\underset{X}{\underset{|}{\dot{C}}}H + ZH \longrightarrow R{-}[CH_2{-}\underset{X}{\underset{|}{C}}H]_n{-}CH_2{-}\underset{X}{\underset{|}{C}}H_2 + Z\bullet$$

The radical $Z\cdot$ then reacts like $R\cdot$.
Chain termination by combination:

$$R{-}[CH_2{-}\underset{X}{\underset{|}{C}}H]_n{-}CH_2{-}\underset{X}{\underset{|}{\dot{C}}}H + R{-}[CH_2{-}\underset{X}{\underset{|}{C}}H]_m{-}CH_2{-}\underset{X}{\underset{|}{\dot{C}}}H$$

$$\longrightarrow R{-}[CH_2{-}\underset{X}{\underset{|}{C}}H]_n{-}CH_2{-}\underset{X}{\underset{|}{C}}H{-}\underset{X}{\underset{|}{C}}H{-}CH_2{-}[\underset{X}{\underset{|}{C}}H{-}CH_2]_m{-}R$$

Chain termination by disproportionation:

$$\begin{array}{l} \mathrm{R{-}[CH_2{-}\underset{\displaystyle X}{\underset{|}{C}}H]_n{-}CH_2{-}\underset{\displaystyle X}{\underset{|}{\dot{C}}}H + R{-}[CH_2{-}\underset{\displaystyle X}{\underset{|}{C}}H]_m{-}CH_2{-}\underset{\displaystyle X}{\underset{|}{\dot{C}}}H} \\ \longrightarrow \mathrm{R{-}[CH_2{-}\underset{\displaystyle X}{\underset{|}{C}}H]_n{-}CH{=}\underset{\displaystyle X}{\underset{|}{C}}H + R{-}[CH_2{-}\underset{\displaystyle X}{\underset{|}{C}}H]_m{-}CH_2{-}\underset{\displaystyle X}{\underset{|}{C}}H_2} \end{array}$$

Let us for the moment disregard chain transfer reactions. Radical polymerization then consists of three component reactions: initiation, propagation of the polymer chains, and termination of chain growth. The rate of primary radical formation, v_i, by decomposition of the initiator I, may be written

$$v_i = k_i[\mathrm{I}] \tag{1}$$

The rate constant k_i contains a factor that allows for the efficiency of initiation; not all the radicals generated by the initiator are capable of starting polymer chains, some being lost by combination or other reactions. The initiator efficiency is defined as the ratio of the number of initiator molecules that start polymer chains to the number of initiator molecules decomposed under the given conditions of polymerization. With most radical initiators the efficiency lies between 0.6 and 0.9; it also depends on the nature of the monomer.

The rate of propagation v_p is given by

$$v_p = \frac{-\mathrm{d}[\mathrm{M}]}{\mathrm{d}t} = k_p[\mathrm{M}]\,[\mathrm{R}\cdot] \tag{2}$$

Here it is assumed that k_p is independent of the number of monomer molecules already added; $[\mathrm{R}\cdot]$ denotes the concentration of radicals in the system.

The rate of the termination reaction v_t is given by

$$v_t = k_t[\mathrm{R}\cdot]^2 \tag{3}$$

According to Bodenstein, for a chain reaction in the steady state, the number of radicals formed and disappearing in a given time must be the same. This applies to most addition polymerizations, at least in the region of low conversion. Under these conditions v_i and v_t may be equated.

$$v_i = v_t$$

$$k_i[\mathrm{I}] = k_t[\mathrm{R}\cdot]^2 \tag{4}$$

$$[\mathrm{R}\cdot] = \left(\frac{k_i}{k_t}\right)^{\frac{1}{2}} [\mathrm{I}]^{\frac{1}{2}} \tag{5}$$

Inserting this value in equation (2) yields

$$v_p = \frac{-\mathrm{d}[\mathrm{M}]}{\mathrm{d}t} = k_p\left(\frac{k_i}{k_t}\right)^{\frac{1}{2}} [\mathrm{M}]\,[\mathrm{I}]^{\frac{1}{2}} \tag{6}$$

v_p is identical with the overall rate of polymerization, R_p, since at sufficiently large chain length it determines the consumption of the monomer M almost completely. Hence the rate of polymerization is proportional to the monomer concentration and the square root of the initiator concentration. For polymerization in bulk at low conversion [M] is nearly constant so that

$$R_p \propto [\mathrm{I}]^{\frac{1}{2}} \tag{7}$$

For initiation of polymerization by light or high-energy radiation, the initiator concentration [I] is replaced by the radiation intensity in the above kinetic equations.

Raising the temperature of a radical chain reaction causes an increase in the overall rate of polymerization, since the main effect is an increase in the rate of decomposition of the initiator and hence the number of primary radicals generated in unit time. At the same time the degree of polymerization falls, since, according to equation (3), the rate of the termination reaction depends on the concentration of radicals (see Example 3–02). Higher temperatures also favour side reactions such as chain transfer and branching, and in the polymerization of dienes the reaction temperature can affect the relative proportions of the different types of CRU in the chains.

Although the above derivations involve certain simplifications they nevertheless represent correctly the kinetics of many addition polymerization reactions.[1,2,3] However, the behaviour is different when the polymerization is conducted under heterogeneous conditions, e.g. in suspension or in emulsion (see literature cited in Section 2.1.5.3).

For radical polymerizations of some monomers in bulk a specific effect can appear when the conversion exceeds a certain value. In these cases the

---

[1] For detailed information on the kinetics of polyreactions see *C.H. Bamford* and *C.F.H. Tipper*, "Comprehensive Chemical Kinetics", Vol. 14A: "Free Radicals", Elsevier, Amsterdam 1977; *A.M. North*, "The Kinetics of Free Radical Polymerization", Pergamon Press, Oxford, New York, Toronto, Paris, Braunschweig 1966; *G.P. Gladyshev* and *V.A. Popov*, "Radikalische Polymerisation bis zu hohen Umsätzen". Akademie-Verlag, Berlin 1978; *Kh. S. Bagdasarŷan*, "Theory of Free-Radical Polymerization", Israel Program for Scientific Translations, Jerusalem 1968.

[2] *G. Henrici-Olivé* and *S. Olivé*, Adv. Polym. Sci. *2* (1961) 496.

[3] *G. Henrici-Olivé* and *S. Olivé*, "Polymerisation: Katalyse-Kinetik-Mechanismen", Verlag Chemie, Weinheim 1970.

viscosity of the reaction mixture increases to such an extent as a result of the formation of macromolecules, that the mobility of the growing macroradicals becomes severely restricted. Bimolecular termination, but not the propagation reaction, is thereby hindered, so that both the degree of polymerization and the reaction rate increase; the system is no longer in a steady state and the radical concentration rises continuously. The increasing reaction rate, coupled with the more difficult heat exchange in the very viscous medium, leads to a rise in temperature as a consequence of the heat evolution, and hence to an auto-acceleration of the reaction, which can become explosive in nature. This phenomenon is called the gel effect or Trommsdorff effect, but does not occur to the same extent in all monomers. It is especially noticeable with methyl methacrylate (see Example 3–14a). Somewhat similar behaviour is observed in the polymerization of some monomers in "bad" solvents, in which the resulting macromolecules are more tightly coiled than in "good" solvents, where the polymer chains are more strongly solvated; (for "goodness" of solvents, see Section 2.3.1).

The interplay of radical formation, propagation and termination of the growing chains determines the overall rate and degree of polymerization, provided there are no chain transfer reactions. When a growing polymer chain undergoes chain transfer its growth is terminated but at the same time a new polymer chain is started; the kinetic chain is therefore uninterrupted (see mechanism on p. 126).

The kinetic chain length is given by the number of monomer molecules consumed per initiation step. Since the efficiency of most initiators is not known quantitatively (see p. 127) it is necessary to compare the rate of the propagation reaction with either the rate of initiation or the rate of termination. If there is no chain transfer, the kinetic chain length ν for termination by disproportionation is equal to the number-average degree of polymerization:

$$\bar{P}_n = \nu = \frac{v_p}{v_i} = \frac{v_p}{v_t} \tag{8}$$

On the other hand for termination by combination the degree of polymerization is equal to twice the kinetic chain length:

$$\bar{P}_n = 2\nu = \frac{2v_p}{v_i} = \frac{2v_p}{v_t} \tag{8a}$$

If chain transfer does take place the rate v_f of the chain transfer reaction ($R\cdot + ZH \rightarrow RH + Z\cdot$) must be taken into account:

$$v_f = k_f[R\cdot][ZH] \tag{9}$$

and one obtains for the degree of polymerization $\bar{P}_n$ in the case of termination by disproportionation:

$$\bar{P}_n = \frac{v_p}{v_i + v_f} \tag{10}$$

and for termination by combination:

$$\bar{P}_n = \frac{v_p}{0.5v_i + v_f} \tag{10a}$$

Taking the reciprocal of $\bar{P}_n$ and inserting the expressions for v_p, v_i and v_f from equations (4) to (6), one obtains

$$\frac{1}{\bar{P}_n} = \frac{v_i}{\alpha v_p} + \frac{v_f}{v_p} = \frac{k_i[\mathrm{I}]}{\alpha k_p \left(\frac{k_i}{k_t}\right)^{\frac{1}{2}} [\mathrm{I}]^{\frac{1}{2}}[\mathrm{M}]} + \frac{k_f[\mathrm{R}\cdot]\,[\mathrm{ZH}]}{k_p[\mathrm{R}\cdot]\,[\mathrm{M}]} \tag{11}$$

where $\alpha = 1$ for termination by disproportionation and $\alpha = 2$ for termination by combination.

Since the chain carrier ZH may be the monomer, the initiator, the solvent (S), an added transfer agent (regulator), or the polymer already formed, a more general form of equation (11) is

$$\frac{1}{\bar{P}_{n,o}} = k\frac{[\mathrm{I}]^{\frac{1}{2}}}{[\mathrm{M}]} + C_M + C_I\frac{[\mathrm{I}]}{[\mathrm{M}]} + C_S\frac{[\mathrm{S}]}{[\mathrm{M}]} + C_{ZH}\frac{[\mathrm{ZH}]}{[\mathrm{M}]} + \cdots \tag{12}$$

where C_X denotes the chain transfer constant k_f/k_p appropriate to the chain transfer agent X (cf. Example 3–14). Since the chain transfer constant C_I for most initiators is approximately zero, equation (12) shows that at moderate conversion the reciprocal of the degree of polymerization is a linear function of the square root of the initiator concentration. Since in turn $[\mathrm{I}]^{\frac{1}{2}}$ is proportional to the overall rate of polymerization R_p, equation (7), the degree of polymerization is lower, the faster the polymerization.

Transfer reactions with solvent and with those compounds termed regulators are especially important because of their marked effect on the molecular weight of the polymer being formed. While the transfer constants for most solvents are not very big (e.g. for benzene reacting with the growing polystyrene radical at 60°C, C_s is of the order of 10^{-5}), there are some with relatively high transfer constants, so that the polymer formed has a correspondingly short chain length. A particularly well investigated case is that of the polymerization of styrene in carbon tetrachloride, where the transfer constant is about 10^{-2}. The resulting polystyrene is of very low molecular weight and consists of a mixture of oligomers:

$$\begin{aligned} R\cdot + CCl_4 &\longrightarrow RCl + \cdot CCl_3 \\ Cl_3C\cdot + M &\longrightarrow Cl_3C\text{—}M\cdot \end{aligned} \tag{13}$$

The polymer thus contains chloro and trichloromethyl end groups. Likewise the polymerization of ethylene in carbon tetrachloride gives oligomers of formula $Cl\text{—}(CH_2\text{—}CH_2)_n\text{—}CCl_3$ ($n = 2$, 3 and 4), which can be converted by hydrolysis to the corresponding ω-chlorocarboxylic acids. Such a process is called telomerization.[1,2,3]

More important, however, is the ability to control the reduction of molecular weight by the use of regulators. The molecular weight of a polymer can only be controlled to a limited extent by the adjustment of monomer concentration, initiator concentration and temperature. Hence in technology it is often the practice to add transfer agents. For this purpose one may use various thiols, which, because of their high transfer constants (e.g. for 1-dodecanethiol in the polymerization of styrene, $C_f = 19$), need only be added in very small amount (about 0.1% with respect to monomer). The simplest way of determining the transfer constant of such a regulator, using equation (12), is by polymerization experiments at constant initiator and monomer concentration, at varying concentrations of transfer agent ZH.

In the absence of regulator one may write:

$$\frac{1}{\bar{P}_{n,o}} = k\frac{[I]^{1/2}}{[M]} + C_M + C_I\frac{[I]}{[M]} + C_S\frac{[S]}{[M]} \tag{14}$$

Then with addition of regulator ZH:

$$\frac{1}{\bar{P}_n} = \frac{1}{\bar{P}_{n,o}} + C_{ZH}\frac{[ZH]}{[M]} \tag{14a}$$

Plotting the reciprocal of the number-average degree of polymerization of polymers obtained at different regulator concentrations against [ZH]/[M] a straight line is obtained which intersects the ordinate at $1/\bar{P}_{n,o}$ and has slope equal to the transfer constant of the regulator C_{ZH} (also see Section 2.1.5.4 and Example 3–14).

[1] *J.W. Breitenbach* and *A. Maschin*, *Z. Phys. Chem.* A*187* (1940) 1975.
[2] *R. Kh. Freidlina* and *Sb. A. Karapetyan*, "Telomerisation and New Synthetic Materials", Pergamon Press, London 1961.
[3] *C.M. Starks*, "Free Radical Telomerization", Academic Press, New York, 1974.

Finally there are many substances that can inhibit polymerization reactions. Amongst these are molecular oxygen, nitric oxide, phenols such as hydroquinone and 4-*tert*-butylpyrocatechol, quinones, certain aromatic amines such as *N*-phenyl-β-naphthylamine, nitrocompounds and some sulfur compounds. The mechanism of action of most inhibitors is not yet fully clarified; the inhibitors react either with the primary radicals or with the growing chains to yield products that are no longer active in propagation. Stable free radicals such as *N*,*N*-diphenyl-*N*′-picrylhydrazyl

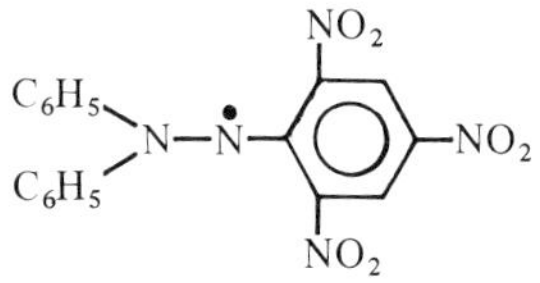

$$(C_6H_5)_2N{-}\dot{N}{-}C_6H_2(NO_2)_3$$

are also effective inhibitors. Since this radical is strongly coloured its consumption during the inhibition period can be followed photometrically[1]; in this case, however, the inhibition reaction does not consist of simple combination of the stable radical with the growing radical end of the polymer chain.

Inhibitors are frequently used to stop a polymerization quickly, for example in kinetic investigations. Another important application is the stabilization of monomers against undesired polymerization during storage. Autoxidation of unsaturated monomers by the action of atmospheric oxygen frequently results in the formation of peroxidic compounds which can generate radicals at relatively low temperatures, so initiating polymerization; inhibitors are added as stabilizers to suppress such undesired and uncontrolled polymerization (cf. Section 2.1.5.4); these must of course be removed before using the monomer for polymerization reactions (see, for example, Example 3–01). The effectiveness of an inhibitor depends, amongst other things, on its structure. Since the inhibitors are consumed by growing polymer chains, the time during which they prevent polymerization (incubation time, induction period) depends on their concentration in the stabilized monomer (cf. Example 3–19).

Molecular oxygen plays a special role in radical polymerizations. It is known to react very rapidly with hydrocarbon radicals with the formation of peroxy radicals:

$$R\cdot + O_2 \rightarrow R{-}O{-}O\cdot$$

---

[1] For summary see *C. Walling*, "Free Radicals in Solution", Wiley, New York, 1957; also see *P.D. Bartlett* and *H. Kwart*, J. Am. Chem. Soc. *72* (1950) 1051; *F. Tüdös*, *T.F. Berezhnikh* and *M. Azori*, Acta Chim. Acad. Sci. Hung. *24* (1960) 91.

It will, therefore, be understood why atmospheric oxygen must be carefully excluded during radical polymerization. Peroxyl radicals are much less reactive than most alkyl (or aryl) radicals, but they can add a further monomer molecule, regenerating an alkyl radical which can react again with oxygen. The rate of consumption of monomer, relative to that in the absence of oxygen, is substantially reduced. An alternating addition of unsaturated monomer and oxygen is observed, resulting in the formation of a polymeric peroxide (copolymerization with molecular oxygen):

$$\cdots\!-\!CH_2\!-\!\underset{\displaystyle X}{\underset{|}{CH}}\!-\!O\!-\!O\!-\!CH_2\!-\!\underset{\displaystyle X}{\underset{|}{CH}}\!-\!O\!-\!O\!-\!\cdots$$

Normal polymerization commences only after complete consumption of the oxygen; this is then accelerated by the formation of additional initiating radicals through the thermal decomposition of the polymeric peroxide. Thus molecular oxygen at first inhibits the polymerization, but, after its consumption, there is an acceleratory action.

Unlike ionic polymerizations, radical chain polymerizations have so far been found to occur only with unsaturated compounds. In some cases they can be induced purely thermally, or by means of light or high-energy radiation; generally, however, radical initiators such as per-compounds, azo-compounds, and redox systems are used.

3.1.1. Polymerization with per-compounds as initiators

Organic and inorganic per-compounds are especially important as initiators of radical polymerizations. Hydroperoxides, dialkyl peroxides, diacyl peroxides and per-esters are typical organic per-compounds. Since they dissolve not only in organic solvents but also in most monomers, they are suitable for solution polymerization as well as bulk or bead polymerization. Their decomposition into radicals can be brought about either thermally, or by irradiation with light, or by redox reactions (see Section 3.1.3). The rate of decomposition of organic per-compounds depends on their structure and on the temperature. For initiation by thermal decomposition of per-compounds an acceptable rate of polymerization is generally attained only above 50°C. Per-esters, for example diethyl peroxydicarbonate, are exceptional in decomposing rapidly at room temperature and, because of the danger of explosion, should be added only in dilute solution.

A per-compound that is frequently used (concentration 0.1–1 wt.% with respect to monomer) is dibenzoyl peroxide (see Examples 3–02, 3–05, 3–06). It decomposes in solution at temperatures of about 50–80°C, mainly into benzoyloxy radicals; at higher temperatures phenyl radicals are formed to an increasing extent by elimination of carbon dioxide, so that the end groups of the resulting polymer are either hydrolyzable benzoic ester groups or non-hydrolyzable phenyl groups:

$$C_6H_5-\underset{\underset{O}{\|}}{C}-O-O-\underset{\underset{O}{\|}}{C}-C_6H_5 \longrightarrow 2\, C_6H_5-\underset{\underset{O}{\|}}{C}-O\bullet \longrightarrow 2\, C_6H_5\bullet + 2\, CO_2$$

Dibenzoyl peroxide, when substituted with chlorine or bromine, decomposes faster than dibenzoyl peroxide itself under comparable conditions; such peroxides provide a means of introducing an analytically identifiable end group which then allows determination of molecular weight (see Section 2.3.2.2). Equations (6) and (11), according to which the rate of polymerization increases and the average degree of polymerization decreases with increasing initiator concentration, are satisfied by most monomers when either unsubstituted or substituted dibenzoyl peroxides are used as initiators.

Thermally activated organic per-compounds are generally used for polymerization in bulk or in organic solvents, as well as for bead polymerization; but inorganic per-compounds are the most suitable for initiating polymerization in aqueous solution, suspension or emulsion. Hydrogen peroxide is mainly used as a component of a redox initiator (see Example 3–22); in contrast, potassium and ammonium peroxodisulfate (concentration 0.1–1 wt.% with respect to monomer) are very frequently used without a reducing agent, since even at 30°C they decompose thermally into radicals that can initiate polymerization.[1]

$$^{-}O-\overset{\overset{O}{\|}}{\underset{\underset{O}{\|}}{S}}-O-O-\overset{\overset{O}{\|}}{\underset{\underset{O}{\|}}{S}}-O^{-} \longrightarrow 2\cdot SO_4^{-}$$

$$2\cdot SO_4^{-} + 2H_2O \longrightarrow 2HSO_4^{-} + 2HO\cdot$$

Ammonium peroxodisulfate is more soluble in water than the potassium salt; furthermore it dissolves in some polar organic solvents (e.g. dimethyl-

[1] Also see *A. Ledwith*, J. Polym. Sci. B*13* (1975) 109.

TABLE 3.2

Per-compounds for initiation of radical polymerization

Per-compound	Formula	Suitable temperature for polymerization (°C)
Hydrogen peroxide	H—O—O—H	30–80
Potassium peroxodisulfate	$KO-S(=O)_2-O-O-S(=O)_2-OK$	30–80
Dibenzoyl peroxide	$C_6H_5-C(=O)-O-O-C(=O)-C_6H_5$	40–100
α,α-Dimethylbenzyl hydroperoxide[a]	$C_6H_5-C(CH_3)_2-O-O-H$	50–120
Di-*tert*-butyl peroxide	$CH_3-C(CH_3)_2-O-O-C(CH_3)_2-CH_3$	80–150

[a] Also called cumyl hydroperoxide.

formamide), so that it is sometimes also used for initiating polymerizations in organic media. In polymerizations initiated by peroxodisulfates the reaction medium is liable to become acidic, so that buffering is generally necessary (see Example 3–20).

In the case of initiation with per-compounds in organic solvents (rather than in bulk or in aqueous medium), chain transfer with the solvent and consequent lowering of the average degree of polymerization have to be taken into consideration; suitable solvents with low transfer constants are benzene and cyclohexane. According to equations (6) and (12), at constant initiator concentration, the rate and average degree of polymerization fall with decreasing monomer concentration.

A list of some per-compounds that generate free radicals is given in Table 3.2; extensive information can be found in the literature.[1,2]

[1] *Houben-Weyl 14/1*, 59 et seq., 209 et seq.
[2] "Polymer Handbook", *J. Brandrup* and *E.H. Immergut* (Eds.), John Wiley & Sons, New York, London, Sydney, Toronto, 2nd Ed. 1975, pp. II–12 to II–39.

It should be emphasized that any monomer that is susceptible to radical polymerization can generally be polymerized with any per-compound (or azo-compound) that decomposes into radicals, i.e. the type of initiator affects only the rate and average degree of polymerization, the nature of the end groups or the number of branches, but not the polymerizability. With redox systems and especially with ionic initiators this is not the case. Hence, it can generally be established whether a (carefully purified) unsaturated compound is polymerizable radically, simply by heating with 1 wt. % dibenzoyl peroxide (or potassium peroxodisulfate, when working in aqueous medium) under nitrogen for an extended time at 50–120° C.

Example 3–01:
Thermal polymerization of styrene in bulk (effect of temperature)

Monomeric styrene is freed from phenolic inhibitors by shaking twice with 10% sodium hydroxide solution, washing three times with distilled water, drying over calcium chloride or silica gel and distilling into a receiver (see Section 2.1.2) under reduced pressure of nitrogen (b.p. 82° C/100 torr, 46° C/20 torr). It is stored in a refrigerator until required.

4 g (38.4 mmol) of destabilized styrene is weighed into each of five thick-walled Pyrex tubes (content 15–20 ml). The tubes, equipped with a suitable adaptor (see Section 2.1.3), are now cooled in a methylene chloride/dry ice cold bath, thereby freezing the styrene (m.p. − 30.6° C); after evacuation with a filter pump and thawing, the tubes are filled with nitrogen. This sequence is repeated twice more. Finally the tubes are sealed off under nitrogen. The samples are polymerized at 80, 100, 110, 120, and 130° C, respectively, by placing in an appropriately adjusted thermostat or vapour bath (CAUTION: the tubes may explode; place behind shield, or cover with cloth!). After exactly 6 h the sealed tubes are rapidly cooled by immersion in cold water (wear safety goggles) and then opened. The contents are each dissolved in 20–30 ml benzene and the solution run slowly from a dropping funnel into 200–300 ml methanol with stirring, thereby precipitating the polystyrene. The polymers are filtered off using sintered glass crucibles (No. 2 porosity) and dried to constant weight in vacuum at 50° C. The observed yield (in %) is plotted as a function of polymerization temperature. Using an Ostwald viscometer (capillary diameter 0.3 mm) the limiting viscosity numbers of all samples are determined in benzene at 20° C, and the average degrees of polymerization derived (see Section 2.3.2.1). These values are plotted as a function of temperature.

Example 3–02:
Bulk polymerization of styrene with dibenzoyl peroxide (effect of initiator concentration)

4 thick-walled tubes carrying a ground joint and attachable stopcock are charged respectively with 4.7 mg (0.019 mmol), 9.3 mg (0.038 mmol), 46.5 mg (0.19 mmol) and 93 mg (0.38 mmol) of dibenzoyl peroxide[1] (BPO), together with 4 g (38.4 mmol) of destabilized styrene (see Example 3–01). The same amount of styrene is also placed in a fifth tube. The BPO is dissolved in the styrene by shaking. The stopcock is attached and fastened with springs, and the air removed as described in Example 3–01. The samples are now kept in a thermostat at 60° C (± 0.1° C) for 4 h, then rapidly cooled in ice-water and worked up as in Example 3–01. The results are plotted as rate of polymerization (% conversion per hour) against the square root of the initiator concentration (in mole %), and as the degree of polymerization against the reciprocal of the square root of the initiator concentration. These values are compared with those for the thermally polymerized sample and with the corresponding series of experiments with 2,2′-azoisobutyronitrile as initiator (Example 3–11).

Example 3–03:
Emulsion polymerization of styrene with potassium peroxodisulfate

A 250 ml three-necked flask, fitted with stirrer, thermometer and nitrogen inlet, is evacuated and filled with nitrogen three times. The following are then added under nitrogen: 122 mg (0.45 mmol) of potassium peroxodisulfate, 50 mg of NaH_2PO_4, 1.0 g of sodium oleate or sodium dodecyl sulfate, and 100 ml of water that has been boiled under nitrogen.[2] When everything has dissolved, 50 ml of destabilized styrene are added with constant stirring; the resulting oil-in-water emulsion is heated at 60° C for 6 h with steady stirring under a slow stream of nitrogen. After cooling the polystyrene latex, 30 ml is pipetted into a beaker; the polymer is precipitated by addition of an equal volume of a concentrated solution of aluminium sulfate, if necessary by boiling; a further 30 ml sample is precipitated by dropping into 300 ml of methanol. Finally the latex remaining in the flask is coagulated by addition of concentrated hydrochloric acid. The samples are washed with water and methanol, filtered and dried in vacuum at 50° C. The total yield, and the limiting viscosity number (degree of polymeriza-

[1] For preparation see *Gattermann-Wieland*, "Die Praxis des organischen Chemikers", de Gruyter, Berlin.
[2] Sodium dihydrogenphosphate is added because styrene polymerizes best in weakly alkaline medium.

tion) of one sample, is determined. The values are compared with those obtained under similar conditions for polymerization conducted in bulk (Examples 3–01 and 3–02) and in solution (Example 3–13).

Example 3–04:
Preparation of foamable polystyrene and of polystyrene foam

4 g of polystyrene (e.g. prepared as in Example 3–02 or 3–03 with 0.1 mole % of initiator) are intimately mixed with a solution of 6 g of styrene, 0.6 g of pentane and 0.08 g of dibenzoyl peroxide in a beaker until homogeneous. The resulting mass is transferred to a thick-walled tube with the aid of a wide-necked funnel until it is about one-quarter full. The tube is cooled under nitrogen in a methylene chloride/dry-ice bath to −78°C, sealed off and kept at 30° C in a water bath for about 8 days. Finally the temperature is raised to 85° C for 6 h. The tube is cooled and the upper end then softened with the aid of a blow-torch so that the internal pressure can escape through the hole which is formed.[1] The foamable polystyrene, containing pentane, is obtained in the form of a transparent solid[2] by breaking open the tube.

The polystyrene foam is obtained from this material as follows. A piece of the polymer (about 1 g) is placed in a 500 ml beaker containing boiling water. At this temperature the polystyrene softens and is foamed by the vaporizing pentane. The test piece is held below the surface of the water for 5 min with the aid of piece of bent wire and is then removed.

The foamed material obtained in this manner is dipped into a graduated cylinder containing methanol in order to determine its approximate volume; it is also weighed after drying in a vacuum desiccator. The density is found to be below 0.1 $g\,cm^{-3}$ (cf. cork, 0.2 $g\,cm^{-3}$).

Example 3–05:
Bulk polymerization of methyl methacrylate with dibenzoyl peroxide

This experiment illustrates a simple method of preparation of a poly(methyl methacrylate) plate. The reaction vessel consists of a closed container made as follows. Two clean glass plates (100 mm × 100 mm × 5 mm), dried at 80° C, are held apart by means of four polytetrafluoroethylene (PTFE) or PVC spacers, about 3 mm thick, placed at the corners, and the whole is bound together by two wrappings of adhesive

---

[1] The technique is similar to that of halogen analysis by the *Carius* method.
[2] Foamable polystyrene is prepared commercially according to the same principle by suspension polymerization (bead polymerization). The polystyrene beads containing the expanding agent are then heated above the softening point by means of steam, whereby they foam and also fuse together at their points of contact.

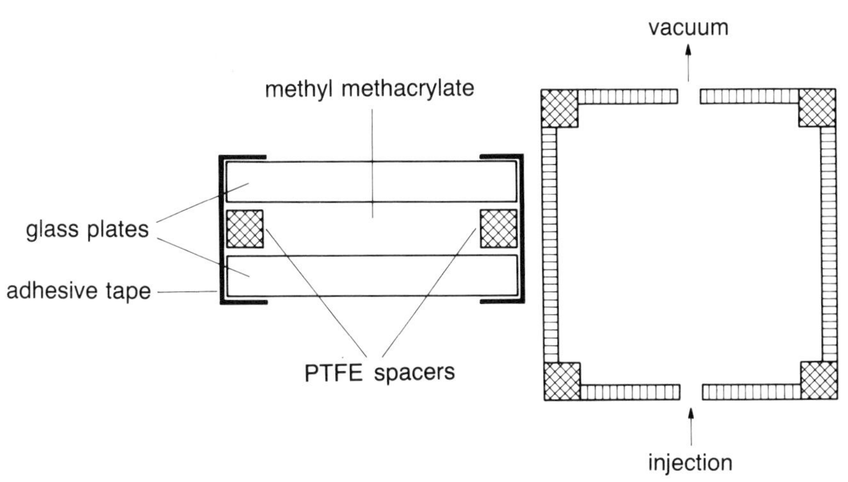

FIGURE 3.1 Cell for making a poly(methyl methacrylate) plate

tape (Figure 3.1). After making a small hole in the wrapping, the cell is placed in a vacuum desiccator.

50 mg of dibenzoyl peroxide are now dissolved in a mixture of 5 g dibutyl phthalate and 70 g of methyl methacrylate (destabilized by distillation under nitrogen), and the solution filtered through a sintered disc into a 250 ml round-bottomed flask. A vertical tube is attached to the flask and the solution then heated on a water bath until it boils gently; the solution begins to polymerize and becomes more viscous. After 25–30 min (this time should not be exceeded, otherwise the mixture becomes too viscous) the cell is filled in a vertical position (see Figure 3.1, right), as follows. A capillary is inserted in the upper opening of the cell and suction applied by means of a filter pump; at the same time the pre-polymerized, bubble-free solution is introduced through the lower opening by means of a hypodermic syringe (without using a needle, because of the high viscosity). The holes in the tape are now resealed. The filled cell is heated for 24 h in an oven at 45°C, the reaction finally being taken to completion by heating for 2 h successively at 60, 80 and 120° C. The cell is now placed in warm water at about 80°C and allowed to cool to room temperature. After removal of the adhesive tape the glass plates are easily removed from the poly(methyl methacrylate) block.

Example 3–06:
Bulk polymerization of vinyl acetate with dibenzoyl peroxide

0.14 g (0.58 mmol) of dibenzoyl peroxide are dissolved in 20 g (0.23 mol) of vinyl acetate that has been destabilized by distillation under nitrogen; this solution is stored in a refrigerator. A 100 ml pear-shaped flask, fitted with reflux condenser, dropping funnel and nitrogen inlet, is evacuated and filled with nitrogen three times; only thick-walled polymerization vessels should be used since the polymer adheres to the wall so tenaciously that thin-walled vessels are easily fractured. The flask is then heated on a water bath to 80°C and the solution of initiator in monomer dropped in from a dropping funnel at such a rate that it does not boil too vigorously; at the same time a slow stream of nitrogen is passed through the apparatus, the reflux condenser being closed by a Bunsen valve (see Section 2.1.1). After all the vinyl acetate has been added the flask is held at 80°C for another 30 min, and at 90°C for a further 60 min, the residual monomer finally being removed by evacuation for 10 min at about 90°C. The flask is now heated to about 170°C and the highly viscous poly(vinyl acetate) pulled out of the flask with a spatula. After cooling, the residue is dissolved in methanol by shaking several times with about 20 ml of solvent until a viscous solution is obtained which is then poured out. A small portion of the polymer (about 1 g) is dissolved in 20 ml of methanol and precipitated by dropping into a 6- to 8-fold excess of water; after filtration it is dried in vacuum at 50°C. Poly(vinyl acetate) is soluble in benzene, toluene, acetone, methanol, ethyl acetate, and methylene chloride. The limiting viscosity number is determined in acetone at 30°C and the average molecular weight derived (see Section 2.3.2.1).

The polymer can be used for the preparation of poly(vinyl alcohol) (Example 5–01).

Example 3–07:
Polymerization of vinyl acetate with ammonium peroxodisulfate in aqueous dispersion

A 500 ml three-necked flask, fitted with stirrer, reflux condenser, thermometer and nitrogen inlet, is evacuated and filled with nitrogen. 5 g of poly(vinyl alcohol) is then placed in the flask (see Example 5–01) and dissolved in 100 ml of distilled water by stirring at 60°C; next are added 2.2 g of oxethylated nonylphenol (e.g. Arkopal N300 from Hoechst A. G.), and 0.4 g (1.8 mmol) of ammonium peroxodisulfate buffered by the addition of 0.46 g sodium acetate in order to prevent the hydrolysis of the monomeric vinyl acetate.

The solution is heated to 72°C and 25 g (0.29 mol) of vinyl acetate (freshly distilled under nitrogen) are added dropwise. The temperature of the water bath is then raised to 80°C. As soon as the internal temperature reaches 75°C, a further 75 g (0.87 mol) of vinyl acetate are added dropwise at such a rate that the internal temperature is maintained between 79 and

83°C at moderate reflux (total time about 20 min); finally a further 0.1 g (0.44 mmol) of ammonium peroxodisulfate in 1 ml of distilled water is added. Refluxing soon abates and the internal temperature rises to about 86°C. The reaction mixture is allowed to polymerize for another 30 min on the water bath at 80°C. On cooling, a creamy dispersion is obtained that contains less than 1% monomer (corresponding to a solid content of about 50%). The polymer can now be precipitated by addition of a 3-fold excess of saturated sodium chloride solution. The poly(vinyl acetate) dispersion can also be spread out as a thin layer on a glass plate; on drying in air the polymer particles coalesce and form a homogeneous, very cohesive film that is resistant to water. These kinds of dispersions are very stable and insensitive to addition of pigments or electrolytes, as well as temperature variations (within certain limits) and are therefore extensively used as paints for wood or plaster surfaces, as well as for cementing wood and for impregnation of leather, paper and textiles.

Example 3–08:
Polymerization of acrylonitrile with ammonium peroxodisulfate in solution

(a) In dimethylformamide
0.025 g (0.11 mmol) of ammonium peroxodisulfate is weighed into a 250 ml flask; using an adaptor (see Section 2.1.3) the flask is evacuated and filled with nitrogen. 90 ml of dimethylformamide (distilled under nitrogen) are then introduced through the adaptor and the ammonium peroxodisulfate dissolved by swirling; to this solution is added 12.5 ml (0.19 mol) of acrylonitrile that has been destabilized by distillation under nitrogen. The flask is now stoppered with a ground joint fitted with a Bunsen valve and heated at 50°C in a thermostat. After 48 h the light yellow, slightly viscous liquid is dropped slowly into about 2 l of water with stirring, whereby the polymer is precipitated in fine flakes. It is filtered off, washed with methanol and dried in vacuum at 50°C. The polyacrylonitrile so obtained is a white powder that is infusible and soluble in only a few solvents (e.g. dimethylformamide). The yield is determined, also the limiting viscosity number in dimethylformamide at 20°C.

(b) In 60% aqueous zinc chloride solution
The air is removed from four 50 ml round-bottomed flasks as described in (a); they are then charged under nitrogen with the following components:

15 ml of a solution of 100 g of very pure anhydrous $ZnCl_2$ in 55 ml of water that has been distilled under nitrogen;
5 ml (76.4 mmol) of acrylonitrile distilled under nitrogen;
0.2 ml of a solution of 5 g (0.05 mmol) of potassium peroxodisulfate in 75 ml of water distilled under nitrogen.

In the second flask 0.1 ml of conc. hydrochloric acid is added after the zinc chloride solution; in the third flask 0.2 ml of a 20% aqueous solution of sodium disulfite (0.21 mmol of $Na_2S_2O_5$) are added, and in the fourth flask 0.1 ml of conc. hydrochloric acid and 0.2 ml of a 20% aqueous solution of sodium disulfite.

The first two flasks are maintained at 50°C overnight; the other two mixtures, which contain sodium disulfite as reducing agent (redox polymerization, see Section 3.1.3), are polymerized at room temperature. The stoppered flasks are well shaken at the beginning of reaction and then allowed to stand. The following morning the first two flasks are observed to contain a pale yellow solution, permeated with bubbles and so viscous that they only flow slowly when inverted. The flasks containing sodium disulfite react very quickly after mixing, with heat evolution, also to give highly viscous solutions. The mixture without the hydrochloric acid is yellow, while that with the acid is colourless. The viscous polymer solution is removed from the flask with the aid of a spatula and mixed with water in a beaker. The lumps of polymer are cut into small pieces with a knife and washed copiously with water in a beaker. The water is changed four or five times until chloride ions are no longer detectable. The polymer is filtered off and dried overnight in vacuum at 50°C. The yields for the two experiments at 50°C are practically 100%, while those for the two experiments using sodium disulfite are about 80%. The limiting viscosity numbers of the individual samples are determined in dimethylformamide and compared with that determined in (a). Should the polymer samples contain an insoluble portion then a weighed sample is extracted with dimethylformamide and the insoluble portion weighed again after drying; the clear solution is diluted with dimethylformamide to the concentration necessary for the viscosity measurements. The polymer can also be isolated by precipitation in water, dried, and weighed out afresh for the viscosity measurements.

Example 3–09:
Bead polymerization of vinyl acetate[1]

0.15 g of a hydrolyzed copolymer of styrene and maleic anhydride (see Example 5–03) are dissolved in 150 ml of hot distilled water to give a 1% solution which is then neutralized with a few drops of ammonia solution. The solution of the ammonium salt is placed in a three-necked flask, fitted with stirrer, reflux condenser, thermometer and dropping funnel (mounted

[1] This experiment need not be carried out under nitrogen.

on the reflux condenser), and heated to about 70°C by means of a water bath set at 80°C. 0.6 g (2.5 mmol) of dibenzoyl peroxide are dissolved in 100 g (1.16 mol) of freshly distilled vinyl acetate and the clear solution allowed to run in through the reflux condenser over a period of about 30 min with vigorous stirring. The water bath temperature is held steady at 80°C and the rate of addition of the vinyl acetate is so regulated that moderate refluxing is maintained. After the addition of monomer is complete the internal temperature rises to about 80°C, the reflux having ceased a few minutes previously.

In order to remove the small amount of unconverted vinyl acetate, steam is blown through the suspension for about 30 min, the flask being fitted with a condenser for distillation. The suspension is finally cooled externally to room temperature and diluted with cold water to about 500 ml. Only now is the stirrer switched off and after settlement of the bead polymer the aqueous layer is drawn off. The product is washed by repeated slurrying with cold water and subsequent decantation until the wash water no longer foams and is therefore free of suspending agent. The moist bead polymer is dried as a thin layer in vacuum at room temperature. The limiting viscosity number is determined in acetone at 30°C and the average molecular weight derived (see Section 2.3.2.1).

Example 3–10:
Polymerization of methacrylic acid with potassium peroxodisulfate in aqueous solution[1]

11 ml of distilled water are heated to 80°C in a 50 ml three-necked flask fitted with a reflux condenser and two dropping funnels. At this temperature 6 g (0.07 mol) of methacrylic acid, purified by vacuum distillation under nitrogen, and a solution of 0.18 g (0.66 mmol) of potassium peroxodisulfate in 4 ml of water are slowly introduced dropwise into the flask over a period of 10–15 min, stirring with a magnetic stirrer. The methacrylic acid polymerizes immediately, as may be seen from the increase of viscosity of the solution. After the additions are complete the temperature is held for another hour at 80°C.

After polymerization the rather viscous solution is diluted with a mixture of 70 ml of acetone and 20 ml of water, filtered and dropped into 1 l of a 4:1(by vol.) mixture of acetone and petroleum ether (b.p. 50–70°C). At first the polymer precipitates in flakes but later sticks together. After precipitation the supernatant liquid is decanted off and another 500 ml of

[1] This experiment need not be carried out under nitrogen.

precipitant mixture are added, causing the polymer to become more solid. The polymer is filtered, if necessary broken up, extracted in a Soxhlet apparatus with petroleum ether for 5 h, and finally dried to constant weight in vacuum at 50°C. The yield is practically quantitative.

Poly(methacrylic acid) is soluble in water, methanol, 1,4-dioxane, and dimethylformamide. The solution viscosity of the polymer is measured in water at 20°C, using concentrations of 0.5, 0.7, 1.0, 1.5, 2.0, 2.5, 3.0, 3.5, and 4.0 g l^{-1} (Ostwald viscometer, capillary diameter 0.3 mm). A plot of η_{sp}/c against c gives a curve that is typical of polyelectrolytes (see Figure 2.13). If, however, the viscosity is measured at the same concentrations in 1 M NaCl solution the behaviour is identical with that for non-electrolyte polymers. It is best to proceed as follows. 30 g of NaCl are dissolved in water and made up to 100 ml in a graduated flask; this solution is 5.13 M. To prepare the solutions for measurement at the aforementioned concentrations, the required amounts of poly(methacrylic acid) are weighed into 10 ml graduated flasks, dissolved in about 5 ml of water, and 2 ml of the 5.13 M NaCl solution added. The solutions are finally made up to the mark with water to give a solution 1.03 M with respect to NaCl.

3.1.2. Polymerization with azo-compounds as initiators

Azo-compounds that are especially suitable as initiators for radical polymerization are those in which the azo-group is bonded on both sides to tertiary carbon atoms that carry nitrile or ester groups in addition to alkyl groups. They are stable at room temperature but decompose thermally above 40°C, or photochemically below 40°C, giving substituted alkyl radicals and liberating nitrogen. This nitrogen can be a nuisance at high initiator concentrations, both in dilatometric measurements (gas bubbles in the measuring capillary) and in bulk polymerizations (the solid polymer is then frequently permeated with minute gas bubbles).

Such azo-compounds decompose in a manner which is essentially independent of solvent and strictly according to a first order rate law so that they are especially suited for kinetic investigations (see Example 3–11); the normal kinetic laws are obeyed (see Section 3.1). The most important azo-compound in this connection is 2,2′-azoisobutyronitrile (AIBN):

$$CH_3-\underset{CN}{\overset{CH_3}{\underset{|}{\overset{|}{C}}}}-N=N-\underset{CN}{\overset{CH_3}{\underset{|}{\overset{|}{C}}}}-CH_3 \longrightarrow 2\,CH_3-\underset{CN}{\overset{CH_3}{\underset{|}{\overset{|}{C}}}}\cdot + N_2$$

The yield of initiating radicals is, however, generally a good deal smaller than would be expected from this equation; this is because a certain amount of tetramethylsuccinic acid dinitrile is formed by combination of the primary radicals, while some methacrylonitrile and isobutyronitrile are formed by disproportionation of the primary radicals. Azo-compounds are especially suited as initiators for polymerizations in bulk or in organic solvents.

Example 3–11:
Bulk polymerization of styrene with 2,2'-azoisobutyronitrile (effect of initiator concentration)

4 thick-walled tubes, having a ground joint and attachable stopcock, are charged with 3.2 mg (0.019 mmol), 6.3 mg (0.038 mmol), 31.5 mg (0.19 mmol), and 63 mg (0.38 mmol) of 2,2'-azoisobutyronitrile[1], respectively, together with 4 g (38.4 mmol) of destabilized styrene (see Example 3–01); 4 g of styrene only are weighed into a fifth tube. The procedure is then as described in Example 3–02. The rate of polymerization (in % conversion per hour) is plotted against the square root of the initiator concentration accordng to equation (6), and the degree of polymerization is plotted against the reciprocal square root of the initiator concentration according to equation (8). The results are compared with those of the thermally polymerized sample and with those obtained with dibenzoyl peroxide as initiator (Example 3–02).

Example 3–12:
Bulk polymerization of styrene with 2,2'-azoisobutyronitrile followed dilatometrically

A polymerization reaction can be followed very conveniently and with great accuracy by observing the resulting contraction of volume in a dilatometer. This contraction results from the considerable difference in density between the monomer and polymer. Knowing the initial volume V and contraction ΔV during polymerization, the percentage conversion U in the absence of a diluent is given by equation (15).

$$U = \frac{100}{K} \frac{\Delta V}{V} \tag{15}$$

[1] For preparation see *J. Thiele* and *K. Heuser*, Justus Liebigs Ann. Chem. *290* (1896) 30.

K is a constant that can be calculated from the specific volumes of the monomer and polymer at the appropriate temperature. One can, however, also determine the relationship between U and ΔV by direct experiment.

(a) Calibration of the dilatometer

In the simplest case the dilatometer consists of a short tube with ground joint which opens out above the joint to an overflow chamber as shown in Figure 2.8. (Section 2.1.3). A capillary, of about 30 cm length, 0.2 ml content and graduated in 0.001 ml, fits into the ground joint of the dilatometer tube.[1]

In order to determine the volume the dilatometer is filled with mercury to the upper rim of the joint and the capillary inserted so that part of the mercury rises into the capillary while the rest runs into the overflow chamber. Care is taken that the capillary is firmly bedded into the tube by attaching small springs. The excess mercury is removed from the overflow chamber by closing the upper end of the capillary with one finger and tilting the whole dilatometer until the mercury has run off. The dilatometer is now brought to thermal equilibrium in a thermostat at 20°C and the meniscus level read off. The capillary is then carefully removed, taking care to leave no mercury behind in the capillary, and the mercury weighed by pouring into a weighed beaker and reweighing to the nearest 10 mg. The volume is calculated from the mass and the density of mercury at 20°C ($\rho_{20} = 13.5457\ \mathrm{g\,cm^{-3}}$). To this must be added the volume from the meniscus level up to the zero mark in order to obtain the total volume V_o of the dilatometer and capillary up to the zero mark. This calibration is performed several times for each dilatometer and the mean values determined.

For very exact measurements the volume of the dilatometer should be determined at the temperature at which polymerization is to be carried out (in the present case at 60°C). However since the coefficient of linear expansion of glass is more than two orders of magnitude smaller than the coefficient of cubic expansion of the liquid to be used, the change of the volume of the dilatometer with temperature can generally be neglected.

(b) Measurement of the coefficient of expansion of styrene

For the later evaluation of the dilatometric experiments it is necessary to be able to convert volumes at 60°C to volumes at 20°C, that is, the coefficient of expansion α must be known. This can be determined using

[1] For other types of dilatometer see "Encyclopedia of Polymer Science and Technology", Interscience Publishers, New York, London, Sydney 1966, Vol. 5, p. 83 et seq; also *D. Braun* and *G. Disselhoff*, Polymer *18* (1977) 963.

the dilatometer. It is filled with stabilized monomeric styrene (containing 0.1 wt. % of hydroquinone); the capillary is inserted and sealed with mercury at the joint. Filling is done at about 10°C above the lowest temperature of measurement. The dilatometer is mounted in a cold bath at about 10°C so that the meniscus falls to near the lower end of the capillary. The whole length of the capillary is then available for measuring the expansion. The temperature of the water thermostat is slowly raised and the position of the meniscus measured as a function of temperature. The volume in ml is plotted against temperature, an approximately straight line being obtained. From its slope the average coefficient of expansion α over this temperature range can be determined:

$$\alpha = \frac{V_h - V_i}{V_i \Delta T}$$

where
V_h = volume at the highest temperature measured,
V_i = volume at initial temperature of measurement,
ΔT = temperature interval between the highest and lowest temperature of measurement,
α = coefficient of cubic expansion.

The coefficient of expansion is determined in the range 20 to 30°C and in the range 50 to 60°C, the mean value being taken as the average coefficient of expansion over the whole range 20 to 60°C.

(c) Dilatometric measurements

The following amounts of 2,2′-azoisobutyronitrile (AIBN) are weighed into 4 graduated 25 ml flasks: 35, 110, 180, and 250 mg (0.21, 0.67, 1.10, and 1.52 mmol). They are filled to the mark with destabilized styrene (at 20°C) and the amount of styrene added determined by weighing. Dividing the amount of styrene by 25 (ml) gives the density of styrene at 20°C (neglecting the partial volume of AIBN).

4 dilatometers are filled with the above solutions of AIBN in styrene, the joints being sealed with mercury, and placed in a large water thermostat at 60°C (±0.05°C) so that the filled parts of the capillaries are completely immersed. The thermostat can easily be constructed from a large glass tank, a powerful stirrer, immersion heater, contact thermometer and relay. The dilatometers are put into a metal test tube rack fitted with suitable mountings to hold them rigidly in the thermostat. After inserting a filled dilatometer in the thermostat the meniscus rises in the capillary until thermal equilibrium is reached (if necessary some of the styrene solution may be withdrawn from the capillaries by means of a syringe or thin glass capillary). It remains steady for a short induction period, because of the

presence of dissolved oxygen, and then falls as polymerization commences. The meniscus level is read every minute after insertion of the dilatometer in the thermostat and plotted against time. When the reaction slows down it is sufficient to take readings every 5 min. Zero reaction time is taken as the intersection of the horizontal line and the initial slope. When the volume has fallen by 0.1 to 0.2 ml the meniscus level is quickly noted and the reaction immediately quenched by immersion of the dilatometer in ice-water.

The dilatometers are emptied as follows. The dilatometer, cooled in ice to below 10°C, is inclined carefully over a small beaker so that the mercury and water can run off from the overflow compartment, which is then washed in turn with methanol and benzene, and finally dried in a stream of air. The capillary is now withdrawn from the dilatometer, the polymer solution poured into a beaker and the capillary and dilatometer bulb washed out several times with small amounts of benzene. The polymer solutions are each added dropwise to an 8- to 10-fold excess of methanol. The amounts of polymer precipitated are determined gravimetrically.

In order to obtain the percentage conversion from the gravimetric data the original amount of styrene in the dilatometer must be known. This is determined as follows. The initial volume V_{60} at 60°C in the dilatometer is converted to the equivalent volume V_{20} at 20°C using the formula:

$$V_{20} = \frac{V_{60}}{1 + 40\alpha} \tag{17}$$

$V_{20}\rho_{20}$ gives the amount of styrene introduced, and the conversion can then be expressed in weight percent.

(d) Evaluation of the dilatometric measurements

The change in volume ΔV is determined from the initial and final dilatometer readings in each experiment, as given by the plot of meniscus level against time. The constant K can now be calculated from equation (15) for each dilatometric measurement and the results averaged. The statistical error is estimated from the scatter of the data (without applying the Gaussian formula).

The rate of polymerization (in % conversion per hour) is plotted against the square root of the initiator concentration (in mole %), according to equations (6) and (7), p. 128. The limiting viscosity numbers and hence the degrees of polymerization are also determined and plotted against the reciprocal square root of the initiator concentration, according to equation (8). The values are compared with those obtained in Example 3–02.

The statistical error determined for K is only a limited measure of the accuracy of the dilatometric measurements. Since the main errors will be similar for each measurement, the accuracy of the method is best determined by estimating the limits of error of each individual measurement. The main source of error and its approximate magnitude should be indicated.

Example 3–13:
Polymerization of styrene with 2,2'-azoisobutyronitrile in solution

(a) Conversion-time curve
100 mg (0.61 mmol) of 2,2'-azoisobutyronitrile (AIBN) are weighed into a 250 ml three-necked flask fitted with stirrer, reflux condenser and nitrogen inlet; it is then evacuated and filled with nitrogen three times. 100 ml of pure toluene (distilled under nitrogen) and 10 ml (0.09 mol) of destabilized styrene are now introduced under nitrogen (best from a receiver, see Section 2.1.2), and the flask immersed in a water bath. Nitrogen is passed gently over the reaction mixture while the water bath is heated to boiling (the outlet is protected against back-diffusion of oxygen as described in Section 2.1.1); the solution is stirred slowly, and gradually becomes more viscous. At intervals of 1 h, 10 ml of the solution are withdrawn by means of a suitable syringe-pipette (during this operation the nitrogen flow is increased) and immediately discharged dropwise into 100 ml of methanol with stirring. After 6 h the polymerization is terminated by cooling the flask; the remaining solution is run from a dropping funnel into 500 ml of stirred methanol. The polymer samples are filtered off, dried in vacuum at 50°C to constant weight and the yield and limiting viscosity number (and degree of polymerization) determined (see Example 3–01). The yield and degree of polymerization are plotted against time.

(b) Effect of monomer concentration
23 mg (0.14 mmol) of AIBN are weighed into each of 7 tubes with ground joints. Using adaptors attached with springs (see Section 2.1.3), they are evacuated and filled with nitrogen three times. Under a flow of nitrogen 0.5, 1.0, 1.5, 2.0, 2.5, and 3.0 ml (4.36, 8.72, 13.07, 17.42, 21.78 and 26.13 mmol) of destabilized styrene are pipetted into the tubes and each diluted to 15 ml with pure tolueme (distilled under nitrogen); the seventh tube is charged with 15 ml of styrene only (bulk polymerization). Using a slight positive pressure of nitrogen the adaptors are removed from the tubes and immediately closed with ground glass stoppers secured with springs. The tubes are placed in a boiling water bath, cooled after 6 h and worked up as before, if necessary after dilution with toluene.

The yield and degree of polymerization are plotted against the monomer concentration. The results are compared with those for the sample polymerized in bulk.

Example 3–14:
Bulk polymerization of methyl methacrylate with 2,2′-azoisobutyronitrile

(a) Observation of the Trommsdorff effect (gel effect)
100 ml of methyl methacrylate are distilled under nitrogen into a graduated receiver or dropping funnel with pressure-equalizing tube (see Section 2.1.2), into which 100 mg (0.61 mmol) of 2,2′-azoisobutyronitrile (AIBN) have previously been weighed. 10 tubes with ground joints and suitable adaptors (see Section 2.1.3) are evacuated, filled with nitrogen, and 10 ml (93.6 mmol) of methyl methacrylate with AIBN introduced to each tube. The tubes are removed from the adaptors under slight positive pressure of nitrogen, immediately closed with glass stoppers secured with springs, and stored until needed in an acetone/dry-ice bath.

To start the experiment all the tubes are placed in a rack at the same time and allowed to warm to room temperature; finally they are placed in a thermostat at 50°C. The tubes are removed at intervals of 1 hour and immediately cooled in an acetone/dry-ice bath. The samples that are still fluid are diluted with approximately 50 ml of chloroform and dropped into about 500 ml of stirred heptane or petroleum ether. For the very viscous or solid samples 1–2 g are dissolved in 50–100 ml of chloroform and the solution added dropwise to 500–1000 ml of heptane or petroleum ether with stirring. The polymers are filtered off and dried to constant weight in vacuum at 50°C. The yield, limiting viscosity number (measured in chloroform at 20°C) and degree of polymerization are plotted against reaction time.

The conversion can also be followed refractometrically, since the change of refractive index during polymerization is directly proportional to the conversion. The measurements can be made with an Abbé refractometer, by placing a drop of the still-liquid sample on the prism by means of a glass rod in the usual way. However, the determination of the refractive index of the highly viscous or solid samples can only be done with special equipment. The conversion corresponding to the measured refractive index is derived from a calibration line connecting the value for the pure monomer (n_D^{20} = 1.4140, 0% conversion) with that for the pure polymer (n_D^{20} = 1.4915, 100% conversion).

(b) Control of the molecular weight; determination of the transfer constant of 1-dodecanethiol
5 tubes with ground joints are filled, as described in (a), with 10 ml (93.6 mmol) of methyl methacrylate (containing 0.1 wt. % of AIBN). 0.1, 0.5, 1.0, and 2.0 mole % 1-dodecanethiol are added as regulator to four of the tubes, while the fifth serves as reference. The tubes are stoppered and stored in an acetone/dry-ice bath until needed. To begin the experiment

the tubes are warmed up to room temperature at the same time and placed in a thermostat at 50°C. After 2 h the tubes are taken out[1], the contents each dissolved in 30 ml of chloroform and the solutions added dropwise to 300 ml of heptane or petroleum ether with stirring. The polymers are reprecipitated from cholorform into heptane or petroleum ether, filtered and dried in vacuum at 50°C. The limiting viscosity number of all samples is determined in chloroform at 20°C and the average degree of polymerization derived (see Section 2.3.2.1). The value for the transfer constant cannot be determined very accurately from these values, since the chain transfer equation (12) is strictly valid only for the number-average degree of polymerization. $1/\bar{P}$ is plotted against the mole ratio of thiol to monomer, [ZH]/[M]. A straight line is obtained, intersecting the ordinate at $1/\bar{P}_o$ (experiment without thiol); the slope gives the transfer constant C_{ZH} (see equation (14a)). On a second graph the yield is plotted against [ZH]/[M]; the rate of polymerization is unaffected by the occurrence of chain transfer.

Example 3–15:
Polymerization of vinyl acetate with 2,2'-azoisobutyronitrile in different solvents

4 ampoules of about 50 ml capacity are each charged with 0.03 g (0.18 mmol) of 2,2'-azoisobutyronitrile (AIBN), 20 g of solvent distilled under nitrogen (toluene, methylene chloride, methanol, and *tert*-butanol, respectively), and 10 g (0.12 mol) of vinyl acetate destabilized by distillation under nitrogen. The air is then removed by evacuating and filling with nitrogen three times; each evacuation is only allowed to proceed until the first appearance of gas bubbles, and nitrogen then admitted. The ampoules are finally sealed off and placed in a water bath at 70°C.

After 4 h the tubes are opened and the polymer solutions poured into Petri dishes; the polymerizations are quenched by addition of 0.5 ml crotonaldehyde to each dish. The solvent and residual monomer are removed by evaporation in vacuum at 60°C. After weighing, about 1 g of each polymer is dissolved in about 25 ml methanol and precipitated by adding the solution dropwise to water. After purifying the polymers in this way and drying in vacuum at 50°C, the limiting viscosity numbers are determined in acetone at 30°C (capillary diameter 0.3 mm), and the molecular weights

---

[1] The conversion should not be much above 10%, otherwise the ratio [ZH]/[M] can no longer be regarded as constant.

derived (see Section 2.3.2.1). The results are tabulated to show the effect of solvent on the conversion, the limiting viscosity number, and the molecular weight.

Example 3–16:
Determination of the molecular weight distribution of poly(vinyl acetate) by fractional extraction

The fractional extraction of poly(vinyl acetate) is carried out according to the method of Fuchs (see Section 2.3.3). For this purpose, a 5–10% solution of 1 g of polymer[1] is prepared in ethyl acetate and poured into a level, rectangular glass trough (about 8 cm × 18 cm). The concentration of solution should be adjusted according to the molecular weight of the polymer; the higher the molecular weight, the lower should be the concentration. The best conditions are easily determined by preliminary experiments in which aluminium foils are dipped into polymer solutions of different concentration and subsequently dried and weighed. The amount of polymer adhering to the surface should be about 100 mg per 100 cm^2; it should not exceed this value by more than 50%, but may be lower. Four pieces of aluminium foil (about 20–50 μm thick) are cut to a size of about 7 cm × 17 cm (total area about 1000 cm^2) and weighed exactly. They are dipped into the polymer solution, taken out, the solvent evaporated by means of a hot-air blower, and the foils dried overnight in vacuum at 40°C. The weight of poly(vinyl acetate) deposited on the foils is determined and the foils then cut into strips of about 1 × 7 cm. In order to prevent them from sticking together during fractionation they are folded into a tight zig-zag form by means of forceps (fold height 0.5–1 cm). The fractionation is best carried out in a special vessel (see Section 2.3.3) using mixtures of methyl acetate and petroleum ether. The methyl acetate and petroleum ether must be previously distilled through a column; the latter should not contain any component boiling above 60°C, otherwise occlusion may occur when drying the fractions. The two liquids must be well mixed before filling the fractionating vessel. The aluminium foils are shaken for 10 min in the solvent mixture; the resulting polymer solution is run off through the stopcock into a weighed conical flask with ground joint. To ensure complete removal of the solution from the kinks in the foils the fractionation vessel is shaken several times. About 10 fractions are separated off in this way after successive changes in the solvent composition; the polymer solutions so obtained are carefully dried at 50°C to constant weight, first at

[1] A typical distribution curve is obtained by using the poly(vinyl acetate) prepared in *tert*-butanol (see Example 3–15).

atmospheric pressure, then in vacuum. The composition of the solvent mixtures for the fractionation steps are adjusted according to the molecular weight of the poly(vinyl acetate); they should be so chosen that the weights of the individual fractions are about the same. The mixtures set out in the table are suitable for a poly(vinyl acetate) having an η_{sp}/c value of 0.12 g dl^{-1} (measured at a concentration, $c = 10\,\mathrm{g}\,l^{-1}$ in acetone at 30°C). If the η_{sp}/c value is more than 5% higher, the corresponding mixtures for fractionation must be somewhat richer in methyl acetate; but if the sample has a lower molecular weight the proportion of petroleum ether must be raised.

Example: Fractionation of poly(vinyl acetate), total sample 1016 mg

Fraction no.	Volume of methyl acetate[a] (ml)	Polymer weight in mg	Polymer weight %	$I(P)$ %	$10^5 \frac{dm_p}{dP}$	η_{sp}/c (g dl^{-1})	$[\eta]$[b] (g dl^{-1})	$\bar{P}$
1	52.0	43	4.2	2.1	5.8	0.020	0.019	400
2	54.0	84	8.3	8.4	11.9	0.041	0.037	1080
3	56.0	146	14.4	19.7	15.6	0.071	0.060	2020
4	57.0	152	15.0	34.4	16.6	0.096	0.076	2790
5	57.5	201	19.8	51.8	11.8	0.142	0.103	4220
6	58.0	220	21.6	72.5	7.4	0.214	0.136	6250
7	58.5	128	12.6	89.6	4.1	0.338	0.177	9030
8	59.0	18	1.8	96.8	—	—	—	—
9	60.0	9	0.9	98.2	—	—	—	—
10	62.0	7	0.7	99.0	—	—	—	—
11	100.0	5	0.5	99.8	—	—	—	—
Totals		1013	99.8					

[a] Volume of petroleum ether = 100 − volume of methyl acetate
[b] Calculated from η_{sp}/c

The amount of polymer in each fraction is determined by weighing the conical flasks after drying. Sufficient acetone is then run in from a microburette so that polymer solutions of concentration 10 g l^{-1} are obtained. The approximation is made that the volume of the solutions is the same as that of the acetone added. The conical flasks are then closed with ground glass stoppers (upper halves lightly greased) and shaken for several hours until the poly(vinyl acetate) is completely dissolved. Finally the specific viscosity is determined in an Ostwald viscometer (capillary diameter 0.3 mm) at 30°C and the limiting viscosity number $[\eta]$ calculated from the formula of Schulz and Blaschke ($K_\eta = 0.27$); hence the molecular weight (see Section 2.3.2.1).

The fractionation results are now tabulated. As explained in Section 2.3.3 the numbers in column 5 are given by the summation of the numbers for the earlier fractions in column 4 and half the value for the fraction in question (see the example tabulated above).

The integral distribution curve of the poly(vinyl acetate) is obtained by plotting $I(P)$ against the degree of polymerization $\bar{P}$. The values for dm_p/dP (column 6) are determined from the slopes at different points (see Section 2.3.3) and the differential distribution obtained by plotting dm_p/dP against $\bar{P}$.

Example 3–17:
Determination of the molecular weight distribution of polystyrene by fractional precipitation

(a) Preparation of polystyrene
300 ml of destabilized, distilled styrene are placed in a three-necked flask, flushed for 15 min with nitrogen, and heated with stirring to 60°C in a thermostatted water bath. A solution of 0.1 g of 2,2′-azoisobutyronitrile in 50 ml of distilled styrene is added and flushed with nitrogen for another 15 min. Polymerization is allowed to proceed at 60°C with stirring under nitrogen atmosphere, the polystyrene (ca. 8 g) finally being isolated by precipitation in methanol, filtration and drying. The limiting viscosity number $[\eta]$ of the polymer is determined in benzene at 20°C.

(b) Fractional precipitation of the polystyrene
7 g of the polymer are dissolved in about 550 ml of toluene in a 2 l three-necked flask. This is placed in a thermostat held at 25.0 ± 0.05°C, and after temperature equilibrium has been reached, methanol is added with vigorous stirring until phase separation (precipitation) just occurs, as evidenced by a weak, persistent turbidity (about 235 ml of methanol required). The solution is warmed under continuous stirring until it is clear again. The stirrer is removed and the solution allowed to cool slowly to 25°C. When the precipitated portion has settled on the bottom of the vessel (overnight), the upper sol phase is carefully decanted into another three-necked flask, the gel phase (coacervate) being dissolved in a little toluene, precipitated in methanol, filtered and dried. This is the first fraction and contains the highest molecular weights. The second fraction is obtained by repeating this cycle after adding a little more methanol (ca. 5 ml) to the stirred sol phase at 25°C so as to produce further phase separation. By continuing this procedure one can isolate a total of about 8 to 12 fractions, depending on the amounts of methanol added in the individual precipitation steps. The last fraction, which contains the lowest molecular weights, is recovered by concentrating the final sol phase and precipitating in methanol.

The weight fraction w_i for fraction i is calculated from the total weight of all fractions. It is arbitrarily assumed that the losses during separation, precipitation, filtration, etc., are proportional to the amounts of polymer of the individual fractions. The number-average molecular weight $(\bar{M}_n)_i$ of each of the fractions is determined from viscosity measurements in benzene at 20°C, or by membrane osmometry. The integral distribution curve is derived as described in Section 2.3.3 and in Example 3–16. (For the conversion of viscosity measurements into degrees of polymerization, see Table 2.3, p. 83).

3.1.3. Polymerization with redox systems as initiators

Numerous redox reactions generate radicals that can initiate polymerization.[1,2] Chief amongst oxidizing agents are organic and inorganic peroxides; reducing agents include either low valency metal ions, or non-metallic compounds that are readily oxidized, for example certain sulfur compounds (cf. Table 3.3). There are also redox systems that consist of a mixture of a per-compound with metal ions (e.g. Fe^{2+}) and a second reducing agent such as a hydrogensulfite. In this case the iron (III) ion produced by the redox reaction between the per-compound and iron (II) compound is reduced again to the iron (II) state by the hydrogensulfite, so that only a very small amount of iron (II) ions is required.

In a redox system consisting of a per-compound and iron (II) salt, the initiating radicals are formed by electron transfer from Fe^{2+} to the per-compound, causing the peroxy link to be cleaved, with simultaneous formation of a radical and an anion:

$$\mathrm{R{-}O{-}O{-}R} + Fe^{2+} \longrightarrow \mathrm{R{-}O^-} + Fe^{3+} + \mathrm{R{-}O\cdot}$$

$$\mathrm{RO\cdot} + \mathrm{CH_2{=}\underset{\displaystyle X}{\underset{|}{CH}}} \longrightarrow \mathrm{RO{-}CH_2{-}\underset{\displaystyle X}{\underset{|}{CH}}\cdot}$$

Table 3.3 lists some suitable oxidizing and reducing agents.

It must be emphasized that, in contrast to the initiation of polymerization with per-compounds or azo-compounds, not every redox system is suitable for initiating polymerization of every unsaturated monomer. Before attempting to polymerize a new compound with a redox system it is, therefore, advisable first to test its radical polymerizability with

---

[1] *W. Kern*, Angew. Chem. *6* (1949) 471.

[2] *D.C. Blackley*, "Emulsion Polymerisation", Applied Science Publishers, London 1975, p. 204.

TABLE 3.3

Some common oxidizing and reducing agents that are suitable for initiating radical polymerization by redox reactions

Oxidizing agents	Reducing agents
Hydrogen peroxide	Ag^{+}, Fe^{2+}, Ti^{3+}
Peroxodisulfates	Hydrogensulfites, sulfites, thiosulfates, thiols, sulfinic acids
Diacyl peroxides	Amines (e.g. *N,N*-dimethylaniline)
	Certain sugars
	Benzoin/Fe^{2+}
	Hydrogensulfite/Fe^{2+}

dibenzoyl peroxide (see Section 3.1.1). Furthermore, the effectiveness of a redox system is influenced by a number of factors, and the redox components must be carefully balanced in order to attain optimum polymerization conditions. The most favorable conditions do not always correspond to a stoichiometric ratio of oxidizing and reducing agents. However, at constant molar ratio of oxidizing agent to reducing agent it is generally the rule that the rate of polymerization increases, and the mean degree of polymerization decreases, with increasing initiator concentration (see Example 3–20). The order of addition of the components can also be important; while it is normal to add the reducing agent first (in order to remove any oxygen which may be present), with subsequent dropwise addition of the oxidizing agent, there are cases where the reverse order must be applied. In aqueous medium the pH value is also important; if it is necessary to work in alkaline medium, iron salts can only be used in combination with complexing agents such as sodium pyrophosphate ($Na_4P_2O_7$).

Redox polymerizations are usually carried out in aqueous solution, suspension or emulsion; seldom in organic solvents (see Example 3–22). Their particular importance lies in the fact that they proceed at relatively low temperatures with high rates and with the formation of high molecular weight polymer. Furthermore, transfer and branching reactions are relatively unimportant. The first large-scale commercial application of redox polymerization was the production of synthetic rubber from butadiene and styrene (Buna S) at temperatures below 5°C (cf. Example 3–47).

Example 3–18:
Polymerization of acrylamide with a redox system in aqueous solution

5 g of pure acrylamide (recrystallized from benzene) are dissolved in 500 ml of water (that has been boiled and distilled under nitrogen) in a 1 l flask

fitted with stirrer, gas inlet and gas outlet. To the solution are added 25 ml of a 0.1 M aqueous solution of iron (II) ammonium sulfate and 25 ml of a 0.1 M aqueous solution of hydrogen peroxide. The solution is gently stirred and nitrogen passed through the flask for 5 min in order to displace the atmospheric oxygen. The polymerization is conducted at room temperature (ca. 20°C). After 30 min the viscous solution is run dropwise with vigorous stirring into 4 l of methanol to which a few drops of concentrated hydrochloric acid have been added. The precipitate, coloured brown by iron (III) hydroxide, is filtered off and dissolved in 50 ml of water. To this solution is added ammonia solution, the precipitated iron (III) hydroxide is filtered off, and the polymer solution is added dropwise to 500 ml methanol. After filtration, the polyacrylamide is dried to constant weight in vacuum at 20°C; yield about 50%. Polyacrylamide is soluble in water but infusible. The limiting viscosity number is determined in water at 25°C (capillary diameter 0.35 mm).

Example 3–19:
Fractionation of polyacrylamide by gel permeation chromatography[1]

10 g of Sephadex[2] G100 are placed in a beaker and swollen with 400 ml water for two days. In a 2 m chromatography column (diameter 1.5 to 2 cm) a layer of glass wool and some glass beads are placed above the outlet tap. The tube is then partially filled with water and the swollen Sephadex is carefully run in so as to avoid trapping air bubbles. The gel particles gradually settle and the supernatant water is drawn off. The gel is now washed with fresh water until the runnings no longer show turbidity when added to methanol. (Fresh Sephadex often contains a small amount of water-soluble material that may have an adverse effect on the fractionation). Finally the water is run off until the liquid meniscus is a few mm above the gel particles; the liquid level must never be allowed to fall below the level of the gel.

A solution of 250 mg of polyacrylamide (from Example 3–18) in 25 ml of distilled water is introduced at the top of the column. At the same time the collection of the eluate is commenced, taking 10 ml portions (measuring cylinder) for each fraction; the flow time under these conditions is about 10 ml h^{-1}. When the meniscus of the polyacrylamide solution reaches the gel layer, distilled water is added to the column as eluting agent.

The fractions collected are each run dropwise into 100 ml of stirred methanol to which two drops of concentrated hydrochloric acid have been

[1] See Section 2.3.3.
[2] Supplied by Deutsche Pharmacia GmbH, Postfach 5480, D-7800 Freiburg, W. Germany.

added. Turbidity or precipitation will be observed from about the 6th fraction to about the 20th fraction. The precipitated fractions are filtered off, washed with methanol and dried to constant weight in vacuum at 20°C. For each fraction the viscosity is measured in water at 25°C using a capillary viscometer (capillary diameter 0.35 mm) and at as high a concentration as possible ($10\,g\,l^{-1}$) in order to minimize errors. The limiting viscosity number, and hence the molecular weight, is estimated by means of the Schulz-Blaschke formula (see Section 2.3.2.1). Adjacent fractions, for which there may be insufficient material for a viscosity measurement, can be combined where necessary.

After the fractionation, the amount of the individual fractions in mg is plotted against the elution volume. One can test whether there has been any loss of polymer by comparing the total mass of all the fractions with the initial amount; the loss should not be more than 3%.

The integral and differential molecular weight distribution curves are finally determined as described in Section 2.3.3 and in Example 3–16. In summing the percentage amounts of the fractions (see Example 3–16), one proceeds from the last fraction having the lowest molecular weight to the first fraction having the highest molecular weight.

Example 3–20:
Polymerization of acrylonitrile with a redox system in aqueous solution (precipitation polymerization)

(a) Effect of the ratio of oxidizing agent to reducing agent
The following solutions are prepared in distilled water;

5% solution of sodium disulfite ($Na_2S_2O_5$),
5% solution of potassium peroxodisulfate ($K_2S_2O_8$),
0.010 g of $FeSO_4 . 7H_2O$ in 100 ml of water + 2 ml of conc. H_2SO_4.

Four 250 ml round-bottomed flasks are evacuated and filled with nitrogen using a suitable adaptor (see Section 2.1.3).[1] The reagents listed in the following table are then introduced; the potassium peroxodisulfate solution is added last to all four samples at about the same time. The flasks are shaken briefly and allowed to stand under nitrogen at 20°C. For each sample the time of appearance of the first turbidity is noted. After 20 min the precipitated polymer is filtered off by suction from the four samples, washed with water, then with methanol, and dried overnight at 50°C in vacuum. The yield, the rate of polymerization and the limiting

[1] This experiment need not be conducted under nitrogen.

Sample	Vol. of water[a] in ml	Vol. of acrylonitrile[b] in ml	$Na_2S_2O_5$ Vol. of soln. in ml	$Na_2S_2O_5$ Amount in mmol	$FeSO_4$ Vol. of soln. in ml	$FeSO_4$ Amount in mmol	$K_2S_2O_8$ Vol. of soln. in ml	$K_2S_2O_8$ Amount in mmol
1	175	15	0.5	0.13	2.5	9×10^{-4}	2.5	0.46
2	173	15	2.5	0.66	2.5	9×10^{-4}	2.5	0.46
3	170	15	5.0	1.32	2.5	9×10^{-4}	2.5	0.46
4	165	15	10.0	2.63	2.5	9×10^{-4}	2.5	0.46

[a] The water is previously boiled under nitrogen.
[b] Destabilized by distillation under nitrogen.

viscosity number (measured in dimethylformamide at 25°C) are plotted against the mole ratio of oxidizing agent to reducing agent.

(b) Effect of initiator concentration at constant ratio of oxidizing agent to reducing agent

Four 250 ml round-bottomed flasks are filled with the following components as described in (*a*).

Sample	Vol. of water[a] in ml	Vol. of acrylonitrile[b] in ml	$Na_2S_2O_5$ Vol. of soln. in ml	$Na_2S_2O_5$ Amount in mmol	$FeSO_4$ Vol. of soln. in ml	$FeSO_4$ Amount in mmol	$K_2S_2O_8$ Vol. of soln. in ml	$K_2S_2O_8$ Amount in mmol
1	181	15	0.5	0.13	0.5	1.8×10^{-4}	0.5	0.09
2	175	15	2.5	0.66	2.5	9.0×10^{-4}	2.5	0.46
3	167	15	5.0	1.32	5.0	1.8×10^{-3}	5.0	0.93
4	152	15	10.0	2.63	10.0	3.6×10^{-3}	10.0	1.85

[a] The water is previously boiled under nitrogen.
[b] Destabilized by distillation under nitrogen.

The samples are allowed to react for 20 min at 20°C and then worked up as described in (*a*). The rate of polymerization (in % conversion per minute) is plotted against the square root of the initiator concentration c_i (in mol $K_2S_2O_8$ per litre); the limiting viscosity number and the degree of polymerization are plotted against $c_i^{-\frac{1}{2}}$.

(c) Inhibition of polymerization

5, 20, 100, and 200 mg of hydroquinone are weighed into four 250 ml round-bottomed flasks which are then evacuated and filled with nitrogen. To each are now added 15 ml of acrylonitrile (destabilized by distillation under nitrogen), 165 ml of deaerated water, 10 ml of $Na_2S_2O_5$ solution[1] and

[1] Concentration of solution as in (*a*).

2.5 ml of $FeSO_4$ solution[1]; finally 2.5 ml of $K_2S_2O_8$ solution[1] are added to all four samples (20°C) at about the same time. The time required for the appearance of the first turbidity (incubation time, induction period) is noted. The induction periods are compared with each other and with that for sample 4 of experiment (*a*).

(d) Solution-spinning of polyacrylonitrile
3.5 g of one of the polyacrylonitriles obtained above are dissolved in 25 ml of dimethylformamide. The viscous solution is poured into a 1 cm wide glass tube which is drawn out to a jet at the lower end and dips into a dish of cold water. The polymer solution flows continuously out of the jet under its own weight, the polyacrylonitrile being precipitated in the form of an endless filament. This is guided through the water bath and wound on to a rotating drum driven slowly by means of a motor. It is also possible to use a hypodermic syringe in place of the drawn-out glass tube; the filament thickness and rate of spinning can then be easily varied.

Example 3–21:
Emulsion polymerization of isoprene with a redox system

Most emulsion polymerizations are performed with water-soluble initiators; however, the following experiment describes a redox polymerization in which one component (dibenzoyl peroxide) is insoluble in water, while the other component is soluble.

A 100 ml three-necked flask, fitted with stirrer and nitrogen inlet, is evacuated and filled with nitrogen three times. The following solutions are prepared: a) 500 mg of sodium oleate (or sodium dodecyl sulfate) in 16 ml of deaerated water; b) 125 mg (0.32 mmol) of iron (II) ammonium sulfate and 125 mg of sodium pyrophosphate (as buffer) in 4 ml of deaerated water. The latter solution is maintained at 60–70°C for about 15 min, with occasional shaking, and is then poured into the flask together with solution a). The mixture is cooled to room temperature and 20 ml (0.2 mol) of isoprene (distilled under nitrogren) containing 50 mg (0.21 mmol) of dissolved dibenzoyl peroxide are added. Vigorous stirring produces a stable emulsion which becomes more viscous as polymerization proceeds. After 6 h at room temperature the isoprene is almost completely polymerized. The polymer is precipitated from the latex in the form of large flakes by adding it dropwise to 500 ml of methanol containing 500 mg of *N*-phenyl-β-naphthylamine as stabilizer for the polyisoprene; flocculation can

[1] Concentration of solution as in (*a*).

be improved by the addition of a few drops of concentrated hydrochloric acid. The solid elastic product is filtered off, washed with methanol and dried in vacuum at 50°C. The solubility is tested in different solvents, the limiting viscosity number determined (toluene, 25°C), also the proportion of 1,2- and 1,4-repeating units, and the *cis/trans* ratio (see Example 3–30). These values are compared with those obtained in the polymerization of isoprene with butyllithium (Example 3–30).

Example 3–22:
Polymerization of styrene with redox systems in an organic solvent

Three 100 ml round-bottomed flasks are evacuated and filled with nitrogen using a suitable adaptor (see Section 2.1.3). In one of the flasks is placed 3 mg of iron (III) benzoate, in another, 3 mg of iron (III) acetylacetonate. To each of the two flasks are now added, under nitrogen, a solution of 121 mg (0.5 mmol) of dibenzoyl peroxide in 5 ml of dry benzene (distilled under nitrogen), and a solution of 106 mg (0.5 mmol) of benzoin in 15 ml of benzene. In the third flask is placed just a solution of 121 mg of dibenzoyl peroxide in 5 ml of dry benzene. Into each flask is then pipetted, under nitrogen, 11 ml (0.1 mol) of dry destabilized styrene. Finally the contents of each flask are diluted to 50 ml with benzene, the adaptors removed under slight positive pressure of nitrogen, and the flasks immediately closed with ground glass stoppers secured with springs. The samples are now heated to 50°C for 48 h, after which the polymerizations are terminated by running the solutions from a dropping funnel into 500 ml of stirred methanol. The polymers are filtered off at sintered glass crucibles and dried to constant weight in vacuum at 50°C. The conversion, limiting viscosity number (benzene, 20°C), and degree of polymerization are determined for each sample (see Section 2.3.2.1).

3.2. IONIC HOMOPOLYMERIZATION[1]

The chain carriers in ionic polymerizations may be cationic or anionic. Like radical polymerizations, ionic polymerizations occur by a chain mechanism. In contrast to radical polymerization the chain carriers are macroions: carbonium ions in the case of cationic polymerization and carbanions

[1] *G. Heublein*, "Zum Ablauf ionischer Polymerisationsreaktionen", Akademie-Verlag, Berlin 1975.

in the case of anionic polymerization of C=C compounds:

cationic $$R^{\oplus} + n(CH_2{=}\underset{\displaystyle X}{\underset{|}{CH}}) \longrightarrow R{+}CH_2{-}\underset{\displaystyle X}{\underset{|}{CH}}{+}_{n-1}CH_2{-}\underset{\displaystyle X}{\underset{|}{CH^{\oplus}}}$$

anionic $$R^{\ominus} + n(CH_2{=}\underset{\displaystyle Y}{\underset{|}{CH}}) \longrightarrow R{+}CH_2{-}\underset{\displaystyle Y}{\underset{|}{CH}}{+}_{n-1}CH_2{-}\underset{\displaystyle Y}{\underset{|}{CH^{\ominus}}}$$

Again in contrast to radical polymerization, there is no chain termination by combination, since the growing chains (macro-ions) repel each other electrostatically as a consequence of their like charge. Chain termination occurs only by reaction of the growing chain ends with substances such as water, alcohols, acids, and amines. The ions produced by reaction of these substances can sometimes initiate new chains (chain transfer). Under certain conditions the ionic propagating species retain their ability to grow over extended periods of time, even after complete consumption of monomer ("living polymers", see Section 3.2.1).

Radical initiators have so far been employed successfully only for the polymerization of compounds containing C=C bonds; but the number of ionically polymerizable monomers is considerably larger, and includes compounds containing C=O groups, C=N groups, and a series of heterocyclic compounds.[1] In some cases migration of an atom or group occurs during polymerization (isomerization polymerization[2]). It is a particular characteristic of some ionic polymerizations that they proceed stereospecifically, leading to tactic polymers.

In contrast to radical polymerizations, ionic polymerizations proceed at high rates even at low temperatures, since the initiation and propagation reactions have only small activation energies. For example, isobutene is polymerized commercially with boron trifluoride in liquid propane at −100°C (see Example 3–23). The polymerization temperature often has a considerable influence on the structure of the resulting polymer.

3.2.1. Ionic polymerization via C=C bonds

The tendency of unsaturated compounds to undergo cationic or anionic polymerization differs greatly according to the type of substituent at the double bond. As well as monomers that polymerize only cationically or only anionically there are some that can polymerize by both mechanisms (see Section 1.1, Table 1.1). Electron-repelling groups (alkyl, phenyl,

---

[1] *A.D. Jenkins* and *A. Ledwith*, "Reactivity, Mechanism and Structure in Polymer Chemistry", Wiley-Interscience, London 1974.

[2] *J.P. Kennedy* and *T. Otsu*, Adv. Polym. Sci. 7 (1970) 369.

alkoxy) cause polarization, such that the unsubstituted carbon atom of the double bond carries a partial negative charge; hence, protons or other suitable cations can attach themselves to this unsubstituted carbon atom, the positive charge being transferred to the substituted carbon atom which can likewise add monomer by the propagation reaction.[1,2,3] In cationic polymerization the propagating chains can be terminated only by addition of reactive anions, since combination of two macro-cations is not possible; β-elimination of a proton from the chain end may also take place in a transfer reaction.

Initiation $$R^{\oplus} + \overset{\delta^-}{C}H_2{=}\overset{\delta^+}{C}H(X) \longrightarrow R{-}CH_2{-}\overset{\oplus}{C}H(X)$$

Propagation $$R{-}CH_2{-}\overset{\oplus}{C}H(X) + \overset{\delta^-}{C}H_2{=}\overset{\delta^+}{C}H(X) \longrightarrow R{-}CH_2{-}CH(X){-}CH_2{-}\overset{\oplus}{C}H(X)$$

Termination $$\ldots{-}CH_2{-}CH(X){-}CH_2{-}\overset{\oplus}{C}H(X) + B^{\ominus} \longrightarrow \ldots{-}CH_2{-}CH(X){-}CH_2{-}CH(X){-}B$$

Transfer $$\ldots{-}CH_2{-}\overset{\oplus}{C}H(X) \longrightarrow \ldots{-}CH{=}CH(X) + H^{\oplus}$$

Nucleophiles, such as water, alcohols, esters, acetals, and ethers, can also act as transfer agents by reacting with the macro-cations, at the same time forming a new cation which initiates the growth of a new chain; the kinetic chain thus remains unbroken.

$$\ldots{-}CH_2{-}\overset{\oplus}{C}H(X) + HOH \longrightarrow \ldots{-}CH_2{-}CH(X){-}OH + H^{\oplus}$$

$$H^{\oplus} + CH_2{=}CH(X) \longrightarrow CH_3{-}\overset{\oplus}{C}H(X)$$

Cationic polymerizations can be initiated with protonic acids (e.g. sulfuric acid, perchloric acid, trifluoroacetic acid), with Lewis acids (see Section 3.2.1.1) and with compounds that form suitable cations (e.g. iodine, acetyl perchlorate). Some monomers are also polymerized by high-energy radiation according to a cationic mechanism.

By choice of appropriate conditions it is sometimes possible to stop the reaction at the dimer stage; for example, in the case of styrene this leads

[1] *P.H. Plesch* (Ed.), "The Chemistry of Cationic Polymerization", Pergamon Press, Oxford 1963.

[2] *J.P. Kennedy* and *A.W. Langer, jr.*, Adv. Polym. Sci. *3* (1961) 508; *J.P. Kennedy*, "Cationic Polymerization of Olefins", John Wiley, New York 1975.

[3] J. Polym. Sci., Polym. Symp. *56* (1977).

to an unsaturated linear dimer, or, by ring closure, to a cyclic dimer (see Example 3–25):

$$H^{\oplus} + CH_2{=}\underset{C_6H_5}{CH} \longrightarrow CH_3{-}\underset{C_6H_5}{\overset{\oplus}{C}H}$$

$$CH_3{-}\underset{C_6H_5}{\overset{\oplus}{C}H} + CH_2{=}\underset{C_6H_5}{CH} \longrightarrow CH_3{-}\underset{C_6H_5}{CH}{-}CH_2{-}\underset{C_6H_5}{\overset{\oplus}{C}H}$$

$$\nearrow \; CH_3{-}\underset{C_6H_5}{CH}{-}CH{=}\underset{C_6H_5}{CH} + H^{\oplus}$$

$$\searrow \; CH_3{-}CH{-}CH_2{-}CH{-}C_6H_5 \text{ (cyclic, ring closure onto the phenyl ring)} + H^{\oplus}$$

For unsaturated compounds with electron-withdrawing substituents (carboxyalkyl, nitro, nitrile, vinyl), polymerization can be initiated anionically (e.g. by OH^-, NH_2^- or carbanions)[1]. The addition of the anion takes place at the unsubstituted carbon atom, which, in this case, carries a partial positive charge. Since the growing chain end is a genuine anion, chain termination can occur by addition of a reactive cation. As in cationic polymerization, combination of two growing ends is not possible. Chain transfer with electrophiles can also occur.

Initiation $\quad R^{\ominus} + \overset{\delta^+}{C}H_2{=}\underset{X}{\overset{\delta^-}{C}H} \longrightarrow R{-}CH_2{-}\underset{X}{\overset{\ominus}{C}H}$

Propagation $\quad R{-}CH_2{-}\underset{X}{\overset{\ominus}{C}H} + \overset{\delta^+}{C}H_2{=}\underset{X}{\overset{\delta^-}{C}H} \longrightarrow R{-}CH_2{-}\underset{X}{CH}{-}CH_2{-}\underset{X}{\overset{\ominus}{C}H}$

Termination $\quad R{-}CH_2{-}\underset{X}{CH}{-}CH_2{-}\underset{X}{\overset{\ominus}{C}H} + H^{\oplus} \longrightarrow R{-}CH_2{-}\underset{X}{CH}{-}CH_2{-}\underset{X}{CH_2}$

Transfer $\quad R{-}CH_2{-}\underset{X}{\overset{\ominus}{C}H} + HOH \longrightarrow R{-}CH_2{-}\underset{X}{CH_2} + HO^{\ominus}$

$$HO^{\ominus} + \overset{\delta^+}{C}H_2{=}\underset{X}{\overset{\delta^-}{C}H} \longrightarrow HO{-}CH_2{-}\underset{X}{\overset{\ominus}{C}H}$$

[1] *J.E. Mulvaney, C.G. Overberger* and *A.M. Schiller*, Adv. Polym. Sci. *3* (1961) 106; *M. Szwarc*, "Carbanions, Living Polymers and Electron Transfer Processes", Interscience Publishers, New York 1968; *L.L. Böhm, M. Chmelir, G. Löhr, B.J. Schmitt* and *G.V. Schulz*, Adv. Polym. Sci. *9* (1972) 1.

For some monomers (e.g. nitroethylene and 2-cyano-2,4-hexadienoic acid ester, $CH_3—CH=CH—CH=C(CN)—COOR$) anionic polymerization can be conducted in aqueous alkaline solution. Other anionic initiators are Lewis bases, e.g. tertiary amines or phosphines, and organometallic compounds (see Section 3.2.1.2). Since the polarizability of unsaturated compounds depends very much on the substituents and on the solvent used, there are considerable differences in the effectiveness of the initiators mentioned.

Another way to initiate anionic polymerization is by electron transfer.[1] The reaction of sodium with naphthalene gives sodium naphthalene (sodium dihydronaphthylide) in which the sodium has not replaced a hydrogen atom but has transferred an electron to the electronic levels of the naphthalene; this electron can be transferred to styrene or α-methylstyrene, forming a radical anion:

$$Na^{\oplus}\left[C_{10}H_8\right]^{\ominus\bullet} + CH_2{=}\underset{\displaystyle X}{\underset{|}{CH}} \longrightarrow \bullet CH_2{-}\underset{\displaystyle X}{\underset{|}{\overset{\ominus}{C}H}} + C_{10}H_8 + Na^{\oplus}$$

Such radical anions combine very quickly forming a dianion which can then add styrene at both ends:

$$2\,\dot{C}H_2{-}\underset{\displaystyle X}{\underset{|}{\overset{\ominus}{C}H}} \longrightarrow \underset{\displaystyle X}{\underset{|}{\overset{\ominus}{C}H}}{-}CH_2{-}CH_2{-}\underset{\displaystyle X}{\underset{|}{\overset{\ominus}{C}H}}$$

Provided the reaction mixture is prepared under stringent conditions, such that reaction of the dianions with impurities (e.g. water) is prevented, the polymer chains can grow until the monomer is completely consumed. If another batch of styrene is added the "living polymer" can grow afresh. If, finally, a chain breaker is added (e.g. a proton donor), a "dead" polymer results.

Living polymers can also be used for the preparation of block copolymers; after the consumption of the first monomer, a second anionically polymerizable monomer is added which then grows on to both ends of the initially formed block. By termination of the living polymer with electrophilic compounds the polymer chains can be given specific end groups; for example living polystyrene reacts with carbon dioxide to give a polystyrene with carboxyl end groups.

---

[1] *M. Szwarc*, "Carbanions, Living Polymers and Electron Transfer Processes", Interscience Publishers, New York 1968; *L.L. Böhm*, *M. Chmelir*, *G. Löhr*, *B.J. Schmitt* and *G.V. Schulz*, Adv. Polym. Sci. *9* (1972) 1; *B.J. Schmitt* and *G.V. Schulz*, Eur. Polym. J. *11* (1975) 119.

Under ideal conditions, two sodium naphthalene molecules give rise to one polymer chain, so that provided initiation is rapid compared with propagation,

$$\bar{P} = 2 \times \frac{\text{amount of monomer consumed (mole)}}{\text{amount of initiator (mole)}}.$$

The polydispersity of polymers prepared in this way is usually very low; for example a value of $\bar{M}_w/\bar{M}_n$ of 1.05 was found for a sample of poly-(α-methylstyrene). This requires that the experiment is performed under extremely pure conditions and that the monomer is mixed with the initiator solution very rapidly and completely (cf. Example 3–27).

In the anionic polymerization of α-methylstyrene with sodium naphthalene the reaction proceeds to an equilibrium position and it is possible to observe the temperature dependence of the equilibrium between monomer and polymer. On addition of the monomer the deep green colour of the initiator solution is transformed into the red colour of the carbanions. At low temperature ($-70°$ to $-40°$C) the living polymer is formed and the solution becomes viscous. On warming, the macro-anions depolymerize again, but reform reversibly on cooling. The temperature at which the equilibrium is established is termed the ceiling temperature[1] for the prevailing monomer concentration; for pure α-methylstyrene this lies at about 60°C, but for most vinyl monomers it occurs above 250°C. For some monomers with polymerizable C=O bonds, or for cyclic monomers, the ceiling temperature is relatively low, for example formaldehyde or trioxane (126°C) and tetrahydrofuran (85°C). In spite of its thermodynamic instability poly (α-methylstyrene) is isolable after termination of the living anionic chain ends; capping the chain ends by reaction with water or carbon dioxide blocks the depolymerization. The thermal degradation of poly(α-methylstyrene) to monomer proceeds with measureable speed only above 200°C where chain fission begins to occur (cf. Example 5–14).

In certain cases of ionic polymerization the chain growth is stereoregular. It was first shown possible to make tactic polymers by stereospecific catalysis using organometallic mixed catalysts (see Section 3.2.1.3), but other initiator systems have since been found that are suitable for stereoregular polymerization. For example under certain conditions styrene can be polymerized with many organo-alkali-metal compounds to

[1] For determination of the ceiling temperature see *R.E. Cook*, *F.S. Dainton* and *K.J. Ivin*, J. Polym. Sci. *29* (1958) 549; *A.D. Jenkins* and *A. Ledwith*, "Reactivity, Mechanism and Structure in Polymer Chemistry", Wiley 1974, Chapter 16.

TABLE 3.4

Structure of polymeric dienes prepared with various initiators (proportions of repeating units in %)

Initiator	Butadiene			Isoprene			
(R = alkyl)	1,4 *cis*	1,4 *trans*	1,2	1,4 *cis*	1,4 *trans*	1,2	3,4
Peroxide in emulsion	28	51	21	22	65	6	7
$AlCl_3$				—	90	—	—
Lithium in heptane				93	—	—	—
Lithium in ether				4	21	6	63
Sodium in heptane	14	23	61	—	46	9	45
Potassium in heptane				—	55	9	36
$AlR_3/TiCl_4$	49	49	2	97	—	—	3
AlR_3/VCl_4	—	99	—	—	99	—	—
AlR_3/chromium acetylacetonate	—	—	90 (isotactic)				

give tactic polystyrene[1,2] (Example 3–28). A certain degree of stereoregulation during chain growth has also been observed in cationic and even in radical polymerization of some monomers.

Stereospecific polymerization has particular significance for the preparation of stereoregular polymeric dienes. In the radical polymerization of butadiene or isoprene the molecular chains always consist of varying proportions of adjacent *cis*- and *trans*-1,4-units as well as 1,2- and 3,4-linked units, depending on the polymerization conditions; but it is now possible, using particular ionic initiation systems to make a "synthetic natural rubber" that contains more than 90% *cis*-1,4-isoprene repeating units (see Examples 3–30 and 3–34).

Table 3.34 shows a series of initiators and initiator systems for diene polymerization and information about the structure of the resulting polymers. It may be seen that both the solvent and the catalyst affect the structure of the polymer produced. For example the structure of the polyisoprene varies markedly with the alkali metal, even when used in the same solvent medium. Experiments with a typical organometallic mixed catalyst, consisting of trialkylaluminium and titanium tetrachloride, show that the same initiator can lead to quite different structures in the products of polymerization of isoprene and of butadiene.

---

[1] *D. Braun* and *W. Kern*, J. Polym. Sci. *C4* (1964) 197.
[2] *W. Kern* and *D. Braun*, Chem.-Ztg./Chem. Appar. *87* (1963) 799.

3.2.1.1. Cationic polymerization with Lewis acids as initiators

The following Lewis acids are suitable for initiating cationic polymerization: BF_3, $AlCl_3$, $TiCl_4$, $SnCl_4$, $SnCl_2$, and $FeCl_3$. So-called co-initiators (cocatalysts[1]) play an essential part in the initiation mechanism. Amongst these are proton-donating compounds (protonic acids, water, alcohols) and compounds such as alkyl halides, that form ionic complexes with Lewis acids, which can then dissociate to give cations capable of initiating polymerization. Thus in the case of water or alcohol as cocatalyst the real initiator is a proton:

$$BF_3 + ROH \longrightarrow \left[F_3B{-}O\begin{smallmatrix} H \\ R \end{smallmatrix} \right] \rightleftharpoons F_3\overset{\ominus}{B}{-}OR + H^{\oplus}$$

and, when alkyl halides are used, an alkyl cation:

$$SnCl_4 + RCl \longrightarrow [RSnCl_5] \rightleftharpoons SnCl_5^{\ominus} + R^{\oplus}$$

The initiator concentration required for cationic polymerization is smaller than for radical polymerization; frequently 10^{-3} to 10^{-5} mol of initiator per mol monomer is sufficient to achieve a high rate of reaction. The effect of initiator concentration on the rate and average degree of polymerization depends on the monomer and a variety of other factors, and does not follow a consistent pattern.

The type and amount of cocatalyst required for optimum polymerization conditions must be determined for every case; generally the amount of cocatalyst required (in mol) is much less than that of the initiator.

In polymerizations of unsaturated compounds with Lewis acids the required reaction temperatures are below room temperature, down to −100°C or lower (see Example 3–23). On the other hand cyclic monomers (see Section 3.2.4) frequently need higher temperatures.

Cationic polymerizations of unsaturated compounds are practically always carried out in solution. In radical polymerization, dilution with a solvent, under otherwise similar conditions, always results in a decrease of rate and molecular weight; but in cationic polymerization, addition of one to four parts by volume of a suitable solvent (with respect to monomer) often causes a significant increase of molecular weight and sometimes also of the rate. The main reasons for this behaviour reside in the effect of solvent polarity and the cocatalytic action of the solvent. The choice of

[1] For summary see *M. Kucera*, J. Macromol. Sci., Chem. 7 (1973) 1611.

solvent is thus extremely important. The rate of polymerization and molecular weight generally increase with increasing polarity and relative permittivity (dielectric constant) of the solvent; some solvents can form complexes with Lewis acids which then initiate polymerization by a carbonium-ion mechanism. The solvent can also interfere with the course of a cationic polymerization through chain transfer reactions. Taking into account the above mentioned limitations the following solvents are suitable for cationic polymerizations: benzene, cyclohexane, methylene chloride, carbon tetrachloride, dichloroethylene, trichloroethylene, chlorobenzene, nitrobenzene, and liquid sulfur dioxide.

In solution polymerizations catalyzed by Lewis acids, the polymerization frequently does not begin immediately after addition of the initiator, and there is an induction period which cannot be completely eliminated even by careful purification of the starting materials. In contrast some cationic polymerizations proceed so quickly (flash polymerization), even after dilution, that conversion is already complete after a few minutes (e.g. isobutene, Example 3–23).

Finally it should be noted that cationic polymerizations are very sensitive to impurities. These can act as cocatalysts, accelerating the polymerization, or as inhibitors (e.g. tertiary amines); they can also give rise to chain transfer or chain termination and so cause a lowering of the degree of polymerization. Since these effects can be caused by very small amounts of impurity (10^{-3} mole % or less), careful purification and drying of all materials and equipment is imperative.

Example 3–23:
Cationic polymerization of isobutene with gaseous BF_3 at low temperatures

Monomeric isobutene is passed from a cylinder through a drying column filled with solid potassium hydroxide pellets and condensed in a dry cold trap.

A dry 100 ml three-necked flask, fitted with stirrer, ground glass stopper and thermometer, is cooled to −80°C in a methanol/dry-ice bath; 10 ml isobutene and 5 g of dry ice are then added while stirring (the dry ice should be taken from the middle of a larger piece in order to be as free from water as possible). A dry 10 ml syringe pipette is filled with gaseous boron trifluoride from a cylinder via an empty wash bottle; this is then injected into the liquid monomer. Polymerization begins immediately and gives a rubber-elastic product. The cold bath is removed after 45 min so that the excess monomer slowly evaporates on warming to room temperature. The polyisobutene obtained in this way is soluble in aliphatic, cycloaliphatic and chlorinated hydrocarbons. The limiting viscosity number is determined in cyclohexane at 24°C and the molecular weight derived (see Section 2.3.2.1).

Example 3–24:
Cationic polymerization of isobutyl vinyl ether with BF_3-etherate at low temperatures

35 ml of isobutyl vinyl ether are washed five times with distilled water, dried over $MgSO_4$ and distilled (b.p. 80°C) under nitrogen over solid NaOH into a dry receiver (see Section 2.1.2). About 80 ml of propane, taken from a cylinder, are passed through a drying column filled with solid KOH pellets and condensed in a dry cold trap.

A dry 500 ml three-necked flask, fitted with stirrer, thermometer and nitrogen inlet, is heated under vacuum with a Bunsen flame in order to remove traces of water. It is then filled with nitrogen and cooled to −70°C in an acetone/dry-ice bath, the gas outlet being protected against back-diffusion (see Section 2.1.1). The propane in the cold trap is transferred to the flask and 20 ml (0.15 mol) of isobutyl vinyl ether added from the receiver. 0.30 ml (0.39 mmol) of freshly distilled BF_3-etherate ($\rho_{20} = 1.125$) are added dropwise with vigorous stirring and the mixture held at −70°C for 30 min; another 0.30 ml of BF_3-etherate is then dropped in and stirring continued at this temperature for a further 90 min. During polymerization, which takes place at the surface of the initiator droplets, a very slow stream of nitrogen is passed through the apparatus. Finally the initiator is destroyed by addition of an excess of cyclohexylamine and the mixture slowly warmed to room temperature; the propane evaporates, leaving the polymer behind in the form of small lumps. It is dried in vacuum at 50°C. Before drying, a small sample is dissolved in benzene and reprecipitated in petroleum ether; this is used for the determination of the limiting viscosity number in benzene at 20°C; also for the determination of the melting point (90°C), the product being crystalline (see Section 2.3.4.3).

Example 3–25:
Dimerization of styrene

A 250 ml three-necked flask, fitted with stirrer, thermometer and reflux condenser, is immersed in an oil bath at 120°C. A mixture of 52 g (0.5 mol) of destabilized styrene (see Example 3–01) and 5 ml of concentrated sulfuric acid in 7.5 ml of water is introduced and stirred vigorously. A milky-white emulsion is first formed, which turns a pale yellow colour after about 10 min. The temperature is held at 120°C and the progress of dimerization followed by measuring the refractive index of the organic phase at 20°C every 15 min. In order to take a sample the stirrer is switched off for a short time and after the two phases have separated a drop is withdrawn from the upper layer containing the monomer and dimer. As soon as the same value is obtained several times running (after about 4 hours), the experiment is stopped, the dimerization then being complete; a graph is plotted of refractive index against reaction time (monomeric styrene, $n_D^{20} = 1.5472$; styrene dimer, $n_D^{20} = 1.5877$).

The reaction vessel is now stoppered and held for 1 h at 50°C on a water bath; the two layers are then separated with the aid of a separating funnel. The organic phase is diluted with 15 ml of benzene and washed three times with a 10% sodium chloride solution and then twice with a saturated sodium carbonate solution; the layers separate rather slowly. The solution is now washed again with saturated calcium chloride solution, the aqueous phase removed and the organic phase dried for several hours over solid calcium chloride. Finally, the liquid is distilled under vacuum (0.5 torr) using an oil pump, the benzene distilling off first. The fraction distilling between 120 and 123°C is the purest product ($n_D^{20} = 1.5877$), but contains about 3% of the cyclic dimer of styrene, as may be shown by gas chromatography. Yield of styrene dimer: 50%.

Example 3–26:
Cationic polymerization of α-methylstyrene in solution

Monomeric α-methylstyrene is freed from phenolic inhibitors by shaking twice with 10% sodium hydroxide. It is then washed three times with distilled water, dried over calcium hydride and distilled before use over calcium hydride under reduced pressure of nitrogen (b.p. 54°C/12 torr). Methylene chloride is refluxed over P_2O_5 (best deposited on a solid carrier) for 1 h and then distilled into a dry receiver.

A carefully dried 100 ml three-necked flask, fitted with stirrer, thermometer and nitrogen inlet, is evacuated and filled with nitrogen; 5.9 g (0.05 mol) of α-methylstyrene and 50 ml of methylene chloride are pipetted in and cooled to −78°C in a methanol/dry-ice bath. 0.040 ml (0.75 mmol) of concentrated sulfuric acid are now added, causing polymerization to begin immediately, as shown by the slight temperature rise. After 3 h the viscous solution is added dropwise to about 500 ml of methanol in order to precipitate the polymer, which is filtered off after settling and washed with methanol; a portion of the polymer is reprecipitated from a 2% benzene solution into methanol. Both samples are dried to constant weight in vacuum at 50°C; yield: about 70%. The reprecipitated sample is used to determine the softening range (150–200°C) and the limiting viscosity number in toluene at 25°C (molecular weight about 100 000).

This experiment can also be carried out with styrene; boron trifluoride or tin tetrachloride may also be used as initiators.

3.2.1.2. *Anionic polymerization with organometallic compounds as initiators*

There are numerous organometallic compounds that are capable of initiating the polymerization of unsaturated monomers. The following are of general importance: organic compounds of the alkali metals (e.g. butyllithium), organic compounds of zinc and cadmium (e.g. diethylzinc, diisobutylzinc), and organomagnesium compounds. Polymerizations with organometallic compounds as initiators are generally carried out in solution. The following can be used as solvents: aliphatic and aromatic hydrocarbons (hexane, heptane, decalin, benzene, toluene), and cyclic ethers (tetrahydrofuran, 1,4-dioxane). Polarity and solvating power of the solvent have a major effect on the course of the polymerization and on the structure of the resulting polymer when stereospecifically acting initiators are used (see Example 3–29).

For the polymerization of unsaturated monomers with organometallic compounds, the initiator concentration must generally lie between 10^{-1} and 10^{-4} mol per mol of monomer; cocatalysts are usually unnecessary. Polymerization frequently occurs at temperatures below 20°C. Raising the temperature increases the rate of polymerization but usually decreases the tacticity or tactic content when stereospecific initiators are used. An induction period is seldom observed.

All the above-mentioned initiators are very sensitive towards substances with active hydrogen. Care must therefore be taken to exclude acids, water, alcohols, thiols, amines, and acetylene derivatives. Oxygen, carbon dioxide, carbon monoxide, carbonyl compounds, and alkyl halides, which can react with the initiator, also interfere with the reaction. Careful purification and drying of the starting materials and apparatus is, therefore, absolutely essential, especially when dealing with "living polymers" (see Example 3–27).

Example 3–27:
Anionic polymerization of α-methylstyrene[1] with sodium naphthalene in solution ("living polymer")

All operations must be carried out under especially careful exclusion of air and moisture, otherwise the experiment will fail.

[1] This experiment can also be carried out with styrene.

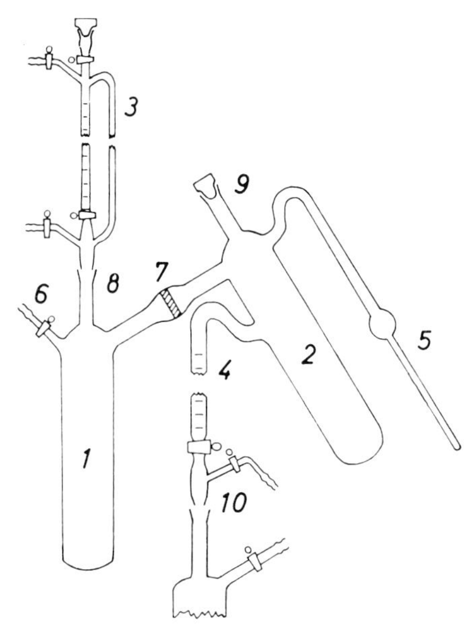

FIGURE 3.2 Apparatus for the preparation of living polymers

(a) Purification of the monomer and solvent
Monomeric α-methylstyrene is destabilized and dried as described in Example 3–25. Pure tetrahydrofuran (THF) (previously distilled through a column) is refluxed over potassium under nitrogen for at least a day. A little pure benzophenone is then added; if the THF is sufficiently pure a blue colour appears, due to the formation of metal ketyl. If the blue colour fails to appear, the purification with potassium must be continued. The THF is distilled from this ketyl solution under nitrogen before use.

(b) Preparation of the initiator solution
The individual parts of the apparatus shown in Figure 3.2 are first carefully cleaned. After assembly the apparatus is evacuated with an oil pump, dried by heating with a flame and filled with pure dry nitrogen.[1]

Very reactive, finely divided sodium, that is easy to handle and relatively safe, can be prepared by mixing 5 g of sodium and 50 g of neutral aluminium oxide powder under nitrogen with vigorous stirring at 150–170°C.[2] A fine powder is obtained that can easily be dosed under nitrogen using the burette shown in Figure 3.2.

[1] The nitrogen is dried either by passing it through a cold trap (−180°C) or over phosphorus pentoxide and through a ketyl solution.
[2] *Houben-Weyl*, *14/1*, p. 635.

To prepare the initiator solution, 1.28 g (10 mmol) of very pure naphthalene are introduced into tube 1 through the joint 8, while passing a slow stream of nitrogen through stopcock 6. After inserting a magnetic stirrer, 100 ml of purified tetrahydrofuran are run in through joint 8 with exclusion of air. The 10 ml burette 3 (bore of the stopcock barrel of this powder burette is 4 mm) is filled with the sodium/aluminium oxide powder and mounted on joint 8. The apparatus is now completely closed to the external atmosphere. 5 ml of the sodium powder are then run into the stirred solution in tube 1; the solution immediately becomes deep green. The burette can now be removed and replaced with a ground glass stopper. The sodium is allowed to react with the naphthalene for 30 min, while stirring is continued; the solution in tube 1 is then transferred to tube 2 by tilting the apparatus and applying a pressure of nitrogen. The aluminium oxide powder is held back by means of the sintered disc 7 (porosity 1). Capillary 5 can now be filled and sealed off after cooling to −78°C; this solution can be used to obtain the electron spin resonance spectrum and thus to demonstrate the radical character of sodium naphthalene. 10 ml of the initiator solution are run from burette 4 into a conical flask containing some distilled water. The content of sodium naphthalene is determined by titration of the hydrolysis product with 0.1 M hydrochloric acid. The solution made by this procedure should contain about 100 mmol of initiator per litre.

(c) Polymerization procedure

A Schlenk vessel is dried by heating under vacuum and is then charged with 50 ml of purified tetrahydrofuran under nitrogen. The vessel is attached through joint 10 to the burette 4 under a gentle stream of nitrogen. To remove traces of impurities that might still be present, a few drops of initiator solution are run into the vessel. As soon as the green colour persists 1 mmol of initiator (about 10 ml depending on the result of the titration) is run into the reaction vessel from burette 4. The vessel is removed from the burette under nitrogen, closed with a stopper and cooled to −78°C. 5.90 g (0.05 mol ≃ 6.5 ml) of α-methylstyrene are now injected in one dose from a hypodermic syringe, the solution being vigorously shaken. The solution immediately turns red as a result of the formation of the carbanions.

After 1 h at −78°C an aliquot of solution is removed from the reaction vessel under a nitrogen stream using a dry syringe filled with nitrogen; this is dropped into a tenfold excess of methanol in order to precipitate the poly(α-methylstyrene). Another 5.9 g of α-methylstyrene are now added to the solution remaining in the reaction vessel, the red colour of which should not be affected by the removal of the sample aliquot. After a further hour another sample is taken and a fresh batch of monomer added. The experiment is finally brought to an end after another hour, the remaining solution being dropped into methanol to precipitate the polymer. The polymers obtained during the experiment are reprecipitated from THF into methanol, filtered and dried in vacuum at 50°C.

(d) Determination of the molecular weight of the poly(α-methylstyrene)s
If the polymerization of α-methylstyrene with sodium naphthalene proceeds without termination according to the above mechanism, the degree of polymerization can be represented by the following simple relation:

$$\bar{P} = 2 \times \frac{\text{amount of monomer consumed (in mole)}}{\text{amount of initiator (in mole)}}.$$

Thus for 0.05 mol of α-methylstyrene and 0.001 mol of initiator the polymer should have a degree of polymerization of 100, corresponding to a molecular weight of 11 800.

The limiting viscosity numbers of the samples are determined in toluene at 30°C and the molecular weights derived. The observed and calculated values are tabulated and compared.

Example 3–28:
Stereospecific polymerization of styrene with pentylsodium

(a) Preparation of pentylsodium[1]
Heptane and petroleum ether (b.p. 110°C) are each refluxed under nitrogen for 4 h and then distilled and stored over sodium.

Pentylsodium is prepared in a dry 250 ml three-necked flask with a sealed-in gas inlet tube. The flask is fitted with a high-speed stirrer (20 000 r.p.m.) and should have indentations on four sides to provide better turbulence during stirring. The second neck contains a protective adaptor with dropping funnel and reflux condenser, and the third neck is closed with a glass stopper. The reflux condenser is fitted with a stopcock and bubbler or non-return valve (see Section 2.1.1). For safety reasons toluene is used as cooling liquid instead of water. Care must be taken to see that the apparatus is assembled in a strain-free condition to avoid the possibility of breakage by vibration during high-speed stirring. It is further recommended that all the joints are firmly secured with spiral springs.

Air is removed from the apparatus by evacuating and filling with nitrogen three times. 100 ml of pure petroleum ether are then introduced

[1] Safety goggles must be worn when handling pentylsodium. All operations must be conducted with the exclusion of moisture and air.

under nitrogen, followed by 4.6 g (0.2 mol) of sodium that has been carefully freed from crusts under petroleum ether. The flask is then heated, using a heating mantle, until the petroleum ether is boiling (no open flames!) After the sodium has melted the stirrer speed is increased to 20 000 r.p.m. over a period of 2 min using a variable transformer. After 5 min the stirrer is switched off and the heater immediately removed to allow the sodium suspension (particle diameter 5–20 μm) to cool to room temperature. The petroleum ether is now exchanged for heptane, as follows. The glass stopper is replaced by a half-bored rubber stopper or another type of self-sealing closure (see Section 2.1.3), and the petroleum ether above the deposited sodium is carefully drawn off with a syringe. 100 ml of pure heptane are then run in from the dropping funnel, swirled round briefly with the stirrer, and drawn off again by syringe after the sodium has settled; this operation is repeated twice more. The glass stopper is put back, 100 ml heptane added through the dropping funnel and finally, with vigorous stirring, a solution of 10.66 g (0.1 mol) of pentyl chloride (distilled under nitrogen) in 20 ml of heptane is added dropwise. After a few drops of solution have been added, the reaction mixture darkens; the flask is now cooled to −25°C in a methanol/dry-ice bath and the remainder of the pentyl chloride solution dropped in within 30 min. The bath temperature should not be allowed to rise above −20°C (if necessary, add dry ice). When all the solution has been added, stirring is continued for a further 30 min at −25°C and the bluish-black initiator suspension (about 120 ml) is transferred by syringe to a dry receiver filled with nitrogen (see Section 2.1.2); it can be kept for several months at −20°C.

Before using the initiator the content of pentylsodium in the suspension is determined as follows.[1]

(b) Determination of the total alkali content

2 ml of the initiator suspension are decomposed in 50 ml of distilled methanol and, after addition of 20 ml of water, titrated with 0.1 M hydrochloric acid using phenolphthalein as indicator (titre V_1).

(c) Determination of free alkali

5 ml of the initiator suspension are added to 15 ml of distilled benzyl chloride and, after addition of 20 ml distilled methanol and 30 ml of water, titrated with 0.1 M hydrochloric acid using phenolphthalein as indicator (titre V_2). From the difference of the two determinations the content of

[1] Cf. *Houben-Weyl 14/1* (1961) 798.

pentylsodium can be derived, as shown in the following example:

Total alkali: $V_1 = 14.53$ ml of 0.1 M HCl, corresponding to 0.7260 mmol of Na per ml of initiator suspension.
Free alkali: $V_2 = 7.33$ ml of 0.1 M HCl, corresponding to 0.1465 mmol of Na per ml of initiator suspension.
Pentylsodium: $0.7260 - 0.1465 = 0.5795$ mmol of Na per ml initiator suspension (79.8% of the total alkali), corresponding to 0.5795 mol of pentylsodium per litre.

(d) Polymerization procedure

250 ml of pure heptane and 26 g (0.25 mol) of dry styrene (distilled under nitrogen) are placed in a dry 500 ml three-necked flask, fitted with stirrer, thermometer and nitrogen inlet, and cooled to −20°C. The air is then displaced by evacuating and filling with dry nitrogen three times. The thermometer is now exchanged for a half-bored rubber stopper or other self-sealing closure. 0.05 mol (0.2 mol per mol styrene) pentylsodium are injected with a syringe while stirring the solution (86.4 ml initiator suspension would be needed in the above example). The thermometer is replaced and the mixture stirred under a slow stream of nitrogen for 6 h at −20 ± 1°C. The polymerization is terminated by injection of about 10 ml of methanol. The polymer is precipitated from the reaction mixture by dropping it into 2.5 l methanol; the polymer is then filtered off, washed with methanol and dried overnight in vacuum at 50°C. In order to remove the inorganic components the dry polymer is dissolved in benzene and centrifuged for 30 min at 4000 r.p.m. The polystyrene is precipitated by running the clear solution into excess methanol, and is then filtered, washed with methanol, and dried to constant weight in vacuum at 50°C (about 20 h). Yield: 25%.

(e) Crystallization and characterization of isotactic polystyrene

The polystyrene prepared above must still be freed from atactic components and induced to crystallize; both are achieved by heating in heptane. 4 g of polymer are refluxed for 4 h in 200 ml of pure heptane; this is filtered hot in order to separate the insoluble (isotactic) portion. For comparison, the same procedure should also be carried out with an atactic polystyrene of not too high molecular weight, prepared by radical polymerization.

The atactic polystyrene is precipitated by dropping the heptane solution into methanol and filtered at a sintered glass crucible; the soluble and insoluble portions are dried in vacuum at 50°C and finally weighed. The solubility of the samples is tested in butanone and their densities (see Section 2.3.7) and limiting viscosity numbers determined in benzene at 20°C. The X-ray diffraction patterns of the two samples are compared with each other and with that of a polystyrene made by radical polymerization; likewise for the infrared spectra (see Sections 2.3.6 and 2.3.9).

The melting range of the isotactic, crystalline sample is determined with the aid of a hot-stage microscope; the following conditions of the sample can be distinguished:

1. clearly defined particles,
2. blurred edges,
3. beginning of sintering,
4. beginning of melting,
5. melt runs together,
6. clear melt.

The melting range is defined by the temperature interval between steps 4 and 5; for the crystalline polystyrene prepared above, this lies between 205 and 215°C.

Example 3–29:
Preparation of isotactic and syndiotactic poly(methyl methacrylate) with butyllithium in solution

(a) Preparation of the initiator solution
0.5 g (72 mmol) of finely cut lithium and 4.64 g (50.1 mmol) of freshly distilled butyl chloride are added to 50 ml of pure benzene under nitrogen in a well-dried, nitrogen-filled, 250 ml three-necked flask, fitted with stirrer and nitrogen inlet. A reflux condenser is then attached and the reaction started by stirring and slowly heating to about 80°C. The mixture is refluxed for a further 4 h and then allowed to stand. The butyllithium solution so obtained is about 1 M. For the exact determination of the butyllithium content, 2 ml of this solution is withdrawn under nitrogen with the aid of a hypodermic syringe, added to 20 ml of pure methanol and titrated with 0.1 M hydrochloric acid using methyl red as indicator.

(b) Polymerization procedure
Methyl methacrylate, benzene, toluene, and 1,2-dimethoxyethane are fractionated using a column, and finally distilled over calcium hydride into a receiver under dry nitrogen (see Section 2.1.2).

Two 250 ml three-necked flasks are baked out under vacuum, fitted with stirrer, nitrogen inlet and self-sealing closure (see Section 2.1.3), and evacuated and filled several times with dry nitrogen. 100 ml of toluene are introduced to one of the flasks and 100 ml of 1,2-dimethoxyethane to the other; to both are added 0.006 mol of butyllithium (about 6 ml 1 M solution) using a hypodermic syringe, and the solutions cooled to −78°C; 10 ml (0.1 mol) of methyl methacrylate are then injected into each. After 30 min the polymerizations are terminated by addition of 10 ml of methanol, and the polymers precipitated by dropping each solution into 1500 ml of low-boiling petroleum ether. The polymers are filtered off, the damp polymer then being dissolved in benzene and centrifuged for about 30 min at 4000 r.p.m. in order to separate insoluble residues (inorganic hydrolysis products and some crosslinked polymer). The polymer is reprecipitated in a 15-fold amount of petroleum ether, filtered and dried in vacuum at 40°C. Yield in toluene as solvent: 60–70% (isotactic polymer); yield in 1,2-dimethoxyethane as solvent: 20–30% (syndiotactic polymer). The limiting viscosity numbers of the two samples are measured in acetone at 25°C (see Section 2.3.2.1); also the infrared spectra in potassium bromide discs (see Section 2.3.9). From the latter, the isotactic and syndiotactic content can be determined both qualitatively and quantitatively.[1]

Example 3–30:
Stereospecific polymerization of isoprene with butyllithium

*(a) Polymerization in bulk: preparation of poly(*cis*-1-methyl-1-butenylene); (*cis*-1,4-polyisoprene)*
Monomeric isoprene is destabilized and dried as for styrene (see Example 3–01); before use it is run, under nitrogen, through a 30 cm column packed with neutral aluminium oxide.

A dry 100 ml two-necked flask, fitted with magnetic stirrer, thermometer and nitrogen inlet, is flamed in vacuum and filled with nitrogen dried by passage over phosphorus pentoxide. 38 g (0.56 mol) of dry isoprene and 0.5 mmol of butyllithium (2.5 ml of 0.2 M benzene solution, see Example 3–29) are then pipetted in under nitrogen. The initiator solution and the monomer are mixed by gentle shaking of the apparatus, at the same time warming on a bath at about 30°C. As soon as the polymerization has begun, as shown by the boiling of the isoprene (b.p. 34°C), the bath is removed; the reaction mixture now becomes gradually more viscous. If polymerization fails to commence it means that the initiator has been destroyed by the presence of too much moisture or other impurities in the system. In this case a further similar amount of butyllithium is added and the above procedure repeated. The polymerization comes to an end after 7 h, the yield being almost quantitative. The highly elastic, transparent product is characterized as described under (*c*). For this purpose about 1 g of the sample is purified by dissolving in 50 ml of benzene and precipitating in 500 ml of ethanol. The polymer is soluble in benzene, toluene, decalin, and carbon disulfide. The limiting viscosity number is determined in toluene at 25°C and the molecular weight derived (see Section 2.3.2.1).

[1] *U. Baumann*, *H. Schreiber* and *K. Tessmar*, Makromol. Chem. *36* (1960) 81.

(b) Polymerization in solution: preparation of poly(1-isopropenylethylene); (3,4-polyisoprene).

Monomeric isoprene is purified as described under (*a*); cyclohexane and 1,2-dimethoxyethane are refluxed over sodium for 6 h and distilled off under nitrogen.

A carefully dried 250 ml two-necked flask, fitted with thermometer, nitrogen inlet and glass-sealed magnetic stirrer, is evacuated and filled with nitrogen (dried over phosphorus pentoxide) three times. 20 ml of 1,2-dimethoxyethane, 100 ml of cyclohexane, and 6.8 g (0.1 mol) of isoprene are pipetted in under a stream of nitrogen. The magnetic stirrer is switched on and 0.1 mmol of butyllithium (0.5 ml of 0.2 M solution in benzene) is added; the solution turns lemon yellow and gradually warms up, a sign that polymerization has been triggered. The reaction may fail to commence on account of impurities present in the system, in which case more initiator solution is pipetted in until the yellow colour of the reaction mixture persists. Polymerization ceases after 3 h; 50 mg of *N*-phenyl-β-naphthylamine are now added to the solution as antioxidant and the polymer precipitated by dropping the solution into ethanol. It is filtered off and dried in vacuum at 50°C. Yield: 90–95%. Before characterizing the polymer as described under (*c*), about 1 g is reprecipitated from a 2% benzene solution into a 10-fold excess of ethanol.

(c) Structural investigations of polymeric dienes by infrared spectroscopy

The type of arrangement of the monomeric units in polymeric dienes can be determined qualitatively and quantitatively by infrared spectroscopy.[1] For this purpose thin films are prepared by dropping an approximately 2% solution in carbon disulfide (spectroscopically pure) on to suitable rock salt plates and allowing the solvent to evaporate at room temperature. The plates are placed in the spectrometer beam and the i.r. spectrum recorded. The different types of chemical linkage are associated with characteristic i.r. bands as summarized in Table 3.5.

The spectra may first be evaluated qualitatively. The polyisoprene prepared in solution shows a pronounced band at 888 cm^{-1}, which indicates a high proportion of 3,4-linkages. For the product of bulk polymerization this band is much reduced in favour of absorptions at 1127 and 1315 cm^{-1}, indicating predominantly *cis*-1,4-linkage of the monomeric units in this case. The polymer made by radical polymerization in emulsion (see Example 3–21) shows the presence of all possible structural units, though the proportion of *cis*-1,4 linkages is low.

[1] 1,2- and 1,4-arrangements can also be differentiated by reaction with perbenzoic acid: double bonds in the chain react more quickly than double bonds in the side chain (see *Houben-Weyl 14/1* (1961) 692).

TABLE 3.5

Characteristic infrared absorption bands for the different possible repeat units in polymeric dienes

Type of linkage		Wave length (μm)	Wave number (cm^{-1})
Polybutadiene			
1,2-	—C—C(—C=C)—	11.0	909
trans-1,4-	—C—C(H)=C(H)—C—	7.4 10.3	1355 971
cis-1,4-	—C—C(H)=C(H)—C—	7.62 13.5–13.8	1311 741–725
Polyisoprene			
1,2-	—C—C(CH_3)(—C=C)—	11.0	909
3,4-	—C—C(—C=C—CH_3)—	11.27	888
trans-1,4	—C—C(CH_3)=C(H)—C—	7.55 8.71	1325 1148
cis-1,4-	—C—C(H_3C)=C(H)—C—	7.60 8.87	1315 1127

3.2.1.3. *Polymerization with organometallic mixed catalysts (Ziegler-Natta catalysts)*[1]

In 1953 Ziegler and coworkers[2] discovered a class of heterogeneous catalysts that allowed ethylene to be polymerized at low pressure and low temperatures (low pressure polyethylene = high density polyethylene HDPE).

[1] *L. Reich* and *A. Schindler*, "Polymerization by Organometallic Compounds", Interscience Publishers, New York 1966; *J. Boor*, J. Polym. Sci., Macromol. Rev. *2* (1967) 115; *Y.I. Ermakov* and *V.A. Zakharov*, Russ. Chem. Rev. (Engl. Transl.) *41* (1972) 203; *J.C.W. Chien*, "Coordination Polymerization", Academic Press, 1975.

[2] *K. Ziegler*, *E. Holzkamp*, *H. Breil* and *H. Martin*, Angew. Chem. *67* (1955) 541.

These catalysts consist of mixtures of compounds of metals of Groups IV-VIII with metal alkyls or hydrides of Groups I-III of the periodic table. Especially effective are combinations of the transition element compounds $TiCl_4$, $TiCl_3$, or $VOCl_3$ with aluminium alkyls or aluminium alkyl halides such as R_3Al,R_2AlHl and $R_3Al_2Hl_3$ (R = alkyl). The transition element compound (catalyst) reacts with the metal alkyl (activator) to form an organometallic complex which generates the active polymerization centre (Cat).

The present view concerning the polymerization mechanism[1,2,3,4] is as follows. The first step is the formation of a π-complex between ethylene and the catalytically active centre Cat, onto which an alkyl group has already been transferred by the activator. The complexed monomer then pushes its way into the transition metal-carbon bond in the sense of an insertion reaction, so lengthening the chain by two carbon atoms. The growth process consists of a series of such insertion steps. Chain termination is a consequence either of β-elimination or of reaction with hydrogen. In both cases the active centre Cat remains intact.

All factors which influence the stability of the transition metal-carbon bond (Mt—R) and/or the stability of the transition metal-ethylene bond ($Mt \leftarrow C_2H_4$) are liable to affect the course of the reaction. Such factors are:

— type and valence state of the transition metal Mt,
— type and number of ligands L on the transition metal Mt,
— type of metal alkyl (activator) R–m,
— catalyst morphology (crystallinity, porosity, external and internal surface).

The polymerization process consists of a series of consecutive and concurrent steps:

— transfer of the monomer and molecular weight regulator, H_2, from the gas phase to the suspension medium,
— diffusion of these components to the active centre Cat; with increasing conversion the diffusion will be affected by the increasing amount of partially crystalline polymer phase,
— adsorption and complex formation of the monomer with the active centre Cat,

[1] *P. Cossee*, J. Catal. *3* (1964) 80, 99.
[2] *G. Henrici-Olivé* and *S. Olivé*, Angew. Chem. *79* (1967) 764; Adv. Polym. Sci. *6* (1969) 421; "Coordination and Catalysis", Verlag Chemie, 1977.
[3] *T. Keii*, "Kinetics of Ziegler-Natta Polymerizations", Chapman and Hall, London 1972.
[4] *L.L. Böhm*, Polymer *19* (1978) 545.

— occurrence of the chemical reactions,
— formation of a steadily growing partially crystalline polymer phase around the active centre Cat; this can result in the catalyst particle breaking off from the polymer (disintegration), whereby under certain circumstances a new active centre can be formed; the diffusion of the reactants (monomer, H_2) can also be hindered by the growing polymer phase.

Formation of the catalyst Cat:

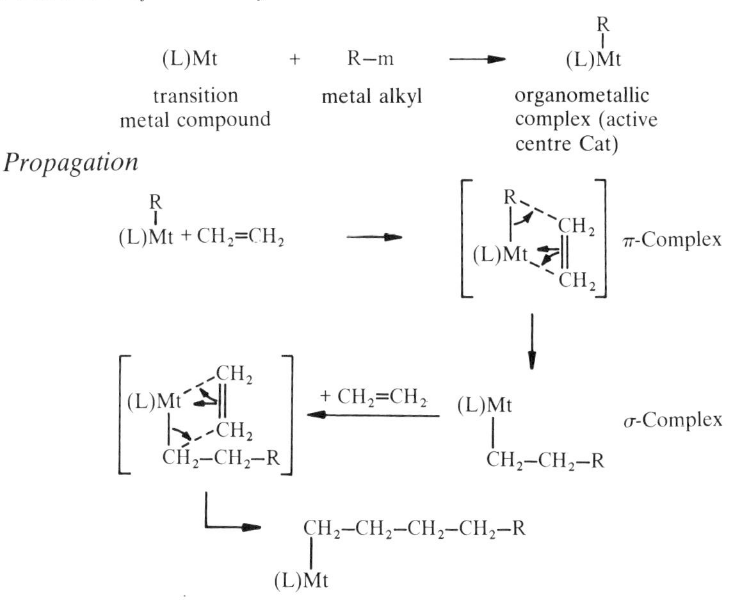

Termination:
(a) by β-elimination

$$\text{(L)Mt}(\text{CH}_2{=}\text{CH}_2)\text{–CH}_2\text{–CHR–H} \longrightarrow \text{(L)Mt–CH}_2\text{–CH}_3 + \text{CH}_2{=}\text{CHR}$$

(b) by reaction with hydrogen

$$\text{(L)Mt–CH}_2\text{–CH}_2\text{–R} \xrightarrow{+\text{H}_2} \text{(L)Mt–H} + \text{CH}_3\text{–CH}_2\text{–R}$$

$$\text{(L)Mt–H} \xrightarrow{\text{CH}_2=\text{CH}_2} \text{(L)Mt–CH}_2\text{–CH}_3$$

The polymerization is carried out with the exclusion of air and water. The organometallic catalyst and the activator are suspended in an aliphatic hydrocarbon, the active centres Cat thereby being formed. Ethylene is passed in and the polymerization allowed to proceed at or slightly above atmospheric pressure, at a temperature below 100° C. The polyethylene precipitates as a swollen powder. As soon as the mixture has developed into a thick, dark-coloured slurry, the reaction is terminated by destroying the catalyst (e.g. with butanol). The polymer is freed from most of the disperse medium by filtration, the remainder being removed by steam distillation. The damp polyethylene is finally dried.

Besides ethylene, higher olefins[1] (propene, 1-butene), dienes and a number of other monomers can be polymerized with organometallic mixed catalysts; the polymerization frequently proceeds stereospecifically, leading to tactic polymers (see Examples 3–32, 3–33).

Recently the activity of organometallic mixed catalysts has been found to be substantially increased by depositing them on suitable carriers.[2,3] In the polymerization of propene a significant enhancement of the stereospecificity of the catalyst has been achieved by the addition of special stereoregulators.

Finally a particular class of organometallic mixed catalysts should be mentioned. These are frequently based on tungsten compounds (e.g. $WCl_6/C_2H_5OH/C_2H_5AlCl_2$) and can bring about the ring-opening polymerization of cycloolefins to linear unsaturated polymers, e.g. of cyclopentene to poly(1-pentenylene) (see Example 3–35):

$$n \; \text{(cyclopentene)} \longrightarrow \ldots \!\left[CH_2{-}CH_2{-}CH_2{-}CH{=}CH \right]_n\! \ldots$$

Such reactions of cycloolefins are called metathesis polymerizations[4,5,6] by analogy with the transalkylidenation reactions of linear olefins[7], which proceed under comparable conditions and in which the double bonds of the olefins are broken and new double bonds are formed (metathesis):

---

[1] *G. Natta*, Angew. Chem. *68* (1956) 393.

[2] For summary see *K. Weissermel*, *H. Cherdron*, *J. Berthold*, *B. Diedrich*, *K.D. Keil*, *K. Rust*, *H. Strametz* and *T. Toth*, J. Polym. Sci., Polym. Symp. *51* (1975) 187.

[3] *L.L. Böhm*, Polymer *19* (1978) 553, 562.

[4] *N. Calderon* and *M. Morris*, J. Polym. Sci., Part A–2, *5* (1967) 1283.

[5] *G. Pampus*, Angew. Makromol. Chem. *14* (1970) 87.

[6] *G. Dall' Asta*, Rubber Chem. Technol. *47* (1974) 511.

[7] *N. Calderon*, *E.A. Ofstead*, *D.P. Ward*, *W.A. Judy* and *K.W. Scott*, J. Am. Chem. Soc. *90* (1968) 4133; *T.J. Katz*, Adv. Organomet, Chem. *16* (1977) 283.

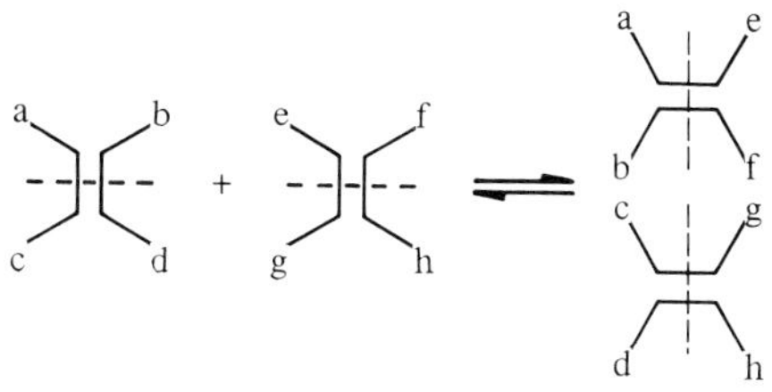

In the polymerization of cycloalkenes, not only linear polymers but also cyclic oligomers are formed. These reactions are thus examples of ring/chain equilibria[1], but their mechanism is not yet fully clarified.

Example 3–31:
Polymerization of ethylene with organometallic mixed catalysts

The greatest possible care must be exercised when working with organoaluminium compounds since they ignite very easily on contact with air and water. All operations must, therefore, be carried out with the complete exclusion of air and moisture and pipettes must be flushed with nitrogen.[2] Moreover, these substances cause wounds that are slow to heal, so that the wearing of safety goggles is mandatory and all contact with the skin must be avoided.

The polymerization apparatus (see Figure 3.3) consists of a 1-litre four-necked flask, fitted with stirrer, thermometer, gas inlet with tap, and gas outlet. On the inlet side the gas stream passes through three wash bottles: one as a safety bottle (A), one for the purification of ethylene, filled with 30 ml of petroleum ether (b.p. 100–140°C) and 5 ml of diethylaluminium chloride (B), and one filled with molecular sieves 4 Å(C). The last of these dries the ethylene further and also serves to trap aluminium hydroxide carried over from B. On the outlet side there are two wash bottles: the first is a safety bottle (D), and the second, (E), is filled with 50 ml of dry bis(2-hydroxyethyl) ether (diglycol), and isolates the apparatus from the external atmosphere.

500 ml of petroleum ether (b.p. 100–140° C) are placed in the flask and the apparatus is flushed with nitrogen. The ethylene cylinder is attached and ethylene passed through the apparatus while stirring the solvent in the flask; at the same time the bath is heated to 50° C. After 15 min the reaction

[1] *H. Höcker, W. Reimann, K. Riebel* and *Z. Szentivany*, Makromol. Chem. *177* (1976) 1707.

[2] The pipettes are cleaned as follows. After all the $(C_2H_5)_3Al$ has been run out, the pipette is filled with petroleum ether and allowed to drain again. It is then washed with acetone, dried and later cleaned with a solution of dichromate in concentrated sulfuric acid.

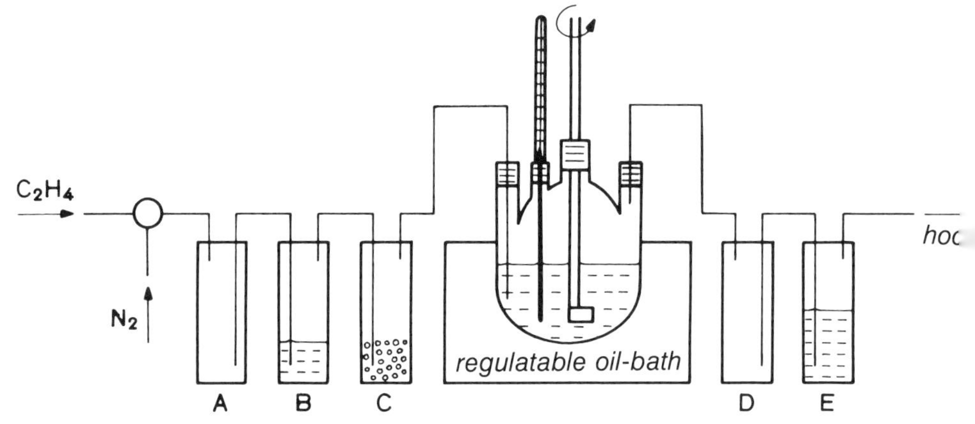

FIGURE 3.3 Apparatus for suspension polymerization of olefins in water-free medium

medium is saturated with ethylene. 2 ml (15 mmol) of $(C_2H_5)_3Al$,[1] followed by 1 ml of $TiCl_4$ (9.1 mmol), are pipetted in, the thermometer being lifted from its neck for this purpose. During this operation the mixture is stirred continuously and the ethylene flow is maintained through the flask and the neck, so preventing the ingress of air.

After the addition of the catalyst components, the reaction mixture turns brown; it can then absorb an increased amount of ethylene. Stirring is continued for 30 min at 700 r.p.m., and then at 900 r.p.m. because of the increased viscosity of the contents of the flask. The ethylene flow is so adjusted that only a small amount escapes through E as a result of the excess pressure; the rate of polymerization can be controlled to some extent by increasing the flow of ethylene or by cutting back until there is some suckback in E. If the ethylene is taken from a home-made graduated gasometer rather than from a cylinder the polymerization can be followed by determining the gas uptake at intervals of 5 min and plotting against time.

The rapid onset of polymerization is marked by a rise of temperature to 55–60° C and the initially clear reaction medium becomes turbid. As the reaction rate falls off, the temperature drops again to 50° C. In the course of 45–55 min the contents of the flask change to a brown pasty mass. After 60 min the polymerization is stopped by the addition of 30 ml of butanol, thereby causing the reaction mixture suddenly to decolorize and become white. Stirring is continued for a further 10 min, and then 150 ml of

[1] Diethylaluminium chloride $(C_2H_5)_2AlCl$ or aluminium sesquichloride $(C_2H_5)_3Al_2Cl_3$ can be used in place of $(C_2H_5)_3Al$.

methanolic hydrochloric acid (2:1) are added with stirring for another 10 min. The polymer is filtered at a Büchner funnel, thoroughly washed with methanol and acetone, and dried to constant weight in vacuum at 50° C. Yield: approximately 50 g.

Depending on the molecular weight, polyethylene is a waxy or hard crystalline substance. The above preparation yields a high-molecular-weight, crystalline product with a melting range around 130° C. It is insoluble in all solvents at room temperature but dissolves at higher temperatures (100–150° C) in aliphatic and aromatic hydrocarbons. Viscosity measurements[1] can be carried out in xylene, tetralin or decalin at 135° C, with the addition of 0.2% of anti-oxidant (e.g. *N*-phenyl-*β*-naphthylamine) to prevent oxidative degradation. Samples of polyethylene can easily be pressed between two metal plates (see Section 2.4.2.1) to give thin films (press for about 2 min at 180–190° C). After quenching with water the films can be detached from the plates. They are convenient for the determination of infrared spectra, from which, for example, the crystallinity (see Section 2.3.6) or the degree of branching (see Section 2.3.9) can be determined.

Example 3–32:
Stereospecific polymerization of propene with organometallic mixed catalysts

(a) Preparation of isotactic polypropene
The polymerization apparatus (Figure 3.3) is flushed with nitrogen and the reaction flask filled with 500 ml of dry petroleum ether (distilled over sodium under nitrogen, b.p. 100–140° C). The flask is then heated to 60° C (±2° C) in a thermostatted oil bath and the petroleum ether is saturated with propene (about 15 min) while stirring (about 500 r.p.m.). 1.26 ml (10 mmol) of diethylaluminium chloride and about 1 g (5 mmol) of $TiCl_3$, $\frac{1}{3}AlCl_3$ (Staufer AA, handled with exclusion of air and moisture) are added through the thermometer neck. The flow of propene is so adjusted that very little escapes through the protective wash bottle E. The polymer separates out as a red-violet powder. The temperature is held steady by thermostatting at 60 ± 2°C. The stirring speed during polymerization is about 700 r.p.m.

After two hours the polymerization is terminated by addition of 20 ml of 2-propanol, and stirring continued for another 30 min at 60° C; the reaction mixture decolorizes and becomes white. After filtration at a Büchner

[1] See *H. Wesslau*, Kunststoffe *49* (1959) 230.

funnel and several washings with warm petroleum ether the polymer is dried to constant weight in vacuum at 70° C. Yield: 29.5 g.

The polypropene so obtained has a high molecular weight and is crystalline. The proportion of isotactic polymer, determined by extracting with heptane for 10 h in a Soxhlet apparatus, is 98.5%. Isotactic polypropene shows similar solubility behaviour to polyethylene, but has a higher melting point (crystalline melting range 165–171°C).

(b) Effect of heterogeneous nucleation on the crystallization of isotactic polypropene

In the crystallization of isotactic polypropene from the melt, the number and size of spherulites (and hence the rate of crystallization) can be influenced by the addition of certain nucleating agents.[1,2] The smaller the spherulites, the greater is the transparency of the polypropene film. The mechanical properties can also be affected in some cases.[3]

The effect of heterogeneous nucleation on the crystallization of isotactic polypropene from the melt can be easily established as follows. A small amount of powdered polypropene is well mixed with about 0.1 wt. % sodium benzoate in a mortar or by means of an analytical mill.[4] Some of the mixture is transferred with a spatula to a microscope slide and melted at about 250°C on a hot block. A cover slip is pressed on to the melt with a cork so as to obtain as thin a film as possible. The sample is held at 200–250°C for some minutes and then allowed to crystallize at about 130°C on the hot stage of the microscope; an unadulterated polypropene sample is crystallized in the same way.[5] Both samples are observed under a polarizing microscope during crystallization, the difference in spherulite size between nucleated and unnucleated polypropene being very clearly seen. An ordinary microscope can also be used by placing polarizers on the condenser and eyepiece, and adjusting these to give maximum darkness.

Example 3–33:

Stereospecific polymerization of styrene with organometallic mixed catalysts

A dry 1 litre three-necked flask, fitted with stirrer, thermometer, nitrogen

---

[1] *C.J. Kuhre, M. Wales* and *M.E. Doyle*, SPE-J. *20* (1964) 1113; *H.N. Beck*, J. Appl. Polym. Sci. *11* (1967) 673.

[2] *F.L. Binsbergen*, Polymer *11* (1970) 253, 309.

[3] *H.J. Leugering*, Makromol. Chem. *109* (1967) 204.

[4] For preliminary experiments it is sometimes sufficient to place some powdered polypropene on a microscope slide, and, after melting, to add a few seeds of sodium benzoate. It can then be observed whether the resulting spherulites are significantly smaller in the neighbourhood of the sodium benzoate particles.

[5] For quantitative investigations, in which the thermal history must be reproduced as closely as possible, the two samples are best cooled to room temperature on a hot plate after switching off the heater.

inlet, and dropping funnel with pressure equalizer, is evacuated and filled with nitrogen three times. 0.9 ml (8.2 mmol) of $TiCl_4$ are added by means of a syringe pipette under a brisk flow of nitrogen, and a mixture of 3.3 ml (24 mmol) of triethylaluminium[1] (or triisobutylaluminium) and 5 ml of pure heptane dropped in from the dropping funnel over a period of 20 min, with stirring. The reaction of the two catalyst components is initially strongly exothermic and the flask must therefore be cooled externally to about 0°C. As a precaution against possible breakage of the flask, the cooling bath must not contain water since it reacts extremely violently with triethylaluminium. A dry-ice/1,2-dimethoxyethane bath can be used. When the additions are complete, stirring is continued for 30 min at room temperature and 400 ml (3.5 mol) of carefully dried styrene (see Example 3–01) are then added quickly from a second dropping funnel. The stirring rate is now increased and the oil bath heated to 50°C. After 1–2 h the contents of the flask become viscous and eventually gel-like after 3–6 h. The hot bath is removed and 50 ml of methanol added over a period of 10 min through the dropping funnel, with vigorous stirring, in order to destroy the catalyst. The addition of methanol must be done very carefully; above all one must ensure immediate and thorough mixing. After the catalyst has been destroyed a further 350 ml of methanol is added, once again with vigorous stirring; the polystyrene then precipitates from the gel-like reaction mixture in the form of fine flakes. Stirring is continued for another 10 min, and the polymer filtered off and washed with methanol. In order to remove catalyst residues the polymer is stirred for 1 h in a mixture of 500 ml of methanol and 5 ml of concentrated hydrochloric acid. It is then filtered off, washed with methanol and dried in vacuum at 60°C. Yield: 5–30%.

For the separation of the amorphous portion the dried polymer is stirred for 3 h in 500 ml of acetone, to which 2 ml of concentrated hydrochloric acid has been added. The insoluble portion is filtered off and dried in vacuum at 60°C. Yield: 85–95% of crystallizable polystyrene.

The crystallization of the acetone-insoluble polystyrene is completed by boiling for 2 h in freshly distilled butanone; it is then allowed to stand overnight at room temperature and finally filtered and dried in vacuum at 60°C. Yield of crystalline isotactic polystyrene: 95–100% of the acetone-insoluble portion. The crystalline melting range (see Example 3–28) and the density (see Section 2.3.7) are determined; also the limiting viscosity number in benzene at 20°C.

[1] Aluminium alkyls must be handled under total exclusion of atmospheric oxygen and moisture.

Example 3–34:
*Stereospecific polymerization of butadiene with organometallic mixed catalysts: preparation of poly (*cis*-1-butenylene); (*cis*-1,4-polybutadiene)*

A trap, to be used for condensing the butadiene (see Figure 3.4), is dried for 1 h at 120°C, evacuated, and filled with pure dry nitrogen. A butadiene cylinder is attached to the three-way tap via a P_2O_5 drying tube. The air in the tubing and drying tube between the three-way tap and cylinder is displaced by flushing with butadiene. The trap is then cooled in a dry-ice/methanol bath and 20 g (33 ml) of butadiene condensed in.

Benzene is refluxed for a day over potassium. 200 ml are then distilled under nitrogen into a dry dropping funnel with pressure-equalizing tube. It is well stoppered for storage. 600 mg (1.68 mmol) of cobalt (III) acetylacetonate are weighed into a dry tube with attached stopcock. This is evacuated, filled with dry nitrogen, and 20 ml of dry benzene are then introduced through the bore of the tap with the aid of a hypodermic syringe. The closed tube is shaken until the cobalt compound has dissolved.

The polymerization is carried out in a 500 ml four-necked flask, fitted with stirrer, thermometer, adaptor and nitrogen inlet (see Figure 3.4). The

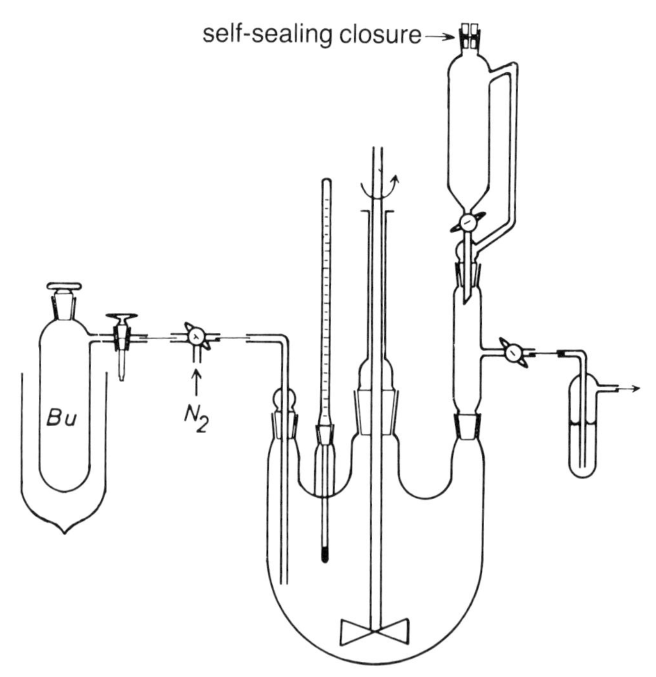

FIGURE 3.4 Apparatus for polymerization of butadiene.

individual parts.of the apparatus are previously dried for 1 h at 120°C, and while still hot are assembled as quickly as possible; the whole is then evacuated and filled with P_2O_5-dried nitrogen three times. Finally the dropping funnel containing the 200 ml of pure benzene is mounted and fitted with a self-sealing closure. Using a hypodermic syringe, that has been dried at 120°C and flushed with dry nitrogen, 1.0 ml (4.6 mmol) of $(C_2H_5)_3Al_2Cl_3$ are injected through the self-sealing closure into the benzene in the dropping funnel; the syringe is washed out with the benzene by several strokes of the piston.[1] The benzene solution of the aluminium sesquichloride[2] is now run into the flask, warmed on a water bath to 20–25°C, and 10 ml of the cobalt (III) salt solution added through the dropping funnel by means of a second hypodermic syringe, prepared as for the first. The colour now changes from green to grey-brown and the temperature climbs to about 40°C. The two components of the catalyst are allowed to react with each other for 10 min. The cold bath round the butadiene is removed and the 20 g (0.37 mol) of butadiene allowed to evaporate into the polymerization flask with stirring; this takes about one hour. The reaction mixture is held at 20°C for another hour and a benzene solution of an anti-oxidant (e.g. 0.2% 2,6-di-*tert*-butyl-4-methylphenol) is then added to prevent crosslinking reactions during work-up. The polymer is precipitated by dropping the highly viscous solution into a five-fold amount of methanol. After settling, the supernatant liquid is decanted from the polymer which is then broken down into small pieces and stirred with fresh methanol for a few minutes; this purification process is repeated twice more, the polymer finally being filtered and dried in vacuum at 40°C. Yield: >40%. About 1 g of the polybutadiene is reprecipitated from benzene solution with methanol. The limiting viscosity number is determined in cyclohexane at 20°C and the molecular weight derived. The polymer may be further characterized as described in Example 3–30.

Example 3–35:
Poly(1-pentenylene) by metathesis polymerization of cyclopentene with an organometallic mixed catalyst in solution

(a) Preparation of $W(OCH_2CH_2Cl)_2Cl_4$
2.64 g (6.67 mmol) of WCl_6 and 50 ml of dry toluene are transferred under nitrogen to a 100 ml three-necked flask, equipped with thermometer,

[1] The piston of the syringe should be smeared with a little paraffin in order to prevent its seizure in the cylinder. It should be washed with a mixture of 2-propanol and decalin (volume ratio 1:1) immediately after use.
[2] The sesquichloride must be handled with total exclusion of atmospheric oxygen and moisture.

nitrogen inlet, dropping funnel with a pressure-equalizing tube, and magnetic stirrer. A solution of 1.07 g (0.9 ml, 13.33 mmol) of 2-chloroethanol in 15 ml of dry toluene are added dropwise at room temperature with stirring, over a period of about 30 min. The temperature should not be allowed to exceed 35°C. Stirring is continued for 1 h. The brown solution of the tungsten compound (0.1 M) is stored under nitrogen.

(b) Preparation of a 0.5 M solution of $(C_2H_5)_2AlCl$ in toluene
30 ml of dry toluene and 2 ml (1.95 g, 16.13 mmol) of diethylaluminium chloride are placed in a 50 ml round-bottomed flask under nitrogen. The solution is well mixed and stored under nitrogen.

(c) Polymerization of cyclopentene
450 ml of pre-dried toluene are placed in a 500 ml flask equipped with stirrer, thermometer, nitrogen inlet, and condenser for distillation. 125 ml of the toluene are distilled off under a gentle stream of nitrogen. The flask is cooled to −15°C under a slight excess pressure of nitrogen, 50 ml (38.6 g, 0.567 mol) of dry cyclopentene are added, and finally 2 ml of the 0.1 M solution of $W(OCH_2CH_2Cl)_2Cl_4$ and 2 ml of the 0.5 M solution of $(C_2H_5)_2AlCl$. The polymerization commences immediately as can readily be seen from the marked increase in viscosity. The reaction temperature is kept at −10°C by external cooling. The polymerization is stopped after 5 h. To deactivate the catalyst a mixture of 0.5 g of 2,6 di-*tert*-butyl-4-methylphenol and 1 ml of ethanol dissolved in 15 ml of toluene are added with stirring; the solution is rapidly decolorized. The poly(1-pentenylene) is isolated by precipitation in about 1 l of propanol and drying to constant weight in vacuum at 50°C. Yield: about 70%. The solution viscosity is determined in toluene and the proportions of *cis* and *trans* double bonds estimated from the i.r. spectrum (see Table 3.5).

3.2.2. Ionic polymerization via C=O bonds

Polymerization by opening of C=O bonds was investigated many years ago in the case of formaldehyde.[1] As may be expected from its polar mesomeric structure formaldehyde can polymerize both anionically and cationically:[2,3]

$$CH_2{=}O \longleftrightarrow \overset{\oplus}{C}H_2{-}\overset{\ominus}{O}$$

[1] *W. Kern* in *H. Staudinger*, "Die hochmolekularen organischen Verbindungen", Springer-Verlag, Berlin-Göttingen-Heidelberg, 1932. New printing 1960 p. 286.
[2] *W. Kern*, *H. Cherdron* and *V. Jaacks*, Angew. Chem. *73* (1961) 177.
[3] *W. Kern*, Chem.-Ztg./Chem. Appar. *88* (1964) 623.

Anionic polymerization can be initiated with tertiary phosphines or amines, with organometallic compounds or with alcoholates. With all of these, initiation occurs by nucleophilic attack on the positive carbonyl carbon atom:

$$R{-}Li + \overset{\delta+}{C}H_2{=}\overset{\delta-}{O} \longrightarrow R{-}CH_2{-}O^{\ominus}\ Li^{\oplus}$$

$$R{-}CH_2{-}O^{\ominus}\ Li^{\oplus} + \overset{\delta+}{C}H_2{=}\overset{\delta-}{O} \longrightarrow R{-}CH_2{-}O{-}CH_2{-}O^{\ominus}\ Li^{\oplus}$$

Cationic polymerization of formaldehyde (which should be carried out under the driest possible conditions to avoid transfer reactions) can be initiated with protonic acids, Lewis acids (see Section 3.2.1.1) or other compounds that yield cations such as acetyl perchlorate or iodine:

$$H^{\oplus} + \overset{\delta-}{O}{=}\overset{\delta+}{C}H_2 \longrightarrow HO{-}CH_2^{\oplus}$$

$$HOCH_2^{\oplus} + \overset{\delta-}{O}{=}\overset{\delta+}{C}H_2 \longrightarrow HO{-}CH_2{-}O{-}CH_2^{\oplus}$$

Polymers of formaldehyde with semi-acetal end groups are thermally unstable; they decompose at temperatures as low as 150° C, splitting out monomeric formaldehyde. Upon acetylation of the hydroxyl end groups, thermal stability up to 220° C is achieved; alkylation also provides stability against alkali, but not against acids since these are capable of splitting the acetal bonds in the polymer chains (see Examples 5–09 and 5–15).

Other carbonyl compounds, such as acetaldehyde or propionaldehyde can also be polymerized to high-molecular-weight products; however their stability is lower than that of polyoxymethylenes with protected end groups.[1,2,3,4,5]

Example 3–36:
Anionic polymerization of formaldehyde in solution (precipitation polymerization)

(a) Preparation of a solution of monomeric formaldehyde
The apparatus for the preparation of monomeric formaldehyde consists of a 1 litre two-necked flask (A), one neck of which is closed with a loosely fitting stopper so that it can be removed quickly in the event of development of excess pressure. To the second neck is fitted a 2 m long, angled

---

[1] *O. Vogl*, J. Polym. Sci. A*2* (1964) 4591.
[2] *J. Furukawa* and *T. Saegusa*, "Polymerization of Aldehydes and Oxides", Polymer Reviews *3* (1963)
[3] Symposium on Polyaldehydes, J. Macromol. Sci., Chem. *1* (1967) 201.
[4] *H. Tani*, Adv. Polym. Sci. *11* (1973) 57.
[5] *P. Kubisa* and *O. Vogl*, Polym. J. *7* (1975) 186.

tube (diameter at least 2 cm), to the end of which is attached another 500 ml two-necked flask (B) filled with Raschig rings. The second neck of flask B serves as gas outlet and is connected to a 1 litre cold trap. The outlet of the trap is fitted with a suitable device (see Section 2.1.1) to prevent the ingress of atmospheric moisture and CO_2, the latter being a powerful inhibitor of polymerization. In the first flask (A), which is not yet connected to the rest of the apparatus, are mixed 100 g of polyoxymethylene (or paraformaldehyde) and 100 g of dry paraffin oil. This is then heated in an oil bath to 130°C until about 10–15% of the polyoxymethylene is decomposed; the resulting gaseous formaldehyde is not passed into the apparatus but into the fume hood; this procedure serves to remove water from the polyoxymethylene.

In the meantime the rest of the apparatus is flamed out under vacuum and filled with dry nitrogen; the cold trap is filled with 500 ml of sodium-dried ether and cooled to − 78°C. Flask A is now attached via the 2 m glass tube. The gaseous formaldehyde generated by further pyrolysis polymerizes partially in the glass tube and in flask B to low-molecular-weight polyoxymethylene, which contains as end groups most of the water formed during depolymerization (thus resulting in purification by pre-polymerization). Care must be taken that the polymer deposited does not block the glass tube; if necessary the pyrolysis must be interrupted so that the tube can be cleaned. Sticking of the ground joints can be prevented by generous greasing. The pyrolysis should be carried out rather quickly (in about one hour), since otherwise too much monomer is lost by pre-polymerization; an oil bath temperature up to 200°C is required. The resulting ether solution contains about 70 g of formaldehyde (about 4 M). Another method of preparing very pure monomeric formaldehyde is from 1,3,5-trioxane, which can be decomposed in the gas phase at 220°C on a supported phosphoric acid catalyst.[1]

(b) Polymerization procedure

150 ml of the ethereal formaldehyde solution are placed in a previously flamed 250 ml three-necked flask cooled to − 78°C, taking care to exclude atmospheric moisture. (For larger quantities care must be taken to make adequate provision for the removal of the heat of polymerization). The flask is fitted with an efficient stirrer, a self-sealing closure, and a pressure equalizer protected with a soda-lime tube. 1 mg of pyridine, dissolved in 5 ml of dry ether, is used as initiator; this is carefully injected into the formaldehyde solution (cooled to − 78°C) over a period of 15 min, with

[1] *A. Giefer* and *W. Kern*, Makromol. Chem *74* (1964) 39; *A. Giefer, V. Jaacks* and *W. Kern*, Makromol. Chem. *74* (1964) 46.

vigorous stirring. After 1 h at − 78°C the conversion attains more than 90% and the smell of formaldehyde has practically disappeared. Should the rate of polymerization be lower because of impurities, the amount of initiator can be raised. The precipitated polyoxymethylene is filtered off, washed with ether and dried in vacuum at room temperature; it melts between 176° and 178°C. The limiting viscosity number is determined on a 1% solution in dimethylformamide at 140°C (η_{sp}/c about 80 cm^3 g^{-1}, corresponding to an average molecular weight of 80 000). The thermal stability can be tested before and after blocking the hydroxyl end groups (see Examples 5–09 and 5–15).

Polyoxymethylene, obtained by the polymerization either of formaldehyde or of 1,3,5-trioxane, is a highly crystalline product that is insoluble in all solvents at room temperature with the exception of hexafluoroacetone hydrate; at higher temperature it dissolves in some polar solvents (e.g. at 130°C in dimethylformamide or dimethyl sulfoxide). If the unstable semi-acetal end groups are blocked (see Example 5–09) polyoxymethylene can be processed as a thermoplastic at elevated temperatures in the presence of stabilizers, without decomposition.

Example 3–37:
Anionic polymerization of chloral in solution (precipitation polymerization)

Chloral can be polymerized cationically (e.g. with $AlCl_3$ or $AlBr_3$) as well as anionically[1,2,3], the latter method giving high molecular weights. Suitable anionic initiators are organometallic compounds and especially tertiary amines, such as triethylamine or pyridine. The polymerization must be carried out at low temperatures on account of the low ceiling temperature.

Chloral is prepared from its hydrate by shaking with approximately four times its weight of warm concentrated sulfuric acid; the resulting layer of monomeric chloral is separated and distilled (b.p. 98°C, m.p. − 57°C).

30 g of chloral and 30 g of dry tetrahydrofuran[4] are placed in an open 100 ml two-necked flask fitted with thermometer and magnetic stirrer. After the contents of the flask have been cooled externally to −70°C the cold bath is removed; 2 ml of pyridine[5] is then stirred in. A white

---

[1] *I. Rosen*, J. Macromol. Sci., Chem. *1* (1967) 267.
[2] *O. Vogl, A.C. Miller* and *W.H. Sharkey*, Macromolecules *5* (1972) 658.
[3] *P. Kubisa* and *O. Vogl*, Polym. J. *7* (1975) 186.
[4] For drying of tetrahydrofuran, see Example 3–27.
[5] Pyridine is distilled from KOH pellets.

precipitate of polychloral immediately forms and the temperature rises. The flask is immersed again in the cold bath and the contents held at −70°C for 15 min. Finally the polymer is filtered off at as low a temperature as possible. The polymer is kept for some hours in a vacuum desiccator. Yield: 30–40%.

The polymer is unstable even at room temperature and slowly depolymerizes to monomeric chloral; at higher temperatures depolymerization proceeds very rapidly and can be observed, for example, with the aid of a melting point microscope at 150°C. As with polyoxymethylene this depolymerization can be suppressed by capping the semi-acetal hydroxyl end groups, for example by acetylation.[1] Polychloral with blocked end groups is stable at 255°C. It is insoluble and does not soften below 400°C.

The ceiling temperature for the anionic polymerization of chloral can be readily determined by a simple experiment. For this purpose 2 ml portions of a mixture of equal parts by weight of chloral and tetrahydrofuran are cooled in test tubes to various temperatures between 0 and −50°C and one drop of pyridine is then added to each. The temperature at which cloudiness, and therefore polymerization, is just noticeable, is the approximate ceiling temperature for the prevailing concentration of monomer.

3.2.3. Ionic polymerization via N=C bonds

The dimerization and trimerization of isocyanates to cyclic compounds has been known for a long time; they can also be polymerized through the N=C bond to give linear macromolecules:[2,3,4]

$$\underset{\text{R}}{\overset{|}{\text{N}}}=\overset{\overset{\text{O}}{\|}}{\text{C}} \longrightarrow \cdots-\underset{}{\overset{\overset{\text{R}}{|}}{\text{N}}}-\overset{\overset{\text{O}}{\|}}{\text{C}}-\cdots$$

Basic compounds are suitable as initiators, for example sodium cyanide in dimethylformamide. The polymers can attain high molecular weight and represent derivatives of the unknown polyamide-1(1-Nylon). Since they do not contain NH groups in the chain, unlike the usual polyamides, they are incapable of forming hydrogen bridges. The thermal stability of polymeric isocyanates is not very great; depolymerization occurs on heating or at room temperature in the presence of basic catalysts.

---

[1] *I. Rosen, D.E. Hudgin, C.L. Sturm, G.H. McCain* and *R.M. Wilhjelm*, J. Polym. Sci., Part A*3* (1965) 1535.
[2] *V.E. Shashoua, W. Sweeny* and *R.F. Tietz*, J. Am. Chem. Soc. *82* (1960) 866.
[3] *G. Natta, J. Di Pietro* and *M. Cambini*, Makromol. Chem. *56* (1961) 200.
[4] *M.N. Berger*, J. Macromol. Sci., Rev. Macromol. Chem. *9* (1973) 269.

Example 3–38:
Polymerization of butyl isocyanate with sodium cyanide in solution

(a) Preparation of the initiator solution
A 100 ml two-necked flask is flamed in vacuum and filled with dry nitrogen. Into this is weighed 1 g of finely powdered sodium cyanide that has been dried over P_2O_5 for 24 h. The flask is fitted with a joint and stopcock. 50 ml of dimethylformamide are then pipetted in, having previously been dried over $CaCl_2$ and distilled three times over P_2O_5. The closed flask is occasionally shaken; after 1 h a saturated solution (≈1%) of sodium cyanide in dimethylformamide is obtained.

(b) Polymerization procedure
A dry 250 ml four-necked flask, fitted with stirrer, thermometer, nitrogen inlet, and self-sealing closure (see Section 2.1.3), is flamed in vacuum and filled with dry nitrogen; the exit from the flask is protected against back-diffusion with a suitable device (see Section 2.1.1). 100 ml of dry dimethylformamide and 10 ml (0.09 mol) of butyl isocyanate, that has been previously distilled three times under reduced pressure of nitrogen, are then pipetted in under a brisk stream of nitrogen and cooled to −60°C. Using a dry hypodermic syringe 5 ml of the initiator solution (corresponding to about 1 mmol of NaCN) are injected over a period of 10 min through the self-sealing closure into the stirred solution. Polymerization sets in immediately with evolution of heat. The mixture is held at −60°C for a further 45 min, causing the polymer gradually to precipitate. 100 ml of methanol are then added in order to terminate the polymerization and to complete the precipitation of the polymer. The polymer is filtered off, washed several times with methanol and dried in vacuum at 50°C. Yield: about 6 g. The butyl-Nylon-1 is of very high molecular weight and only slightly soluble in benzene or tetrahydrofuran; transparent films can be cast from these dilute solutions. The softening point of the polymer is 180°C and the melting point about 209°C. Depolymerization occurs on prolonged heating at 200°C. The limiting viscosity number is determined in benzene at 20°C.

3.2.4. Ring-opening polymerization[1]

Many heterocyclic compounds can be polymerized by ring opening under certain conditions with ionic initiators, to produce linear macromolecules.

[1] *K.C. Frisch, S.L. Reegen* (Eds.), "Ring-opening Polymerization", Dekker, New York 1969; *H.L. Frisch*, "Cyclic Monomers", High Polymers Vol. 26.

Amongst these are cyclic ethers, cyclic sulfides, cyclic acetals, cyclic esters (lactones), cyclic amides (lactams) and cyclic amines. Ring-opening polymerizations are carried out under similar conditions, and frequently with similar initiators, to those used for ionic polymerization of unsaturated monomers (see Section 3.2.1); they are likewise sensitive to impurities.

3.2.4.1. Ring-opening polymerization of cyclic ethers

The ring-opening polymerization of cyclic ethers having 3-, 4- and 5-membered rings (e.g. epoxides, oxetane, tetrahydrofuran) yields polymeric ethers.[1,2,3] 6-membered rings (1,4-dioxane) are incapable of polymerization.

Epoxides[3,4] such as epoxyethane (ethylene oxide) can be polymerized cationically (e.g. with Lewis acids) and anionically (e.g. with alcoholates or organometallic compounds), but not radically.

cationic

$$R^{\oplus} + \underset{\diagdown O \diagup}{CH_2{-}CH_2} \longrightarrow \underset{\underset{R}{|}}{\underset{\diagdown \overset{\oplus}{O} \diagup}{CH_2{-}CH_2}} \longrightarrow R{-}O{-}CH_2{-}\overset{\oplus}{C}H_2$$

$$R{-}O{-}CH_2{-}\overset{\oplus}{C}H_2 + \underset{\diagdown O \diagup}{CH_2{-}CH_2} \longrightarrow R{-}O{-}CH_2{-}CH_2{-}O{-}CH_2{-}\overset{\oplus}{C}H_2$$

anionic

$$RO^{\ominus} + \underset{\diagdown O \diagup}{CH_2{-}CH_2} \longrightarrow RO{-}CH_2{-}CH_2{-}O^{\ominus}$$

$$RO{-}CH_2{-}CH_2{-}O^{\ominus} + \underset{\diagdown O \diagup}{CH_2{-}CH_2} \longrightarrow RO{-}CH_2{-}CH_2{-}O{-}CH_2{-}CH_2{-}O^{\ominus}$$

Polymers with extremely high molecular weights result from the polymerization of ethylene oxide initiated by the carbonates of the alkaline earths, e.g. strontium carbonate, which must, however, be very pure indeed.[5] Poly (ethylene oxides) having molecular weights up to about 600 are viscous liquids; above that they are wax-like or solid, crystalline products that are readily soluble not only in water but also in organic solvents such as benzene or chloroform. Polymers of substituted ethylene

---

[1] *Houben-Weyl*, 14/2 (1963) 425.
[2] *J. Furukawa* and *T. Saegusa*, "Polymerization of Aldehydes and Oxides", Polymer Reviews *3* (1963).
[3] *H. Tani*, Adv. Polym. Sci. *11* (1973) 57.
[4] *A.E. Gurgiolo*, Rev. Macromol. Chem. *1* (1966) 39.
[5] *F.N. Hill*, *F.E. Bailey* and *J.T. Fitzpatrick*, Ind. Eng. Chem. *50* (1958) 5.

oxides can be produced in both atactic amorphous and isotactic crystalline forms.[1,2] Optically active poly(propene oxide)s can be obtained from L-propene oxide.[3]

Polymerization of 4-membered cyclic ethers is also brought about by cationic initiators (e.g. Lewis acids) and by anionic initiators (e.g. organometallic compounds). The polymer of 3,3-bis(chloromethyl)oxetane is distinguished by its very high softening point and by its unusual chemical stability.[4]

Tetrahydrofuran can be polymerized only with cationic initiators, for example boron trifluoride or antimony pentachloride.[5,6] The initial step consists of the formation of a cyclic oxonium ion; one of two activated methylene groups in the α-position to the oxonium ion is then attacked by a monomer molecule in an S_N2 reaction, resulting in the opening of the ring. Further chain growth proceeds again via tertiary oxonium ions and not, as formerly assumed, via free carbonium ions:

$$R^{\oplus} + \begin{array}{c} CH_2—CH_2 \\ | \quad\quad | \\ CH_2 \quad CH_2 \\ \diagdown \; \diagup \\ O \end{array} \longrightarrow \begin{array}{c} CH_2—CH_2 \\ | \quad\quad | \\ CH_2 \quad CH_2 \\ \diagdown \; \diagup \\ O^{\oplus} \\ | \\ R \end{array}$$

$$\begin{array}{c} CH_2—CH_2 \\ | \quad\quad | \\ CH_2 \quad CH_2 \\ \diagdown \; \diagup \\ O^{\oplus} \\ | \\ R \end{array} + O\begin{array}{l} \diagup CH_2—CH_2 \\ \\ \diagdown CH_2—CH_2 \end{array} \longrightarrow \begin{array}{c} CH_2—CH_2 \\ | \quad\quad | \\ CH_2 \quad CH_2—O^{\oplus}\begin{array}{l} \diagup CH_2—CH_2 \\ \\ \diagdown CH_2—CH_2 \end{array} \\ \diagdown \\ O \\ | \\ R \end{array}$$

Deviations from this mechanism are observed when so-called super-acids (e.g. fluorosulfonic acid FSO_3H) are used as initiators (macro-ion/macroester equilibrium).[7,8,9]

Example 3–39:

Bulk polymerization of tetrahydrofuran with antimony pentachloride

Tetrahydrofuran is purified as described in Example 3–27 and distilled shortly before use into a dry receiver (see Section 2.1.2).

---

[1] *M. Osgan* and *C.C. Price*, J. Polym. Sci. *34* (1959) 153.
[2] *J. Furukawa* et al., Makromol. Chem. *32* (1959) 90.
[3] *C.C. Price* et al., J. Am. Chem. Soc. *78* (1956) 690, 4787.
[4] *G.M. Taylor* and *E.C. Wenger*, Ind. Eng. Chem. *53* (1961) 46A.
[5] *H. Meerwein*, *D. Delfs* and *H. Morschel*, Angew. Chem. *72* (1960) 927.
[6] *P. Dreyfuss* and *M.P. Dreyfuss*, Adv. Polym. Sci. *4* (1967) 528.
[7] *K. Matyjaszewski*, *P. Kubisa* and *S. Penczek*, J. Polym. Sci., Polym. Chem. Ed. *12* (1974) 1333, 1905.
[8] *S. Kobayashi*, *K. Morikawa* and *T. Saegusa*, Macromolecules *8* (1975) 387.
[9] *M.O. Mahmud*, *G. Wegner*, *W. Kern* and *B. Lando*, J. Macromol. Sci., Chem. *11* (1977) 2233.

A dry 250 ml round-bottomed flask, fitted with an adaptor (see Section 2.1.3) is flamed under vacuum and filled with nitrogen. 15 g (0.21 mol) of tetrahydrofuran are then added by connecting the receiver to the adaptor and applying a slight vacuum to the flask while admitting nitrogen to the receiver through a side arm and stopcock. The flask is cooled to −30°C and 0.6 g (2.0 mmol) of freshly distilled antimony pentachloride are added under nitrogen. The flask is now detached from the adaptor under a slight positive pressure of nitrogen and immediately closed with a ground glass stopper, secured with springs. The mixture is swirled around and allowed to stand at room temperature for 24 h, during which time the mixture becomes viscous. It is then treated with 2 ml of water and 150 ml of tetrahydrofuran, and refluxed for about 30 min until a homogeneous solution is produced. The still viscous solution is diluted with a further 100 ml of tetrahydrofuran and filtered to remove the insoluble portion (hydrolysis product of the initiator). The polymer is precipitated by dropping the solution into 3 l of water with vigorous stirring; it is filtered off and dried in vacuum at 20°C. Poly (oxytetramethylene), (polytetrahydrofuran), may be a viscous oil, a wax or a crystalline solid (melting range around 55°C), depending on the molecular weight. The polymer obtained in the above preparation is a somewhat sticky solid material with a melting range around 40°C. It is soluble in benzene, carbon tetrachloride, chlorobenzene, tetrahydrofuran, 1,4-dioxane, and acetic acid; it is insoluble in water, methanol, and acetone. The limiting viscosity number is determined in benzene at 20°C.

3.2.4.2. *Ring-opening polymerization of cyclic acetals*

Like tetrahydrofuran, cyclic acetals (e.g. 1,3-dioxolane[1,2] and 1,3,5-trioxane) are polymerizable only with cationic initiators. The ring-opening polymerization of 1,3,5-trioxane[3,4,5] (cyclic trimer of formaldehyde) leads to polyoxymethylenes (see Example 3–40), which have the same chain structure as polyformaldehyde (see Example 3–36). They are thermally unstable unless the semi-acetal hydroxyl end groups have been capped in a suitable way (see Example 5–09). Like the cyclic ethers, the polymerization of 1,3,5-trioxane proceeds via the addition of the initiator cation to a ring oxygen atom, with the formation of an oxonium ion which is

---

[1] *Y. Firat* and *P.H. Plesch*, Makromol. Chem. *176* (1975) 1179.

[2] *S. Penczek* and *P. Kubisa*, Makromol. Chem. *165* (1973) 121; *A. Stolarcyk*, *P. Kubisa* and *S. Penczek*, J. Macromol. Sci., Chem. *11* (1977) 2047.

[3] *W. Kern*, *H. Cherdron* and *V. Jaacks*, Angew. Chem. *73* (1961) 177.

[4] *W. Kern*, Chem.-Ztg./Chem. Appar. *88* (1964) 623.

[5] *K. Weissermel*, *E. Fischer*, *K. Gutweiler*, *H.D. Hermann* and *H. Cherdron*, Angew. Chem. *79* (1967) 512.

transformed to a carbenium ion by ring opening; the chain propagates by the addition of further 1,3,5-trioxane molecules:

$$R^{\oplus} + \text{(1,3,5-trioxane)} \longrightarrow \text{(trioxane ring with } O^{\oplus}\text{–R)} \longrightarrow R{-}O{-}CH_2{-}O{-}CH_2{-}O{-}CH_2^{\oplus}$$

$$R{-}O{-}CH_2{-}O{-}CH_2{-}O{-}CH_2^{\oplus} + \text{(1,3,5-trioxane)}$$

$$\longrightarrow R{-}O{-}CH_2{-}O{-}CH_2{-}O{-}CH_2{-}\overset{\oplus}{O}\text{(trioxane ring)}$$

$$\longrightarrow R{-}O{-}CH_2{-}O{-}CH_2{-}O{-}CH_2{-}O{-}CH_2{-}O{-}CH_2{-}O{-}CH_2^{\oplus}$$

The solubility of polyoxymethylene is very poor so that the ring-opening polymerization of 1,3,5-trioxane proceeds heterogeneously both in the bulk (melt) and in solution. 1,3,5-Trioxane can also be readily polymerized in the solid state; the polymerization can be initiated both by high-energy radiation and by cationic initiators[1,2] (see Example 3–40).

Example 3–40:
Polymerization of 1,3,5-trioxane with BF_3-etherate as initiator

(a) Polymerization in the melt
150 g of commercial 1,3,5-trioxane are refluxed (b.p. 115°C) with 9 g of sodium or potassium under nitrogen for 48 h, using an air condenser with suitable protection against ingress of atmospheric moisture[3]. 20 g of the purified monomer are distilled into a 50 ml flask that has previously been flamed in vacuum. The flask contains a glass-sealed magnetic stirrer and is equipped with a self-sealing closure (see Section 2.1.3). The contents are heated to 70°C and 0.05 ml (0.4 mmol) of BF_3-etherate ($\rho_{20} = 1.125$) are

[1] *M. Dröscher, K. Hertwig, H. Reimann* and *G. Wegner*, Makromol. Chem. *177* (1976) 1695.
[2] *A. Munoz-Escalona*, Makromol. Chem. *179* (1978) 219, 1083.
[3] Monomer which is sufficiently pure for polymerization can be obtained by refluxing commercial 1,3,5-trioxane overnight in the presence of 5 wt. % of 1,5-naphthalene diisocyanate and 0.2 wt. % of dibutyltin didodecanoate, 3,6-diazaoctamethylenediamine (triethylenediamine), or bismuth nitrate.

injected as an approximately 10% solution in nitrobenzene. Especial care must be taken that the molten 1,3,5-trioxane is mixed thoroughly with the initiator and a homogeneous mixture obtained as quickly as possible. Immediately after the addition of initiator to the molten monomer the polyoxymethylene begins to precipitate; after only 10 s the whole reaction mixture has congealed. The polymerization is terminated with acetone and the polymer filtered off at a glass sinter after thorough mixing and, if necessary, grinding. The polymer is boiled twice with acetone for 20 min, filtered and dried in vacuum at room temperature. Yield: about 50%; melting range 177–180°C. The viscosity number is determined for a 1% solution in dimethylformamide at 140°C ($\eta_{sp}/c \approx 60\,cm^3\,g^{-1}$, corresponding to an average molecular weight of 60000). The thermal degradation can also be studied before and after blocking the hydroxyl end groups (see Examples 5–09 and 5–15).

(b) Polymerization in the solid state

Gaseous boron trifluoride is required for this experiment. Suitable gas cylinders are available commercially. If necessary boron trifluoride can be prepared as follows. 15 g of powdered sodium tetrafluoroborate are mixed in a 100 ml flask with 2.5 g of boron trioxide and 15 ml of concentrated sulfuric acid. The flask is fitted with a gas delivery tube, to which is attached a piece of PVC tubing, pinched tight at the open end and having a slit in the side to allow pressure equalization. The gaseous BF_3 is generated by heating to 160–170°C and can be withdrawn by means of a hypodermic syringe through the PVC tubing.

Commercial 1,3,5-trioxane is purified by sublimation under normal pressure at 50°C, followed by recrystallization of 60 g from 500 ml dry cyclohexane (yield: about 36 g). Especially well-formed crystalline needles are obtained, which, after filtering off the solvent, can be used without further drying.

20 g of 1,3,5-trioxane are placed in a 300 ml conical flask that is then flushed with nitrogen and sealed with a polyethylene film. 10 ml of BF_3 gas are injected through the film. After a short time the trioxane crystals, which initially have a glassy appearance, become cloudy. After 1 min the polymerization has advanced so far that a sample is nearly all insoluble in methanol. The conical flask is kept for another hour at 80°C to allow the polymerization to die away. The product can be worked up and treated further as described under (*a*). Yield: >50%

(c) Polymerization in solution (precipation polymerization)

90 g of 1,3,5-trioxane, purified as in (*a*), are distilled into a 500 ml flask (that has previously been flamed under vacuum) containing 200 ml of 1,2-dichloroethane (that has been dried over P_2O_5); atmospheric moisture must be excluded. The flask is closed with a self-sealing closure (see

Section 2.1.3)..0.06 ml (0.5 mmol) of BF_3-etherate dissolved in 7 ml of 1,2-dichloroethane are now injected while shaking the flask, which is then warmed to 45°C. After an induction period of about 1 min, solid polyoxymethylene begins to separate, until finally the contents of the flask are completely congealed. After 1 h the product is well ground with 200 ml of acetone, filtered, boiled well with acetone as under (*a*) and dried. In order to remove occluded initiator residues the polymer is boiled with 1 l of ether containing 2 wt % of tributylamine; it is finally filtered and dried in vacuum at room temperature. Yield: 90–95%; melting range, 176–178°C. The limiting viscosity number is determined in dimethylformamide at 140°C (molecular weight ≈ 60 000). The thermal stability can also be measured; see under (*a*).

3.2.4.3. *Ring-opening polymerization of cyclic esters (lactones)*

Cyclic esters of ω-hydroxycarboxylic acids can be polymerized by ring opening to give linear aliphatic polyesters. According to the type of initiator and monomer the polymerization occurs either by alkyl or by acyl fission:[1,2]

$$R_3N + \begin{matrix} CH_2-C=O \\ | \quad\quad | \\ CH_2-O \end{matrix} \longrightarrow R_3\overset{\oplus}{N}-CH_2-CH_2-\underset{\underset{O}{\|}}{C}-O^{\ominus} \qquad \text{alkyl fission}$$

$$R^{\oplus} + \begin{matrix} CH_2-C=O \\ | \quad\quad | \\ CH_2-O \end{matrix} \longrightarrow \begin{matrix} CH_2-C=O \\ | \quad\quad | \\ CH_2-O^{\oplus} \\ | \\ R \end{matrix} \longrightarrow R-O-CH_2-CH_2-\underset{\underset{O}{\|}}{\overset{\oplus}{C}} \qquad \text{acyl fission}$$

The polymerizability depends on the ring size and on the number, size and position of the substituents.

The ring-opening polymerization of dilactide[3,4] (dimeric cyclic ester of lactic acid) allows the preparation of high-molecular-weight, optically active polyesters of lactic acid. The configuration of the asymmetric carbon atoms of the monomer is retained when polymerization is initiated with $SnCl_4$ or Et_2Zn[5,6], for example:

$$\text{dilactide (cyclic: } H(CH_3)\overset{*}{C}-O-C(=O)-\overset{*}{C}H(CH_3)-O-C(=O)\text{)} \longrightarrow \left[-O-\overset{\overset{O}{\|}}{C}-\underset{\underset{CH_3}{|}}{\overset{*}{C}H}- \right]_n$$

---

[1] *H. Cherdron, H. Ohse* and *F. Korte*, Makromol. Chem. *56* (1962) 179, 187.
[2] *E. Bigdeli* and *R.W. Lenz*, Macromolecules *11* (1978) 493.
[3] *J. Kleine* and *H.H. Kleine*, Makromol. Chem. *30* (1959) 23.
[4] *T. Tsuruta, K. Matsuura* and *S. Inoue*, Makromol. Chem. *75* (1964) 211.
[5] *R.C. Schulz* and *J. Schwaab*, Makromol. Chem. *87* (1965) 90.
[6] *E. Lillie* and *R.C. Schulz*, Makromol. Chem. *176* (1975) 1901.

Example 3–41:
Ring-opening polymerization of dilactide with cationic initiators in solution

5 g of L-dilactide[1], 50 ml of pure toluene and a magnetic stirrer are placed in a reaction vessel that has been flamed under vacuum and flushed with nitrogen. The vessel is closed with a pressure-tight rubber cap and heated to 110°C in an oil bath. 0.5 ml of initiator solution (3.32 g of $SnCl_4$/100 ml of toluene) are then injected through the rubber cap with a hypodermic syringe. After about 3 h the viscous solution is cooled and added dropwise to 300 ml of vigorously stirred methanol. The filtered polymer is dissolved in 50 ml of 1,4-dioxane and reprecipitated in methanol. After drying in vacuum at about 70°C one obtains 4.2 g (yield: 74%) of a white polymer with a crystalline melting point (DTA) of 153°C. It is soluble in 1,4-dioxane, chloroform and acetonitrile, and insoluble in methanol, ether and hexane. The solution viscosity is determined in 1,4-dioxane at 25°C using an Ostwald viscometer (capillary diameter = 0.4 mm). The polymer can be characterized by i.r. spectroscopy, and by ORD- and CD- measurements (cf. references 5,6 on p. 203).

3.2.4.4. Ring-opening polymerization of cyclic amides (lactams)

The ring-opening polymerization of cyclic amides[2,3] gives linear polyamides. This very important reaction can be initiated ionically. There is a pronounced dependence of polymerizability of lactams on the ring size and on the number and position of the substituents.[4] The 5-membered lactam (γ-butyrolactam) can be polymerized anionically at low temperature; the polyamide depolymerizes again to monomer in the presence of the initiator at 60–80°C.[5] The corresponding 6-membered ring, δ-valerolactam, is likewise polymerizable.[3] The 7-membered ring, ε-caprolactam, can be polymerized both cationically and anionically to high-molecular-weight polyamides.

Polymers made from ε-caprolactam in the presence of anionic catalysts at high temperature have average molecular weights which are initially high but decrease on long heating of the molten reaction mixture, finally attaining an equilibrium value. This change of molecular weight distribution is caused by a transamidation reaction of the growing chains with the dead polyamide molecules.

---

[1] L-Lactide (obtainable from Fa. Boehringer, Ingelheim, W. Germany) is hygroscopic and should be stored over P_2O_5.
[2] *H. Rinke* and *E. Istel*, Houben-Weyl *14/2* (1963) 111.
[3] *O. Wichterle, J. Sebenda* and *J. Kralicek*, Adv. Polym. Sci. *2* (1961) 578.
[4] *R.C.P. Cubbon*, Makromol. Chem. *80* (1964) 44.
[5] *H.K. Hall jr.*, J. Am. Chem. Soc. *80* (1958) 6404.

The polymerization of ε-caprolactam with anionic initiators is to be thought of as a true addition polymerization reaction:

$$R^{\ominus} + \underbrace{-\underset{\underset{O}{\|}}{C}-(CH_2)_5-NH-}_{\text{ring}} \longrightarrow R-\underset{\underset{O}{\|}}{C}-(CH_2)_5NH^{\ominus}$$

$$R-\underset{\underset{O}{\|}}{C}-(CH_2)_5NH^{\ominus} + \underbrace{-\underset{\underset{O}{\|}}{C}-(CH_2)_5-NH-}_{\text{ring}}$$

$$\longrightarrow R-\underset{\underset{O}{\|}}{C}-(CH_2)_5-NH-\underset{\underset{O}{\|}}{C}-(CH_2)_5-NH^{\ominus}$$

but the reaction initiated by catalytic amounts of water, ε-aminocaproic acid, or benzoic acid, may be conceived as an addition polymerization involving migration of hydrogen atoms (see Section 1.1). The reaction begins by the addition of water or acid to the ε-caprolactam; the resulting NH_2 or NH_3^+ end groups then add further lactam:

$$H_2O + \underbrace{-NH-(CH_2)_5-\underset{\underset{O}{\|}}{C}-}_{\text{ring}} \longrightarrow NH_2-(CH_2)_5-\underset{\underset{O}{\|}}{C}-OH$$

$$NH_2-(CH_2)_5-\underset{\underset{O}{\|}}{C}-OH + \underbrace{-NH-(CH_2)_5-\underset{\underset{O}{\|}}{C}-}_{\text{ring}}$$

$$\longrightarrow NH_2-(CH_2)_5-\underset{\underset{O}{\|}}{C}-NH-(CH_2)_5-\underset{\underset{O}{\|}}{C}-OH$$

The molecular weight increases with increasing conversion. Regulation of the molecular weight can be achieved by adding small amounts of substances (e.g. benzoic acid) which can react with the polyamide chains by transamidation. As a result of the transamidation reaction and hydrolysis of amide bonds, an equilibrium molecular weight distribution is finally attained (see Section 4.1).

Polymers prepared at 250–270°C contain an equilibrium concentration of up to 10% of cyclic monomer and higher oligomers, partly cyclic; after cooling, the monomer and oligomers can be recovered by extraction with water or lower alcohols. The oligomers can be separated and identified chromatographically (see Example 4–08).

Example 3–42:
Bulk polymerization of ε-caprolactam with anionic initiators (flash polymerization)

ε-Caprolactam is recrystallized twice from cyclohexane and dried in vacuum at room temperature for 48 h; m.p. 68–69°C.

(a) Preparation of N-acetyl-ε-caprolactam[1]

A mixture of 67.8 g (0.6 mol) ε-caprolactam and 67 g (0.665 mol) of acetic anhydride are heated for 4 h in a 250 ml flask fitted with reflux condenser and drying tube. Finally, most of the excess anhydride, as well as the acetic acid, is distilled off and the remainder submitted to fractional distillation at low pressure, using an oil pump; b.p. 134–136°C/26–27 torr; $n_D^{25} = 1.4885$; yield: 77.5 g (83.5%).

(b) Polymerization procedure[2]

25 g (0.22 mol) of ε-caprolactam and 0.6 g (0.025 mol) of sodium hydride are placed in a flask which is then evacuated and filled with nitrogen several times. The sample is melted to allow the sodium hydride to react. When the evolution of hydrogen has ceased, 0.33 g (0.002 mol) of *N*-acetyl-ε-caprolactam are added, the flask is well shaken, and then heated in an oil bath to 140°C. The contents rapidly solidify and after 30 min can be cooled and ground up.

The limiting viscosity number is determined in *m*-cresol or concentrated sulfuric acid. The polyamide (Nylon-6) so obtained has a crystalline melting point around 216°C. It still contains low-molecular-weight cyclic oligomers, which can be removed by extraction with water or lower alcohols. These oligomers can be separated and identified chromatographically (see Example 4–08).

3.3. COPOLYMERIZATION[3]

3.3.1. Random copolymerization

By copolymerization[4] one understands the mutual polymerization of two or more monomers, whereby the macromolecules of the resulting copolymers contain monomeric units of all the participating monomers.[5,6,7,8]

---

[1] *R.E. Benson* and *T.L. Cairns*, J. Am. Chem. Soc. *70* (1948) 2115.

[2] *S.R. Sandler* and *W. Karo*, "Polymer Syntheses," Vol. 1, p. 104, Academic Press, New York, London 1974.

[3] In technical jargon and in the patent literature the term "mixed polymerization" is frequently used to mean copolymerization; it must not be confused with a mixture of homopolymers.

[4] *G.E. Ham*, "Copolymerization", High Polymers *18* (1964), Interscience Publishers, New York.

[5] *H.G. Elias*, "Makromoleküle", 3rd Edn., Hüthig and Wepf Verlag. Basel, Heidelberg 1975, p. 632 et seq.

[6] *J.M.G. Cowie*, "Polymers: Chemistry and Physics of Modern Materials", Intertext Books, Aylesbury 1973.

[7] *B. Vollmert*, "Polymer Chemistry", Springer Verlag, Berlin, Heidelberg, New York 1973, p. 94.

[8] *W. Ring*, Makromol. Chem. *101* (1967) 145.

However, monomers that homopolymerize are not necessarily able to copolymerize well with one another. Thus, in the radical copolymerization of vinyl acetate with styrene, only a very small amount of vinyl acetate is incorporated into the resulting polymer chains. Conversely there are some compounds that are not homopolymerizable but which can be induced to copolymerize.

Copolymerization of two monomers has been very thoroughly investigated; but copolymerization of three or more compounds[1,2,3] presents considerable difficulties on account of the multiplicity of variables; however, terpolymers (from three monomers) are of technical importance.

We limit ourselves here to the case of mutual polymerization of two monomers; there are then essentially four different possible propagation reactions: monomer M_1 can react with a polymer chain whose growing chain end (radical or ionic) has been formed either from monomer M_1 or from monomer M_2; similarly for monomer M_2:

$$\cdots\!-\!M_1^* + M_1 \xrightarrow{k_{11}} \cdots\!-\!M_1 - M_1^*$$

$$\cdots\!-\!M_1^* + M_2 \xrightarrow{k_{12}} \cdots\!-\!M_1 - M_2^*$$

$$\cdots\!-\!M_2^* + M_1 \xrightarrow{k_{21}} \cdots\!-\!M_2 - M_1^*$$

$$\cdots\!-\!M_2^* + M_2 \xrightarrow{k_{22}} \cdots\!-\!M_2 - M_2^*$$

k_{11}, k_{12}, k_{21}, k_{22} represent the rate constants of these propagation reactions, the first number in the subscript indicating the type of active centre, and the second, the nature of the monomer that is adding to it. So long as the chain length is relatively large, the propagation reactions are rate-determining; as in homopolymerization, a quasi-stationary state is set up. It can then be shown that at low conversion ($< 10\%$), the relative rates of consumption of the two monomers, and thus their relative amounts (in mole) in the copolymer, m_1/m_2, can be described by the following copolymerization equation:

$$\frac{d[M_1]}{d[M_2]} = \frac{m_1}{m_2} = \frac{[M_1]}{[M_2]}\left(\frac{r_1[M_1] + [M_2]}{[M_1] + r_2[M_2]}\right) \tag{18}$$

$r_1 = k_{11}/k_{12}$, and $r_2 = k_{22}/k_{21}$, are termed the reactivity ratios (copolymerization parameters). $[M_1]$ and $[M_2]$ represent the molar concentrations of monomer in the reaction mixture. The composition of a copolymer thus

[1] *G.E. Ham*, "Copolymerization", High Polymers *18* (1964), Interscience Publishers, New York.
[2] *A. Valvassori* and *G. Sartori*, Adv. Polym. Sci. *5* (1967) 28.
[3] *W. Ring*, Eur. Polym. J. *4* (1968) 413.

depends on the monomer feed ratio in the polymerization mixture. Since the parameters r_1 and r_2 are ratios of rate constants, they express the tendency of the growing chains to add either the same monomer or the other monomer. If r is close to 1 it follows that a particular chain end adds molecules of monomers M_1 and M_2 at random with approximately equal facility; $r > 1$ means that the addition of a monomer to a chain with the same end unit is strongly preferred. Reactivity ratios, being quotients of two rate constants, are not very temperature-dependent, but of course are strictly valid only for a particular polymerization temperature, which must, therefore, always be indicated.

By rearranging equation (18) and inserting known values of the reactivity ratios r_1 and r_2, one can calculate the molar ratio of monomers that must be used in order to arrive at a copolymer with a chosen composition ($f = m_1/m_2$):

$$\frac{[M_1]}{[M_2]} = \frac{1}{2r_1}\{(f-1) + [(f-1)^2 + 4r_1r_2f]^{1/2}\} \tag{18a}$$

The dependence of the composition of the copolymer on the proportions of the monomers in the initial mixture can be portrayed graphically in a so-called copolymerization diagram (Figure 3.5). The mole fraction of one of the two monomeric units in the resulting copolymer is plotted against the mole fraction of this monomer in the original reaction mixture; the curve can also be calculated from the reactivity ratios by means of equation (18).[1]

As can be seen from Figure 3.5 it is very rare for the polymer composition to correspond to that of the monomer mixture. For this reason the composition of the monomer mixture, and hence also that of the resulting polymer, generally changes as the copolymerization proceeds. Therefore, for the determination of the reactivity ratios one must work at the lowest possible conversion. In practical situations where, for various reasons, one is forced to polymerize to higher conversions, chemical non-uniformity of the copolymers[2] is superimposed on the usual non-uniformity of molecular weights. By direct[3] or numerical[4] integration of the copolymerization equation (18) a general idea can be obtained concerning the chemical non-uniformity of a copolymer as a function of the

[1] For the graphical determination of the copolymer composition and the sequence length distribution see *M. Izu* and *K.F. O'Driscoll*, J. Appl. Polym. Sci. *14* (1970) 1515.

[2] *M. Buck*, Angew. Makromol. Chem. *11* (1970) 63.

[3] *F. Lewis* and *F. Mayo*, J. Am. Chem. Soc. *66* (1944) 1594.

[4] *I. Skeist*, J. Am. Chem. Soc. *68* (1946) 1781.

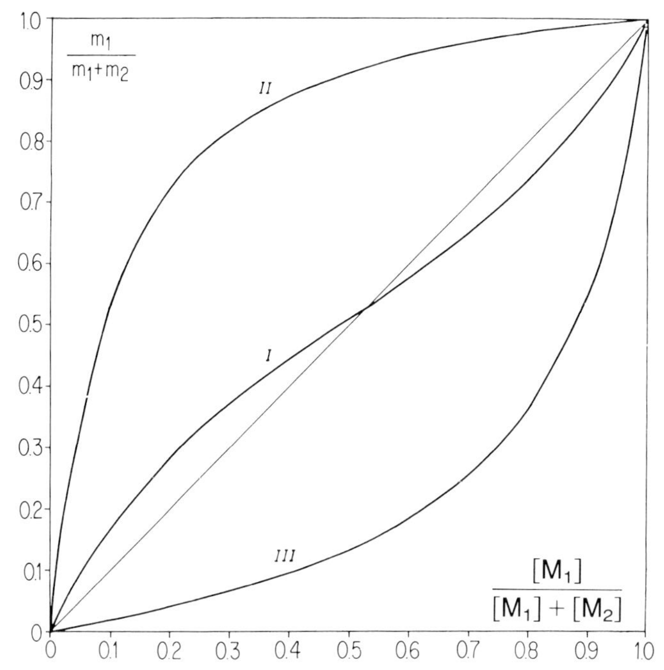

FIGURE 3.5 Copolymerization diagram for the system styrene (M_1)/methyl methacrylate (M_2). I: radical copolymerization at 60°C; $r_1 = 0.52$, $r_2 = 0.46$. II: cationic copolymerization initiated by $SnBr_4$ at 25°C; $r_1 = 10.5$, $r_2 = 0.1$. III: anionic copolymerization initiated by Na in liquid NH_3 at −50°C; $r_1 = 0.12$, $r_2 = 6.4$.

monomer composition and conversion, which can be represented graphically.[1]

From the copolymerization equation (18) and the copolymerization diagram (Figure 3.5) some special cases can be derived (cf. Table 3.6). When the compositions of the monomer feed and the resulting copolymer are the same, one speaks of "azeotropic" copolymerization[2]; equation (18) then reduces to:

$$\frac{m_1}{m_2} = \frac{[M_1]}{[M_2]} = \frac{1 - r_2}{1 - r_1} \tag{19}$$

In such cases the polymerization can be taken to relatively high conversion without change in composition of the copolymer formed (see Example

[1] See for example *L. Küchler*, "Polymerisationskinetik", Springer-Verlag, Berlin, Göttingen, Heidelberg 1951, p. 165; cf. also *E. Gruber* and *W.L. Knoll*, Makromol. Chem. *179* (1978) 733.

[2] Also see *P. Wittmer*, *F. Hafner* and *H. Gerrens*, Makromol. Chem. *104* (1967) 101.

TABLE 3.6

Special cases of the copolymerization equation

r_1	r_2	$r_1 r_2$	Copolymerization equation	Remarks
>0	0	0	$\frac{d[M_1]}{d[M_2]} = 1 + r_1 \frac{[M_1]}{[M_2]}$	M_2 does not homopolymerize
0	0	0	$\frac{d[M_1]}{d[M_2]} = 1$	Neither monomer homopolymerizes; alternating copolymerization
r_1	$\frac{1}{r_1}$	1	$\frac{d[M_1]}{d[M_2]} = r_1 \frac{[M_1]}{[M_2]}$	Ideal copolymerization with no azeotrope
1	1	1	$\frac{d[M_1]}{d[M_2]} = \frac{[M_1]}{[M_2]}$	Ideal copolymerization; azeotrope at all compositions
<1	<1	<1	equation (18)	Non-ideal copolymerization; azeotrope at $\frac{[M_1]}{[M_2]} = \frac{1-r_2}{1-r_1}$
>1	>1	>1		Tendency to block copolymerization; not observed up to now

3–46b). In the copolymerization diagram the azeotrope corresponds to the intersection point of the copolymerization curve with the diagonal. For example, from Figure 3.5 it may be seen that in the radical copolymerization of styrene and methyl methacrylate the azeotropic composition corresponds to 53 mole % of styrene.

For the case where $r_2 = 0$, $k_{22} = 0$, M_2 does not add to growing chains having an M_2 end unit, so that homopolymerization of M_2 is also impossible. Equation (18) then reduces to:

$$\frac{m_1}{m_2} = 1 + r_1 \frac{[M_1]}{[M_2]} \tag{20}$$

The resulting copolymer can contain at most 50 mole % M_2 units, even at high concentration of M_2 in the monomer mixture. This applies for example to maleic anhydride, and especially to such "monomers" as molecular oxygen or sulfur dioxide where, independent of the comonomer used, essentially alternating copolymers are obtained, with almost equal amounts of the two components.

Finally there are also cases in which two "monomers" copolymerize, both "monomers" being themselves incapable of hompolymerization

($k_{11} = k_{22} = 0$); this results in strict alternation of monomeric units of M_1 and M_2 in the copolymer (cf. Example 3–53).

If radical copolymerization is conducted in the presence of complexing agents it is sometimes possible to force monomers into forming alternating copolymers when they would otherwise give random copolymers. Lewis acids such as zinc chloride are especially suitable for this, also organo-aluminium compounds. The reaction mechanism is not yet fully clarified; it is assumed that the additives form electron donor/acceptor complexes with the monomer or the active chain ends, leading to alternating insertion of the two monomers into the growing polymer chains.[1]

It is important to note that the tendency of a monomer towards polymerization and therefore also towards copolymerization is strongly dependent on the nature of the growing chain end. In radical copolymerization the composition of the copolymer obtained from a given monomer feed is independent of the initiating system for a particular monomer pair, but for anionic or cationic initiation this is normally not the case. One sometimes observes quite different compositions of copolymer depending on the nature of the initiator and especially on the type of counter ion.

In the radical copolymerization of styrene and methyl methacrylate the reactivities are about the same, but for anionic initiation of an equimolar mixture of the two monomers the incorporation of methyl methacrylate is much preferred, while for cationic initiation of the same mixture the copolymer contains mostly styrene (see Figure 3.5 and Example 3–43). Conversely monomer pairs whose copolymerization behaviour is well known can be used to test new initiator systems in order to draw conclusions about the polymerization mechanism.

As already indicated, values of reactivity ratios apply only to a given pair of monomers. There have been many attempts, especially for radical copolymerization, to derive parameters from the reactivity ratios, representing individual constants for each monomer[2], which can be related to the structure of the monomers, and used to make predictions.

The basis of the scheme developed particularly by Alfrey and Price[3] is the assumption that the activation energies of the propagation reactions, and hence the related rate constants and reactivity ratios, are governed

[1] *M. Hirooka*, J. Polym. Sci., Part B*10* (1972) 171; *J. Furukawa*, J. Macromol. Sci., Chem. *9* (1975) 867; *Y. Shirota* and *H. Mikawa*, J. Macromol. Sci., Rev. C*16* (2) (1978) 129; for examples see "Macromolecular Syntheses", Coll. Vol. I, p. 517.

[2] *H.G. Elias*, "Makromoleküle", 3rd Edn., Hüthig and Wepf Verlag, Basel, Heidelberg 1975, p. 653.

[3] *T. Alfrey* and *C.C. Price*, J. Polym. Sci. *2* (1947) 101; *C.C. Price*, J. Polym. Sci. *3* (1948) 772; *T.C. Schwan* and *C.C. Price*, J. Polym. Sci. *40* (1959) 457.

primarily by resonance effects and by the interaction of the charges on the double bonds of the monomers with those in the active radicals. Accordingly the rate constant of the reaction between a radical and a monomer is represented by

$$k_{12} = P_1 Q_2 e^{-e_1 e_2}$$

where P_1 denotes the reactivity of a radical having an M_1 unit at the reactive end, Q_2 is the reactivity of the monomer M_2, and e_1 and e_2 are proportional to the charges on the corresponding species. It is assumed that the e-value of the monomer is the same as that of the corresponding radical. Hence, it follows that

$$r_1 = \frac{Q_1}{Q_2} e^{-e_1(e_1 - e_2)}$$

$$r_2 = \frac{Q_2}{Q_1} e^{-e_2(e_2 - e_1)}$$

whence

$$e_1 = e_2 \pm \sqrt{-\ln (r_1 r_2)}$$

$$Q_1 = \frac{Q_2}{r_2} e^{\pm e_2 \sqrt{-\ln(r_1 r_2)}}$$

On this basis, values of Q and e can be calculated for each monomer[1], so long as two arbitrary reference values are assumed. For this purpose Price took the values for styrene as $Q = 1.0$ and $e = -0.8$. Q- and e-values can then be obtained for all monomers that are copolymerizable with styrene. These monomers in their turn can serve as reference compounds for further determinations with other monomers that do not copolymerize with styrene. One of the main advantages of the so-called Q,e-scheme is that the data can be presented in the form of a diagram instead of very comprehensive tables of reactivity ratios.

Obviously the precision of this procedure is not very great, since the assumptions underlying the calculations of Q- and e-values can be regarded at best as semi-quantitative. However it has been shown that when the reactivity ratios are back-calculated from the Q,e-values, quite good agreement is obtained with the experimental values, so that it is possible to make useful predictions of reactivity ratios for monomer pairs not previously investigated. On the other hand it is questionable whether any

[1] *R.Z. Greenley*, J. Macromol. Sci., Chem. *9* (1975) 505.

theoretical conclusions should be drawn from the Q- and e-values concerning the behaviour of different monomers. Thus, the e-value of a monomer as a measure of the charge should be dependent on the nature of the solvent used for polymerization; but it has been shown, for example, that the copolymerization of styrene with methyl methacrylate in solvents of different dielectric constant (benzene 2.28, methanol 33.7, acetonitrile 38.8) give the same reactivity ratios, which should not really be the case if the foregoing assumptions are correct. A broadening of the basis of the Q, e-scheme has therefore been suggested by various people. These further considerations permit somewhat deeper theoretical conclusions to which here we can do no more than draw attention.[1]

In most cases one of the two monomers is preferentially incorporated into the copolymer, so that the composition of the monomer mixture changes with increasing conversion, as also does the composition of the polymer chains. Therefore it is important, when determining reactivity ratios, to work at low conversions, so that at the end of the experiment the ratio of monomer concentrations is essentially the same as at the beginning. However, if high conversions are needed for preparative purposes, constant composition can be achieved by adding the more reactive monomer in a programmed manner (see p. 216).

The determination of the reactivity ratios requires a knowledge of the composition of the copolymers made from particular monomer mixtures; numerous analytical methods are available (cf. Section 2.3.11). In principle it is possible to calculate r_1 and r_2, using equation (18), from the composition of two copolymers that have been obtained from two different mixtures of the two monomers M_1 and M_2. However it is better to determine the composition of the copolymers from several monomer mixtures and to calculate for each individual experiment, values of r_2 that would correspond to arbitrarily chosen values of r_1 from the rearranged copolymerization equation:

$$r_2 = \frac{[M_1]}{[M_2]}\left\{\frac{m_2}{m_1}\left(1 + \frac{[M_1]}{[M_2]}r_1\right) - 1\right\} \tag{21}$$

r_2 is then plotted against r_1 for each experiment to obtain a series of lines intersecting at the actual values of r_2 and r_1. In practice the lines do not intersect precisely at a point so that r_1 and r_2 are taken as the centre of the smallest area that is cut or touched by all the lines. The size of this area allows an estimate of the limits of error.

[1] *T. Alfrey* and *C.C. Price*, J. Polym. Sci. *2* (1947) 101; *C.C. Price*, J. Polym. Sci. *3* (1948) 772; *T.C. Schwan* and *C.C. Price*, J. Polym. Sci. *40* (1959) 457.

A simpler method for determining the reactivity ratios is that of Fineman and Ross[1], whereby the copolymerization equation is rearranged to:

$$(F/f)(f-1) = (F^2/f)r_1 - r_2 \quad (22a)$$

or

$$(f-1)/F = -(f/F^2)r_2 + r_1 \quad (22b)$$

where $f = m_1/m_2$ and $F = [M_1]/[M_2]$.

$(F/f)(f-1)$ is plotted against (F^2/f) to give a straight line (22a), of slope r_1 and intercept $-r_2$ on the ordinate. Alternatively a plot of $(f-1)/F$ against (f/F^2) according to (22b) gives a line of slope $-r_2$ and intercept r_1 on the ordinate. The limits of error or r_1 and r_2 can be determined from the scatter of the points by the method of least squares. Besides these two classical methods a number of further publications have recently appeared concerning the evaluation of copolymerization data. Here should be mentioned the method of Kelen and Tüdös[2] which is a graphical linear procedure, the method of Braun et al[3] which involves curve fitting, and perhaps the most ambitious but also the most exact method of Tidwell and Mortimer.[4] The last two methods can only be applied with the aid of a computer. Comprehensive collections of reactivity ratios for numerous monomer pairs can be found in the literature.[5]

Copolymers often differ significantly from the corresponding homopolymers in their physical properties. Thus, by incorporation of small amounts of vinyl acetate into poly(vinyl chloride), internal plasticization is achieved (see Section 1.4). The ability of synthetic fibres to take up dyes can be improved by incorporation of small amounts of comonomers with functional groups that confer a greater affinity towards organic dyestuffs. Significant differences in solubility between copolymers and the corresponding homopolymers are usually observed (see Example 3–43). Copolymers containing approximately equal proportions of randomly distributed comonomer units are frequently quite different in their properties from the homopolymers; thus, while polyethylene and isotactic polypropylene are crystalline polymers with the characteristic properties of a thermoplastic, copolymers of ethylene and propylene (70:30) are amorphous, rubber-elastic products.

---

[1] *M. Fineman* and *S.D. Ross*, J. Polym. Sci. *5* (1950) 259.
[2] *T. Kelen* and *F. Tüdös*, J. Macromol. Sci., Chem. A*9* (1975) 1.
[3] *D. Braun*, *W. Brendlein* and *G. Mott*, Eur. Polym. J. *9* (1973) 1007.
[4] *P.W. Tidwell* and *G.A. Mortimer*, J. Polym. Sci., Part A-1, *3* (1965) 369.
[5] *J. Brandrup* and *E.H. Immergut*, "Polymer Handbook", John Wiley and Sons, New York, London, Sydney, Toronto 1975, II-105.

Crosslinking copolymerization is also of great practical importance. Such copolymerization is achieved by taking compounds with two or more polymerizable double bonds and copolymerizing them with simple unsaturated monomers to form three-dimensional network materials. For example crosslinked polystyrene is prepared by copolymerizing styrene with small amounts of divinylbenzene (Example 3–50). These crosslinked polymers are insoluble and infusible; however, depending on the degree of crosslinking, they swell to a limited extent in organic solvents and find application, for example, in the preparation of ion-exchange resins (see Section 5.2). The copolymerization of unsaturated polyesters (made from maleic or fumaric acid) with styrene is also a crosslinking copolymerization (cf. Example 4–05). Some non-conjugated dienes can polymerize radically via a cyclization propagation step to yield linear chains containing cyclic repeat units (so-called cyclopolymerization). Thus in the polymerization of acrylic anhydride this tendency to ring formation is so great that the intramolecular reaction occurs almost to the exclusion of the intermolecular crosslinking reaction:[1]

$$CH_2{=}CH(C{=}O){-}O{-}(O{=}C)CH{=}CH_2 \longrightarrow {-}CH_2{-}CH{-}CH_2{-}CH{-} \text{ (six-membered ring: } CH{-}C(=O){-}O{-}C(=O){-}CH \text{ bridged by } CH_2)$$

The rates and degrees of polymerizations in radical copolymerizations conform essentially to the same laws as for radical homopolymerization (see Section 3.1). Raising the initiator concentration causes a rise in the rate of polymerization and at the same time a lowering of molecular weight; a temperature rise has the same effect. However these assertions are valid only for a given monomer composition; in many cases the copolymerization rate depends very much on monomer composition and can pass through a pronounced minimum[2] or maximum.[3]

As already discussed for homopolymerization, radical copolymerizations can be carried out in bulk, in solution and in dispersion (see Sections 2.1.5 and 3.1). The composition of the copolymer obtained in suspension or emulsion may be different from that obtained by polymerization in bulk or solution if one of the monomers is more soluble in water than the other. In such a case the composition of the monomer mixture in the organic phase, or in the micelles where the copolymerization takes place, is not the same as the original composition.

---

[1] For an example see "Macromolecular Syntheses", Coll. Vol. I, p. 441.

[2] *K. Matsuo* and *W.H. Stockmayer*, Macromolecules *10* (1977) 658.

[3] *M. Rätzsch*, *M. Arnold* and *R. Hoyer*, Plaste Kautsch. *24* (1977) 731.

There are no essential differences in experimental technique required for ionic copolymerizations, as compared with ionic homopolymerizations. However the type of initiator and the solvent have a potential influence on the course of ionic copolymerizations as well as on the composition of the copolymers so that the optimum conditions for each monomer pair must be individually determined.

The composition of the copolymers generally changes with increasing conversion. If therefore, for preparative purposes one wishes to attain high conversion at constant composition, the more reactive monomer must be added in a programmed manner.[1] The procedure is as follows. From the copolymerization diagram (or from the reactivity ratios) one obtains the monomer composition that will lead, at low conversion, to the desired copolymer composition. A conversion/time curve is drawn up for this system and the composition of the copolymer determined from time to time. From this one can find how much of the more reactive monomer is to be added at given times during the polymerization in order to maintain an approximately constant composition.

Qualitative and quantitative analysis of copolymers frequently requires special methods (see Section 2.3.11)

Example 3–43:
Copolymerization of styrene with methyl methacrylate

Styrene and methyl methacrylate are destabilized and dried as described in Examples 3–01 and 3–29 respectively. A mixture of 8.00 g (76.8 mmol) of styrene and 7.68 g (76.8 mmol) of methyl methacrylate is then prepared in a dry receiver (see Section 2.1.2) under nitrogen and kept in a refrigerator until required.

(a) Radical copolymerization
16 mg (0.1 mmol) of 2,2′-azoisobutyronitrile are weighed into a 50 ml flask and the air displaced by evacuation and filling with nitrogen, using a suitable adaptor (see Section 2.1.3). 5 ml of the prepared monomer mixture are then pipetted in, the flask removed under a slight positive pressure of nitrogen and immediately closed with a ground glass stopper secured with springs. The flask is now placed in a thermostat at 60°C, cooled quickly in ice after 4 h, and the reaction mixture diluted with 25 ml of benzene. The solution is added dropwise to 200–250 ml of petroleum ether; the copolymer precipitates in the form of small flakes that tend to stick together somewhat (for this reason methanol is less suitable as

[1] *W. Ring*, Makromol. Chem. 75 (1964) 203.

precipitant). The copolymer is filtered off at a sintered glass crucible, washed with petroleum ether and dried to constant weight in vacuum at 50°C. Yield: 10–20% with respect to the monomer mixture.

(b) Anionic copolymerization

The initiator solution is prepared as follows. Phenylmagnesium bromide is prepared from 12 g (76.5 mmol) of bromobenzene and 1.86 g of magnesium in 30 ml of pure dry tetrahydrofuran (see Example 3–27). The apparatus must be carefully dried (all openings being protected by drying tubes); exclusion of atmospheric oxygen is not absolutely necessary. The magnesium should be converted as completely as possible by gentle warming towards the end of the reaction.

The copolymerization is carried out as follows. 5 ml of the prepared monomer mixture (see above) are pipetted into a 50 ml round-bottomed flask that has previously been flamed under vacuum and filled with nitrogen using an adaptor. The flask is closed with a self-sealing closure (see Section 2.1.3) and, after cooling to −50°C, 2 ml of freshly prepared phenylmagnesium bromide solution are injected by means of a hypodermic syringe. After 90 min at −50°C the mixture is diluted with 25 ml of benzene and the copolymer precipitated by dropping the solution into 200 ml of methanol containing about 10 ml of 2 M hydrochloric acid. Further treatment is as described under (*a*). Yield: 10–20% with respect to the monomer mixture.

The anionic polymerization can only be carried out at low temperature, since at higher temperature the Grignard compound can also react with the ester group of methyl methacrylate.

(c) Cationic copolymerization

A 100 ml flask is fitted with an adaptor (see Section 2.1.3), flamed under vacuum using an oil pump, and filled with nitrogen. 5 ml of the prepared monomer mixture (see above) is pipetted in, followed by 40 ml of an initiator solution prepared from 50 ml of pure dry nitrobenzene (see Example 3–50) and 300 mg (2.25 mmol) of anhydrous aluminium trichloride. The flask is now removed from the adaptor under slight positive pressure of nitrogen and immediately closed with a ground glass stopper. The flask is briefly shaken and allowed to stand at room temperature for 1 h. The solution is then dropped into methanol and the copolymer worked up as described above. Yield: 40–50% with respect to the monomer mixture.

Cationic copolymerization proceeds under the above conditions with such speed that even at much lower initiator concentration almost half the monomer mixture is polymerized in less than 1 h. The reaction comes to an end after 50% conversion since with cationic initiators, the copolymer consists almost entirely of styrene units.

(d) Characterization of the copolymers

From the behaviour on precipitation of the copolymers prepared by the three different methods, it may already be suspected that, in spite of the common starting mixture, one is dealing with different kinds of polymer. In order to prove this the following solubility tests are carried out.

50 mg of each of the three copolymers, as well as a mixture of equal parts polystyrene and poly(methyl methacrylate), are warmed with about 5 ml of the following solvents:

1. mixture of acetone and methanol (volume ratio 2:1),
2. acetonitrile,
3. mixture of cyclohexane and benzene (volume ratio 4:1).

If any material remains undissolved, the supernatant liquid is decanted and dropped into methanol in order to precipitate the dissolved portion.

Solvents 1 and 2 are known to be good solvents for poly(methyl methacrylate); solvent 3 readily dissolves polystyrene. The solubility tests show that the radically polymerized sample is insoluble in all three solvents. The solubility is thus different from that of both poly(methyl methacrylate) and polystyrene. The anionically polymerized product dissolves on warming in the acetone/methanol mixture and also in acetonitrile; it is insoluble in cyclohexane/benzene. The solubility is thus similar to that of poly(methyl methacrylate). For the cationically initiated polymerization the product is only slightly soluble in acetone/methanol, is insoluble in acetonitrile, but very readily soluble in cyclohexane/benzene. The solubility thus resembles that of polystyrene.

Example 3–44:

Radical copolymerization of styrene with 4-chlorostyrene (determination of the reactivity ratios)

20 mg (0.12 mmol) of 2,2′-azoisobutyronitrile are placed in each of 9 test tubes fitted with joints, together with the approximate amounts of destabilized styrene and 4-chlorostyrene (cf. Example 3–45) indicated in the following table, weighed accurately to three decimal places. If the samples cannot be polymerized on the same day they must be stored in a refrigerator.

The vessels are cooled in a methanol/dry-ice bath, evacuated through an adaptor (see Section 2.1.3) and, after thawing, filled with nitrogen; this procedure is repeated twice more. The tubes are now withdrawn from the adaptor under slight positive pressure of nitrogen, immediately closed with ground glass stoppers (secured with springs), and brought to 50°C in a thermostat. After 8 h the tubes are quickly cooled, the contents diluted with 5 ml of benzene and dropped into about 80 ml of stirred methanol. The copolymers are filtered off, reprecipitated twice and dried to constant weight in vacuum at 50°C. The yield under these conditions is about 300 mg (10%).

Expt. no.	Styrene weight in g	Styrene amount in mmol	4-Chlorostyrene weight in g	4-Chlorostyrene amount in mmol
1	3.0	29	0	0
2	2.6	25	0.4	3
3	2.3	22	0.7	5
4	1.6	15	1.4	10
5	1.1	11	1.9	14
6	0.7	7	2.3	17
7	0.3	3	2.7	19
8	0.1	1	2.9	21
9	0	0	3.0	22

The chlorine content of the dried copolymers can be determined gravimetrically according to the method of Wurtzschmitt[1], and their composition derived. The copolymerization diagram is drawn and the reactivity ratios calculated. The values are compared with those for the cationic copolymerization of the two monomers (see Example 3–45).

Example 3–45:
Cationic copolymerization of styrene with 4-chlorostyrene (determination of the reactivity ratios)

Monomeric styrene and 4-chlorostyrene are shaken three times with 2 M sodium hydroxide solution to remove the stabilizers, washed three times with water, dried over anhydrous magnesium sulfate, and finally distilled under nitrogen at reduced pressure (filter pump); 4-chlorostyrene, b.p. 81°C/21 torr.

(a) Preparation of the polymerization vessels
One may use as polymerization vessels either 50 ml flasks or wide-mouthed bottles with screw caps that can be fitted with self-sealing closures as described in Section 2.1.3; they should be as dry as possible. Nine of these

[1] See *Houben-Weyl 2* (1953) 41.

vessels are thoroughly cleaned and rinsed out with pure acetone. After drying for 1 h in vacuum at 110°C they are placed while hot in a vacuum desiccator over P_2O_5 and evacuated. After 2 h the desiccator is filled with dry nitrogen, the cover opened while a strong flow of nitrogen is passed through the desiccator, and the vessels closed as quickly as possible with self-sealing closures.

(b) Preparation of the initiator solution
For the preparation of the initiator solution the following distillation apparatus is set up under a hood. A two-necked flask with nitrogen inlet is fitted with a short condenser (which serves initially as a reflux condenser), the upper end of which is connected via a still-head to a condenser for distillation. A dropping funnel serves as receiver and is fitted to the condenser via a vacuum adaptor; the exit tube from the adaptor is closed with a bubbler or suitable isolating device (see Section 2.1.1). About 20 g of $SnCl_4$ are refluxed over P_2O_5 in this apparatus under a gentle flow of nitrogen. After 1 h the cooling water in the reflux condenser is turned off and the $SnCl_4$ distilled over into the dropping funnel acting as receiver (the first fraction is run off through the stopcock). After the distillation is complete the nitrogen flow is increased, the dropping funnel removed, and the neck immediately closed with a half-bored rubber stopper. Using a dry syringe filled with nitrogen, 1.36 ml (11.5 mmol) of $SnCl_4$ (density = 2.23 g cm^{-3}) are withdrawn through the rubber stopper and injected through the self-sealing closure into a 250 ml flask that has been pretreated as described above. Sufficient dry carbon tetrachloride (distilled over P_2O_5) is injected to make up the total volume to 125 ml (1 ml solution contains 24 mg $SnCl_4$).

(c) Polymerization procedure
Styrene and 4-chlorostyrene are transferred with a syringe to the prepared reaction vessels in quantities three times those given in Example 3–44 and weighed to three places of decimals; 12 ml of dry carbon tetrachloride are then injected into each. These solutions are now cooled in an ice/salt mixture to below 0°C; 18 ml of initiator solution (corresponding to 1.7 mmol $SnCl_4$) are then injected into each vessel. After briefly swirling the contents, the polymerizations are allowed to proceed at 0°C, the flasks being placed in an ice-filled Dewar vessel. The vessels are opened after 6 h, the contents diluted with about 40 ml of carbon tetrachloride and the copolymers precipitated by dropping into a tenfold amount of methanol. After filtering off, the copolymers are reprecipitated twice from benzene or butanone solution into methanol and dried to constant weight in vacuum at 60°C. The chlorine content of the dry samples is determined and the results evaluated as in Example 3–44.

Example 3–46:
Radical copolymerization of styrene with acrylonitrile

(a) Determination of the reactivity ratios
6 tubes with ground joints are filled with the amounts of dibenzoyl peroxide (corresponding to 0.1 mole % BPO) given in the following table. The approximate amounts of destabilized styrene and acrylonitrile indicated (see Examples 3–01 and 3–08) are also weighed in to three decimal places.

Expt. no.	BPO[a] weight in mg	Styrene weight in g	Styrene amount in mmol	Acrylonitrile weight in g	Acrylonitrile amount in mmol	Polymerization time in min
1	24.2	1.0	10	4.7	90	105
2	24.2	3.1	30	3.7	70	120
3	12.1	2.0	20	1.5	30	135
4	12.1	3.1	30	1.0	20	180
5	12.1	4.1	40	0.5	10	270
6	9.7	3.7	36	0.2	4	360

[a] Dibenzoyl peroxide

The samples are then frozen in a methanol/dry-ice bath, evacuated through an adaptor and, after thawing, filled with nitrogen; this procedure is repeated twice more. The tubes are now placed in a thermostat at 60°C. After the times indicated in the table the contents are dissolved in about 40 ml of dimethylformamide and dropped into a tenfold amount of stirred methanol. The copolymers are filtered off, precipitated twice more from dimethylformamide into methanol and dried to constant weight in vacuum at 60°C. The conversion should not be much more than 10%, otherwise in the calculation of the reactivity ratios the proportions of the two monomers during polymerization can no longer be regarded as constant.

The composition of the copolymers is determined by nitrogen analysis according to the Kjeldahl method[1]; the value determined for the homopolymer of acrylonitrile can be used as a standard. Spectroscopic methods can also be employed (see Section 2.3.9). The results are used to calculate the reactivity ratios. The solubility and softening ranges (use Kofler hot block) are also determined and compared with those of the homopolymers.

[1] See *Houben-Weyl 2* (1953) 191.

(b) Azeotropic copolymerization of styrene with acrylonitrile

The following experiment is designed to show the independence of the composition of a copolymer on the yield in the copolymerization of an "azeotropic" mixture. 3.23 g (31 mmol) of styrene, 1.01 g (19.0 mmol) of acrylonitrile, and 12.1 mg (0.05 mmol) of dibenzoyl peroxide are weighed into each of 6 tubes with joints, and polymerized as described in (*a*), under nitrogen at 60°C. The tubes are removed successively from the thermostat after 2, 4, 6, 8, 12, and 20 h, and immediately quenched in a cold bath. The contents are dissolved in dimethylformamide. This can best be done as follows. The samples that are still fluid are washed out with 50 ml of dimethylformamide to which a little hydroquinone has been added; where the samples have solidified the tubes are broken open and the polymer dissolved in 100 ml of dimethylformamide with vigorous stirring. The copolymers are precipitated by dropping into 500 ml or 1000 ml of methanol respectively, filtered and dried. The yield in wt. % is plotted against polymerization time; the reasons for the deviation from linearity should be considered. The copolymers are reprecipitated twice more from dimethylformamide into methanol, washed several times with methanol and dried. The compositions are determined as above by nitrogen analysis.

Example 3–47:
Radical copolymerization of styrene with butadiene in emulsion

Monomeric styrene is destabilized (see Example 3–01) and distilled into a suitable receiver (see Section 2.1.2) under nitrogen. Butadiene from a cylinder is condensed under nitrogen atmosphere into a trap cooled in a methanol/dry-ice bath. A 500 ml pressure bottle is evacuated and filled with nitrogen.[1] A solution of 5 g of sodium oleate (or sodium dodecyl sulfate) is then made up in 200 ml of deaerated water, 0.5 g of 1-dodecanethiol is added as regulator and 0.25 g (0.93 mmol) of potassium peroxodisulfate as catalyst; the mixture is shaken until everything has dissolved. The pH is adjusted to a value of 10–10.5 with dilute sodium hydroxide, 30 g (0.29 mol) of styrene and 70 g (1.30 mol) of butadiene are added under nitrogen, and the bottle sealed. The best procedure is to cool the pressure bottle in an ice/salt mixture, place it on a balance and then, under a hood, pour in a small excess of butadiene from the cold trap; the excess butadiene is allowed to evaporate on the balance until the correct

[1] All work is to be undertaken behind a protective screen; the hands should also be protected with asbestos gloves in order to avoid cuts should one of the flasks explode (internal pressure 3 bar).

weight is reached. The sealed bottle is allowed to warm to room temperature behind a safety screen, wrapped with a towel and vigorously shaken to emulsify the contents. It is then placed in a bath at 50°C and should be shaken or rotated continuously. (If a suitable apparatus is not available, the bottle can be shaken again vigorously after 1 h). The bottle is held at 50°C for 15 h behind a safety screen, then allowed to cool to room temperature and finally cooled in ice. It is weighed again to check whether any butadiene has escaped. The latex is then slowly poured into about 500 ml of stirred ethanol contained in a beaker under a hood; to the ethanol has previously been added 2 g *N*-phenyl-β-naphthylamine, being the amount required to protect the copolymer against oxidation. Most of the unconverted butadiene evaporates and the copolymer is obtained as loosely coherent crumbs that are dried for 1–2 days in vacuum at 50–70°C. The composition of the copolymer can be calculated from the analytically determined double bond content or from the spectroscopically determined content of styrene units (see Section 2.3.9); the arrangement of the butadiene monomeric units can be found from the infrared spectrum (see Example 3–30). By vulcanization (see Example 5–21) the copolymer can be converted into an insoluble, highly elastic product.

Example 3–48:
Radical copolymerization of butadiene with acrylonitrile in emulsion

Monomeric acrylonitrile is distilled under nitrogen into a suitable receiver (see Section 2.1.2). Butadiene from a cylinder is condensed under nitrogen atmosphere into a trap cooled in a methanol/dry-ice bath.

As described in Example 3–47, a mixture of 10 g (0.19 mol) of acrylonitrile and 25 g (0.46 mol) of butadiene is polymerized in 50 ml of a 5% aqueous solution of sodium oleate (or sodium dodecyl sulfate) containing 0.25 g (0.93 mmol) of potassium peroxodisulfate as initiator, and 0.1 g 1-dodecanethiol as regulator. After 18 h the pressure bottle is allowed to cool to room temperature and is finally cooled in ice. It is now weighed again to confirm that no butadiene has been lost by leakage. The bottle is carefully opened, the latex poured into a beaker and 0.5 g of *N*-phenyl-β-naphthylamine stirred in as anti-oxidant. A solution of 5 wt. % of sodium chloride in 2% sulfuric acid is now added dropwise with stirring until the copolymer flocculates. It is filtered off, stirred vigorously with water several times in a beaker, filtered again and dried in vacuum at 65–70°C. The composition of the copolymer is determined from the nitrogen content (see Example 3–46) or from the double bond content. The arrangement of the butadiene monomeric units is found by infrared spectroscopy (see Example 3–30).

Copolymers of butadiene and acrylonitrile (Buna N) are insoluble in aliphatic hydrocarbons and are therefore oil-resistant; however they dissolve in aromatic or chlorinated hydrocarbons.

Example 3–49:
Radical copolymerization of vinyl chloride with vinyl acetate (internal plasticization)

(a) Preparation of the copolymers
Vinyl chloride from a cylinder is passed through a wash bottle containing 50% potassium hydroxide, then through a 40 cm long tube filled with calcium chloride, before being condensed in a cold trap at −70°C. Vinyl acetate is destabilized by distillation under nitrogen.

Into each of four, thick-walled, 30 ml tubes, which have been somewhat constricted a few cm from the upper end, are weighed 0.5 g of sodium dodecyl sulfate (or sodium oleate), 25 mg (0.1 mmol) of ammonium peroxodisulfate, and 10 mg (0.1 mmol) of sodium hydrogensulfite. The tubes are then evacuated and filled with nitrogen three times via an adaptor (see Section 2.1.3); 10 ml of deaerated water are added to each from a graduated dropping funnel. The adaptor is removed under a slight positive pressure of nitrogen, the tubes immediately closed with rubber stoppers and cooled to −70°C in a non-inflammable cold bath (methylene chloride/dry-ice). The stoppers are then removed and the amounts of pre-cooled vinyl acetate and vinyl chloride shown in the table are weighed in. If too much vinyl chloride is added, it can be allowed to evaporate on the balance. (Caution: do not inhale the vapour!).

	Vinyl acetate		Vinyl chloride	
Expt. no.	weight in g	amount in mmol	weight in g	amount in mmol
1	0	0	5.0	80.0
2	0.5	5.8	4.5	72.0
3	1.0	11.6	4.0	64.0
4	1.5	17.4	3.5	56.0

The filled tubes are then quickly closed with bored rubber stoppers fitted with Bunsen valves (see Section 2.1.1) and sealed off with the pointed flame of a torch. During this process the tubes should be kept well immersed in the cold bath to prevent the very volatile vinyl chloride from coming into contact with the sealing-off region. Should it do so the decomposition of the vinyl chloride will cause the glass to become brittle, making the seal weak and short-lived.

The sealed tubes are placed in protective metal sleeves, and are then mounted horizontally on a shaking apparatus in a thermostat at 60°C. (Use a safety screen!). After agitating the tubes for 3 h they are cooled to −10°C, opened carefully, and the contents thawed by immersing the tubes in water. The emulsions (about 50 ml each) are poured into 250 ml beakers, warmed slowly to 80°C, and 50 ml of a 10% aluminium sulfate solution added to each with stirring, in order to precipitate the copolymer. The polymers are filtered off, suspended again in cold water and refiltered; this operation is repeated four times and the products finally dried in vacuum at 50°C.

(b) Casting of films from solution

In order to demonstrate in a simple manner the effect of internal plasticization, films of the above samples are prepared as follows. 2 g samples of the homopolymers and copolymers are each dissolved in 30 ml of tetrahydrofuran and the resulting solutions poured on to horizontal glass plates fitted with metal frames (0.5 × 10 × 10 cm). The solvent is allowed to evaporate slowly under a hood and after some hours the films can be peeled away from the glass plates. They are allowed to dry in the air for a further period of time (about 2 days) and their flexibility compared by bending back and forth between the fingers; also their hardness by testing with the finger nail. For comparison an "externally plasticized" film is prepared by dissolving 1.6 g of poly(vinyl chloride) with 0.4 g of dioctyl phthalate[1] in 30 ml of tetrahydrofuran and proceeding as described above. The films modified by external or internal plasticization are softer and more flexible than unplasticized films of poly(vinyl chloride).[2]

Example 3–50:

Radical copolymerization of styrene with 1,4-divinylbenzene in aqueous suspension (crosslinking copolymerization)

Styrene and 1,4-divinylbenzene (the latter as 50–60% solution in ethylbenzene) are destabilized and distilled as described in Example 3–01.

A three-necked flask, fitted with stirrer (preferably with revolution counter), thermometer, reflux condenser and nitrogen inlet, is evacuated and filled with nitrogen three times. 250 mg of poly(vinyl alcohol) are placed in the flask (see Example 5–01) and dissolved in 150 ml of deaerated water at 50°C. A freshly prepared solution of 0.25 g (1.03 mmol) of

[1] For example Vestinol AH from Chemische Werke Hüls or Palatinol A from BASF.
[2] For explanation see Fig. 2.22.

dibenzoyl peroxide in 25 ml (0.22 mmol) of styrene and 2 ml (7 mmol) of 1,4-divinylbenzene is added with constant stirring so as to produce an emulsion of fine droplets of monomer in water. This is heated to 90°C on a water bath while maintaining a constant rate of stirring and passing a gentle stream of nitrogen through the reaction vessel. After about 1 h (about 5% conversion) the crosslinking becomes noticeable (gelation). Stirring is continued for another 7 h at 90°C, the reaction mixture then being allowed to cool to room temperature while stirring. The supernatant liquid is decanted from the beads which are washed several times with methanol and finally stirred for another 2 h with 200 ml of methanol. The polymer is filtered off and dried overnight in vacuum at 50°C. Yield: practically quantitative. The crosslinked copolymer of styrene and 1,4-divinylbenzene so obtained can be used for the preparation of an ion-exchange resin (see Examples 5–11 and 5–13).

The swellability is determined by placing 1 g of the dried polymer in contact with toluene for three days in a closed 250 ml flask. The swollen beads are collected on a sintered glass filter (porosity Gl), suction is applied for 5 min and the filter immediately weighed. The percentage increase in the weight of the beads can then be calculated. The size of the swollen and unswollen beads can be compared under the microscope.

Example 3–51:
Cationic copolymerization of 1,3,5-trioxane with 1,3-dioxolane (ring-opening copolymerization[1]*)*

1,3,5-Trioxane and 1,3-dioxolane are purified as in Example 3–40 or by refluxing for a day over calcium hydride followed by fractional distillation. Nitrobenzene is refluxed over P_2O_5 and distilled.

90 g (1 mol) of 1,3,5-trioxane and 9 g (0.12 mol) of 1,3-dioxolane are dissolved in 300 ml of nitrobenzene in a 500 ml flask that has been flamed under vacuum and filled with dry nitrogen or air. 0.18 ml (1.4 mmol) of boron trifluoride etherate (dissolved in 10 ml of nitrobenzene) are injected through a self-sealing closure (see Section 2.1.3). The mixture is now warmed to 45°C. The copolymer begins to precipitate within a few minutes. After 2 h the product is stirred with acetone, filtered off and washed well with acetone. In order to remove initiator and nitrobenzene residues, the polymer is boiled for 30 min with 1 l of ethanol, containing 1 wt. % of tributylamine. It is then filtered off, washed with acetone and sucked dry. About 90 g of a pure white copolymer are obtained containing

---

[1] *Y. Yamashita*, Adv. Polym. Sci. *28* (1978) 2.

about 30 oxymethylene units for every oxyethylene unit.[1] It melts at 156–159°C and has a molecular weight of about 30 000 (corresponding to an η_{sp}/c value of about 40 $cm^3 g^{-1}$ in dimethylformamide at 140°C). The thermal stability at 190°C (see Example 5–15) is compared with that of a homopolymer of 1,3,5-trioxane (see Example 3–40).

Example 3–52:
Radical copolymerization of styrene with maleic anhydride (alternating copolymerization)

300 ml of distilled benzene, 10.4 g (0.1 mol) of destabilized styrene (see Example 3–01), 9.8 g (0.1 mol) of pure maleic anhydride[2], and 0.1 g (0.4 mmol) of dibenzoyl peroxide are placed in a 500 ml three-necked flask, fitted with stirrer, thermometer and reflux condenser, and stirred at room temperature until a clear solution is obtained.[3] The reaction mixture is continuously stirred and heated to boiling on a water bath; the copolymer gradually precipitates. After 1 h the mixture is cooled, and the solid polymer filtered off and dried to constant weight in vacuum at 60°C. Yield: 19–20 g.

The 1:1 copolymer so obtained has alternating monomeric units of styrene and maleic anhydride and can be hydrolyzed to a polymeric acid (see Example 5–03). It is insoluble in carbon tetrachloride, chloroform, benzene, toluene, and methanol, but soluble in tetrahydrofuran, 1,4-dioxane, and dimethylformamide.

Example 3–53:
Radical copolymerization of cyclohexene with sulfur dioxide (alternating copolymerization)

Cyclohexene is distilled under nitrogen into a receiver (see Section 2.1.2); sulfur dioxide is condensed from a cylinder into a cold trap at −30°C, under nitrogen atmosphere.

Two clean pressure bottles (mineral water or beer bottles can be used), filled with nitrogen, are cooled to −20° to −30°C and charged as follows:

Bottle 1: 30 ml (0.3 mol) of cyclohexene, 20 ml (0.45 mol) liquid SO_2, 5 ml of ethanol (distilled under nitrogen), and 3 ml of a 2% aqueous solution of H_2O_2 (1.76 mmol).

[1] Pyrolytic gas chromatography is especially suited for the analysis of copolymers of 1,3,5-trioxane; see *K.H. Burg*, *E. Fischer* and *K. Weissermel*, Makromol. Chem. *103* (1967) 268. Also n.m.r. spectroscopy; see *D. Fleischer* and *R.C. Schulz*, Makromol. Chem. Suppl. *1* (1975) 235.

[2] If the maleic anhydride is not pure, the solution becomes turbid.

[3] This experiment need not be conducted with exclusion of air.

Bottle 2: 10 ml (0.1 mol) of cyclohexene, 40 ml (0.89 mol) of liquid SO_2, 5 ml of ethanol (distilled under nitrogen), and 3 ml of a 2% aqueous solution of H_2O_2 (1.76 mmol).

The two bottles are sealed and allowed to stand under a hood behind a safety screen for 7 h at room temperature. The bottles are then placed carefully[1] in a metal pot filled with methanol or acetone and the temperature lowered to −20°C by addition of dry ice. After waiting for several minutes the seals on the bottles are opened and 50 ml of chloroform added to each of the bottles which are then taken out of the cold bath. When they have warmed to room temperature the contents are poured into 500 ml beakers. 300 ml of ether are rapidly added to these solutions with vigorous stirring, causing the copolymer to precipitate. After settling, the copolymer is filtered off, dissolved in about 50 ml of chloroform, and reprecipitated by dropping into ether to which 10% methanol has been added. After filtering, the copolymer is dried to constant weight in vacuum at 50°C.

Cyclohexene/sulfur dioxide copolymers are soluble in chloroform and carbon tetrachloride. They also dissolve in concentrated sulfuric acid and can be recovered unchanged by precipitation with water; on the other hand they are decomposed by hot alkali. They are thermally stable to about 200°C.

The composition of the above samples is calculated from the sulfur content, which may be determined by the method of Schöninger. The limiting viscosity number is determined in chloroform at 20°C.

3.3.2. Block and graft copolymerization

Conventional copolymerizations yield random or alternating copolymers. Special procedures must be used to make block or graft copolymers, which have been extensively investigated in recent times.[2,3,4,5,6]

[1] All work is to be carried out behind a protective screen; the hands should be protected with asbestos gloves to avoid cuts should the bottles explode (internal pressure 3 bar).

[2] *S.L. Aggarwal*, "Block Copolymers", Plenum Press, New York 1970.

[3] *D.C. Allport* and *W.H. Janes*, "Block Copolymers", Applied Science Publishers, London 1973.

[4] *R.J. Ceresa*, "Block and Graft Copolymerization", Wiley and Sons, London, New York, Sydney, Toronto, Vol. 1, 1973; Vol. 2, 1976.

[5] *A. Noshay* and *J.E. McGrath*, "Block Copolymers, Overview and Critical Survey", Academic Press, New York, San Francisco, London 1977.

[6] *M. Szwarc*, Makromol. Chem. *35* (1950) 151.

Block copolymers consist of successive series (blocks or segments) of A and B repeating units:

—AAAAA—BBBBBBBBBB—AAAAAAA—

For the preparation of block copolymers one can start from a ready-made centre block with reactive end groups, to which can be added blocks of the second monomer by radical or ionic polymerization. Especially suitable as starting polymers are "living polymers" (see Section 3.2.1.2), having anions at each end capable of adding a second monomer[1] (see Example 3–54).

Butadiene-styrene block copolymers are of particular technical interest since they combine the properties of an elastomer and a thermoplastic. For their preparation one can make use of the different reactivities of the two monomers in anionic polymerization. Thus when butyllithium is added to a mixture of butadiene and styrene, the butadiene is first polymerized almost completely. After its consumption styrene adds on to the living chain ends, which can be recognized by a colour change from almost colourless to yellow to brown (depending on the initiator concentration). Thus, after the styrene has been used up and the chains are finally terminated, one obtains a two-block copolymer of butadiene and styrene (Example 3–55). Oxidative degradation by splitting of the double bonds in the butadiene blocks allows the styrene blocks to be isolated. The degradation of the chains can be followed by molecular weight determinations or viscosity measurements (Example 3–55b).

Block copolymers are linear, but graft copolymers are branched, with the main chain generally consisting of a homopolymer or a random copolymer, while the grafted side chains are composed of another monomer or several monomers:

```
—AAAAAAAAAAAAAA—
   |          |
   B          B
   B          B
   B          B
   B          B
   B          B
              B
              B
```

There are numerous ways of making graft copolymers; an appropriate method must be selected for each individual case. For example one can generate on the main chain of a macromolecule, a radical which can initiate the polymerization of a second monomer. In order to create such a radical

[1] *B.R.M. Gallot*, Adv. Polym. Sci. *29* (1978) 85.

centre one may irradiate with ultraviolet radiation or with high energy radiation. Autoxidation leads to hydroperoxide groups whose decomposition can also lead to suitable radical centres. Chain transfer reactions can sometimes be exploited for the preparation of graft copolymers. However, this method frequently results in a high proportion of the homopolymer of the monomer that forms the side chains. Hence, of particular interest are certain ionic graft copolymerizations in which the polymerization reaction is initiated only on the macromolecular framework and no homopolymer is formed. An example is provided by the formation of polymer carbonium ions from chloride-containing polymers, such as poly (vinyl chloride), in the presence of diethylaluminium chloride:

$$-CH_2-\underset{\displaystyle Cl}{\underset{|}{CH}}- + (C_2H_5)_2AlCl \longrightarrow -CH_2-\overset{\oplus}{CH}- + (C_2H_5)_2\overset{\ominus}{Al}Cl_2$$

The polymer cations can then initiate the polymerization of cationically polymerizable monomers such as styrene or isobutene.[1]

The characterization of block or graft copolymers[2,3] is generally very much more difficult than that of random copolymers (cf. Section 2.3.11). The properties of these types of copolymer depend markedly on the number and length of the blocks and side chains, as well as on the structure of the monomeric units and their mole ratio. In general, graft and block copolymers combine additively the properties of the corresponding homopolymers, while random copolymers normally exhibit the average of the properties of the two homopolymers. There is, therefore, the possibility of preparing copolymers with desired combinations of properties. This can generally not be achieved by blending the corresponding homopolymers, since chemically different polymers are rarely compatible with one another. This incompatibility of different CRU's means that block and graft copolymers often form multiphase systems[4] that are distinguished by special properties. Examples are the well known ABS-polymers made from acrylonitrile, butadiene and styrene; also impact-resistant thermoplastics, and the so-called thermoplastic elastomers which contain both thermoplastic and rubber-elastic segments.

Finally it should be mentioned that graft and block copolymers can be prepared by both condensation and addition polymerization.

---

[1] *J.P. Kennedy* (Ed.), "Cationic Graft Copolymerization", J. Appl. Polym. Sci., Polym. Symp. *30* (1978).

[2] *Y. Ikada*, Adv. Polym. Sci. *29* (1978) 47.

[3] *B.R.M. Gallot*, Adv. Polym. Sci. *29* (1978) 85.

[4] See Angew. Makromol. Chem. *58/59* (1977).

Example 3–54:
Preparation of a block copolymer of 4-vinylpyridine and styrene by anionic polymerization

The block copolymer is prepared by anionic polymerization. As in Example 3–27 the greatest care must be taken to exclude air and moisture.

Monomeric styrene is destabilized as in Example 3–01 and pre-dried with calcium chloride. The monomer is allowed to stand over calcium hydride for 24 h and then distilled under reduced pressure of nitrogen into a previously flamed out Schlenk tube. Pure 4-vinylpyridine is distilled twice over KOH pellets in vacuum. It is then vacuum-distilled under nitrogen through a column packed with Raschig rings into a previously flamed-out Schlenk tube (b.p. 62°C/12 torr). The closed Schlenk tubes containing the monomers are stored in a refrigerator until required. The preparation of the initiator solution (sodium naphthalene) is described in Example 3–27.

The polymerization is carried out as follows. A Schlenk tube, that has been flamed out and filled with nitrogen, is charged with 50 ml of pure tetrahydrofuran and 1 mmol of sodium naphthalene (see Example 3–27). Using a nitrogen-filled syringe, 4.6 ml (40 mmol) of styrene is added to this solution at room temperature with vigorous agitation, while nitrogen is passed through the tube. The closed tube containing the red polymerizing mixture is allowed to stand for 15 min at room temperature.

5.3 g of 4-vinylpyridine are made up to 50 ml with tetrahydrofuran; 5 ml of this solution (containing 5 mmol of 4-vinylpyridine) are added in the same way to the above solution containing the "living" polystyrene, with vigorous agitation. After 15 min another 40 mmol of styrene are added, followed 15 min later by another 5 mmol of 4-vinylpyridine; this operation is repeated once more. 15 min after the last addition of monomer the block copolymer is precipitated by dropping the solution into a mixture of 300 ml of diethyl ether and 300 ml of petroleum ether. The polymer is filtered, washed with ether, filtered again and dried in vacuum at room temperature.

The blocks of 4-vinylpyridine must not be too long, otherwise the polymer is no longer completely soluble in tetrahydrofuran. This can easily be observed when 20 ml lots of 4-vinylpyridine are used instead of 5 ml; the solution then becomes cloudy, but after a fresh addition of styrene it goes clear again.

For comparison, polystyrene (Example 3–27) and poly(4-vinylpyridine) are prepared by anionic polymerization with sodium naphthalene as initiator. Poly(4-vinylpyridine) precipitates from tetrahydrofuran; the mixture is poured into 200 ml of diethyl ether and the polymer filtered off. The polymer is then reprecipitated from pyridine solution into a tenfold amount of diethyl ether and dried in vacuum.

The infrared spectra of all three polymers are recorded and compared with one another. The incorporation of monomeric units of 4-vinylpyridine can also be demonstrated by nitrogen analysis of the block copolymer. The solubility behaviour is also determined. Poly(4-vinylpyridine) is soluble in pyridine, methanol and chloroform, but insoluble in benzene and diethyl ether; it swells considerably in water. On the other hand the block copolymer, like polystyrene, is soluble in pyridine, chloroform, and benzene; but unlike polystyrene, it swells significantly in methanol.

Example 3–55:
Preparation of a butadiene-styrene block copolymer

(a) Preparation
A 500 ml two-necked flask, fitted with ground glass stopper and tap, is dried for 2 h at 110°C. Into the still warm flask are distilled 200 ml of toluene under nitrogen, a moderate stream of nitrogen ($3\,l\,min^{-1}$) then being passed through the solvent for 3 min, using a delivery tube with a sintered glass disc. The flask is immediately closed with a self-sealing rubber cap and the pressure reduced to 0.3–0.5 bar nitrogen. With the aid of a dry hypodermic syringe, 6 g (6.7 ml) of dry, stabilizer-free styrene are added, together with 14 g (23 ml) of dry, stabilizer-free butadiene from a vessel provided with an injection needle (the materials must be cooled to a suitable temperature, cf. Example 3–34). The flask is well shaken and placed in a suitable wire basket. Finally, as initiator, 8–10 ml of a 0.1 M solution of butyllithium in hexane are injected with a hypodermic syringe. The necessary initiator concentration depends on the water content of the reaction mixture and must be determined by trial and error.

The flask in its wire basket is placed in a water bath at 50°C. After 5 h the polymerization has finished and the colour of the solution is then yellow. After cooling to room temperature, 1 ml 2-propanol is injected and the excess pressure released by insertion of an injection needle. The flask is opened, 0.5 g of 2,6-di-*tert*-butyl-4-methylphenol added as stabilizer, and the polymer precipitated in a threefold volume of methanol. The viscous product is dried in vacuum at 50°C. The yield is quantitative. The limiting viscosity number is determined in toluene at 25°C.

(b) Oxidative degradation of the two-block copolymer of butadiene and styrene
6.0 g of the polymer are swollen or dissolved in 600 ml of 1,2-dichlorobenzene at room temperature, heated for 10 min at 120°C and then cooled to 95°C. 90 ml of *tert*-butyl hydroperoxide (80%) are added, the temperature brought back to 95°C, and 15 ml of a solution of 0.08 wt.% of osmium tetraoxide in distilled toluene then added. The temperature rises a

few degrees. After 20 min the reaction mixture is cooled to room temperature and shaken three times in a separating funnel with 600 ml of a mixture of 1500 ml of methanol and 500 ml of water. The lower phase is the 1,2-dichlorobenzene solution. The 1,2-dichlorobenzene is distilled off under vacuum at 60°C (oil pump) and the residue weighed. About 2.4 g of a deep yellow, highly viscous mass are obtained. This can be decolorized by taking it up in benzene and warming with activated charcoal. The polymer then precipitates well in methanol. The reaction products of the degraded butadiene sequences, and also the styrene sequences having molecular weight less than 500, are removed by this treatment. The limiting viscosity number of the residue is now determined in toluene at 25°C and the molecular weight of the styrene sequences determined. The infrared spectrum of the residue from degradation corresponds very closely to that of polystyrene. The amount of styrene sequences is only slightly less than the amount of styrene in the block copolymer since only a small proportion of the styrene is present within short chain lengths containing both monomeric units, which form the junction points of the two-block sequences.

Example 3–56:
Graft copolymerization of styrene on polyethylene

A dry, weighed polyethylene film (length 50 mm, breadth 25 mm, thickness 0.1–0.2 mm) is placed in a tube (about 70 ml capacity) fitted with a ground glass joint and stopcock. After addition of some crystals of benzophenone (as sensitizer) the sample is heated for 1 h on a water bath at 60°C. The outside of the tube is dried, the stopcock closed, and the tube exposed to 15 min irradiation from a mercury lamp.

While the sample is being irradiated a second tube is filled with 50 ml of destabilized styrene (see Example 3–01) and 0.1 ml of 3,6,9-triazaundecamethylenediamine (tetraethylenepentamine)[1], the tube then being evacuated and filled with nitrogen. With the aid of a large syringe 40 ml of this solution is injected through the bore of the stopcock into the tube containing the film. The tube is now evacuated, filled with nitrogen, and the closed vessel maintained at 60°C in a thermostat for 2 h. The grafted polyethylene foil is then extracted in a Soxhlet apparatus for 1 h with ethyl acetate, dried to constant weight in vacuum at 50°C and weighed. The extract is added dropwise to about 400 ml of methanol; the

[1] Available from Merck.

precipitated homopolymer of styrene is filtered off at a sintered glass filter, dried in vacuum at 50°C and weighed.

The amount of grafted styrene is given by the increase in weight of the film. It may also be calculated quite well from the densities of polyethylene (d_1), polystyrene (d_2) and the graft copolymer (d_3) according to the following equation:

$$\frac{100 - x}{d_1} + \frac{x}{d_2} = \frac{100}{d_3}$$

x is the weight percentage of grafted polystyrene, relative to the weight of the grafted film. The density of polystyrene (d_2) may be taken as 1.05 g cm^{-3}; the densities d_1 and d_3 can be determined on small pieces (0.2–0.5 cm^2) of the polyethylene and grafted films by the flotation method using ethanol/water mixtures (see Section 2.3.7). In addition, the ratio of grafted polystyrene and homopolymer of styrene is determined.

4. Synthesis of Macromolecular Substances by Condensation and Addition Polymerization

4.1. CONDENSATION POLYMERIZATION (POLYCONDENSATION)

Condensation polymerizations (polycondensations) are stepwise reactions between bifunctional or polyfunctional components, with elimination of simple molecules such as water or alcohol and the formation of macromolecular substances. For the preparation of linear condensation polymers[1] from bifunctional compounds there are basically two possibilities: one either starts from a monomer which has two unlike groups suitable for polycondensation (type I), or one starts from two different monomers, each possessing a pair of identical reactive groups that can react with each other (type II). An example of type I is the polycondensation of hydroxycarboxylic acids:

$$n\,HO{-}(CH_2)_x{-}COOH \xrightarrow{-(n-1)H_2O} HO{-}\left[(CH_2)_x{-}\underset{\underset{O}{\|}}{C}{-}O\right]_{n-1}{-}(CH_2)_x{-}COOH$$

An example of type II is the polycondensation of diols with dicarboxylic acids:

$$n\,HO{-}(CH_2)_x{-}OH + n\,HOOC{-}(CH_2)_y{-}COOH \xrightarrow{-(2n-1)H_2O}$$

$$H{-}\left[O{-}(CH_2)_x{-}O{-}\underset{\underset{O}{\|}}{C}{-}(CH_2)_y{-}\underset{\underset{O}{\|}}{C}\right]_n{-}OH$$

The formation of a condensation polymer is a stepwise process. Thus, the first step in the polycondensation of a hydroxycarboxylic acid is the formation of a dimer that possesses the same end groups as the initial monomer:

$$2HO{-}(CH_2)_x{-}COOH \xrightarrow{-H_2O} HO{-}(CH_2)_x{-}\underset{\underset{O}{\|}}{C}{-}O{-}(CH_2)_x{-}COOH$$

[1] The same considerations apply to polyfunctional compounds which then lead to branched or crosslinked condensation polymers (see, for example, Section 4.1.1.5).

The end groups of this dimer can react in the next step either with the monomeric compound or with another dimer molecule, and so on. The molecular weight of the resulting macromolecules increases continuously with reaction time, unlike many addition polymerizations, e.g. radical polymerizations. The intermediates, which are formed in independent, individual reactions, are oligomeric and polymeric molecules with the same functional end groups as the monomeric starting compound; in principle these intermediates can be isolated without losing their capability for further growth. Thus, all reactions

$$M_i + M_j \longrightarrow M_{i+j} \qquad i \geq 1,\ j \leq x$$

can occur, where M_i, M_j denote oligomeric or polymeric species containing i, j monomeric units respectively. The formulation of a reaction scheme that takes into account the multitude of possible reactions between i-mers and j-mers, with different rate constants k_{ij}, and with the prevailing concentrations $[M_i]$ and $[M_j]$, would be extraordinarily complicated. If, however, one assumes that the reaction between the end groups takes place in such a way that it is independent of i or j, the kinetic treatment of polycondensation is considerably simplified. Thus it is assumed that the reactivity is independent of molecular weight. This "principle of equal reactivity" holds true for both condensation and addition polymerizations. It means that there is no difference in the reactivity of the end groups of monomer, dimer etc., and therefore that the rate constant is independent of the degree of polymerization over the total duration of reaction.

On this basis the kinetics of polycondensation were worked out a long time ago;[1,2] here the following points are of particular interest:

— dependence of the average molecular weight on conversion,
— dependence of the average molecular weight on the mole ratio of reactive groups,
— dependence of the conversion and average molecular weight on the condensation equilibrium,
— exchange reactions such as transesterification or transamidation.

The progress of a polycondensation can be followed in a simple manner by analysis of the unreacted functional groups. If the reactive groups are present in equivalent amount, which is generally desired (see below), it is sufficient to analyze for one of the two groups, for example the carboxyl groups in polyester formation. If the number of such functional groups initially present is N_0, and the number at time t is N, the extent p of

---

[1] *G.V. Schulz*, Z. Phys. Chem., *A182* (1938) 127.
[2] *P.J. Flory*, "Principles of Polymer Chemistry", Ithaca, New York 1953.

condensation is defined as the fraction of functional groups that have already reacted at that time:

$$p = \frac{N_0 - N}{N_0} \tag{1}$$

Multiplying p by 100 yields the conversion in %.

In polycondensations the average degree of polymerization $\bar{P}_n$ is defined as the ratio of the number of monomer molecules originally present N_0, to the total number of molecules N at the appropriate stage of reaction (including the as yet unconverted monomer molecules). Hence from equation (1) one obtains:

$$\bar{P}_n = \frac{N_0}{N} = \frac{1}{1-p} \tag{2}$$

$\bar{P}_n$ is thus dependent on the percentage conversion ($100p$) as illustrated by the numerical example in Table 4.1. From the table it is clear that to attain the commercially interesting molecular weights of 20 000–30 000, the reaction must be driven to conversions of more than 99%.

The degree of polymerization $\bar{P}_n$ is affected not only by the yield but also by the molar ratio of the reacting functional groups A and B. If N_A and N_B are the number of such groups originally present, and r is their ratio (N_A/N_B), then equation (2) is modified to:

$$\bar{P}_n = \frac{1 + r}{2r(1 - p) + 1 - r} \tag{3}$$

TABLE 4.1

Degree of polymerization $\bar{P}_n$ and number-average molecular weight $\bar{M}_n$ (assuming a structural element of molecular weight 100) as a function of conversion in stepwise condensation and addition polymerizations

Conversion in %	$\bar{P}_n$	$\bar{M}_n$
50	2	200
75	4	400
90	10	1000
95	20	2000
99	100	10 000
99.5	200	20 000
99.7	300	30 000
99.95	2000	200 000
99.97	3000	300 000

When $r = 1$, equation (3) reduces to equation (2). If all the A groups have reacted (i.e. $p = 1$), equation (3) simplifies to:

$$\bar{P}_n = \frac{1 + r}{1 - r} \tag{4}$$

In general the effect of an excess of one component is greater, the higher the conversion: at 99% conversion the molecular weight is depressed to half its value by a 2% excess of one component; at 99.5% conversion the same effect is produced by a 1% excess. Some numerical examples are given in Table 4.2 for the effect of non-equivalence of components on the degree of polymerization (at 100% conversion of the lesser component).

Table 4.2 shows how important it is in polycondensation reactions to ensure the exact equivalence of functional groups, since even a 1 mol % excess of one of the two groups limits the maximum attainable degree of polymerization $\bar{P}_n$ to less than 200. For polycondensations of type I (e.g. hydroxycarboxylic acids and aminoacids) this equivalence is automatic since the monomer contains both groups. On the other hand for polycondensations of type II (e.g. between diols and dicarboxylic acids) a small excess of one component causes the reaction to come to a halt when only the end groups of the component present in excess are left, these being unable to react with each other.

Even if both functional groups are present in equivalent amount at the beginning of reaction, this equivalence can be disturbed during the course of reaction by evaporation, sublimation or side reactions of one of the reaction partners. A monofunctional compound that can react with one of the bifunctional reaction partners acts in the same way as an excess of a bifunctional component. Therefore, high purity of the monomers is

TABLE 4.2

Degree of polymerization $\bar{P}_n$ as a function of the ratio of the functional groups A and B in stepwise condensation and addition polymerizations (conversion 100% with respect to A, i.e. component A has completely reacted).

Excess of component B mol %	$r = \frac{N_A}{N_B}$	$\bar{P}_n$
10	0.9	19
1	0.99	199
0.1	0.999	1999
0.01	0.9999	19 999

absolutely essential in the preparation of high molecular weight polymers by stepwise condensation or addition polymerization; on the other hand this means that it is possible to regulate precisely the average molecular weight of the polymer formed either by using a controlled excess of one component or by addition of a monofunctional compound.

Besides the effects described by equations (1) to (4), which apply to both stepwise addition and condensation polymerizations, two further factors must be considered. In the first place the condensation equilibrium limits the conversion and hence the average molecular weight. As in the case of esterification of monofunctional compounds, the corresponding polycondensations are to be treated as equilibrium reactions[1], governed by the law of mass action. For example in the case of polyesterification, if one mol of hydroxyl groups (0.5 mol of diol) reacts with 1 mol of carboxylic acid groups (0.5 mol of dicarboxylic acid), this takes the form:

$$K = \frac{p n_{\mathrm{w}}}{(1-p)^2} \tag{5}$$

according to Flory[2] and Schulz.[3]

As before, p is the fraction of functional groups that have reacted, (cf. equation (1)), and n_{w} is the mole fraction of water present in the reaction mixture. Solving this equation for p we obtain the upper limit of conversion as a function of the ratio $\beta = K/n_{\mathrm{w}}$:

$$p = \frac{1}{2\beta}\left[1 + 2\beta - (1 + 4\beta)^{\frac{1}{2}}\right] \tag{6}$$

Since the number-average degree of polymerization $\bar{P}_{\mathrm{n}}$ is given by equation (7):

$$\bar{P}_{\mathrm{n}} = \frac{1}{1-p} \tag{7}$$

the upper limit of degree of polymerization, when governed by the condensation equilibrium, is given by equation (8).

$$\bar{P}_{\mathrm{n}} = \frac{2\beta}{(1 + 4\beta)^{\frac{1}{2}} - 1} \tag{8}$$

Since, in practically all cases, one aims at as high a conversion as possible, i.e. a value of p tending towards unity, equations (6) and (8) may be

---

[1] In principle, equilibria can also occur in stepwise addition polymerizations, e.g. the back reaction may give rise to fission of urethane groups into isocyanate and hydroxyl groups; however, such reactions are generally only significant at high temperatures.

[2] *P.J. Flory*, "Principles of Polymer Chemistry", Ithaca, New York 1953.

[3] *G.V. Schulz*, Z. Phys. Chem. *A182* (1938) 127.

simplified to obtain:

$$p \approx 1 - 1/\beta^{\frac{1}{2}} \tag{9}$$

and

$$\bar{P}_n \approx \beta^{\frac{1}{2}} = (K/n_w)^{\frac{1}{2}} \tag{10}$$

If the equilibrium constant K has a value between 1 and 10, less than a thousandth of the total amount of water formed in the reaction mixture is sufficient to prevent the formation of really high-molecular-weight condensation polymers. Hence it follows that it is extremely important to remove as completely as possible the low-molecular-weight reaction products, for example water, eliminated during polycondensation.

In polycondensations one must also take account of exchange reactions which can occur between free end groups and junction points in the chain, for example between hydroxyl end groups and ester groups of a polyester (transesterification):

$$—O—(CH_2)_x—O—\overset{\overset{O}{\|}}{C}—(CH_2)_y—\overset{\overset{O}{\|}}{C}— \quad + \; HO—(CH_2)_x—O—\underset{\underset{O}{\|}}{C}— \longrightarrow \; \begin{matrix} —O—(CH_2)_x \\ | \\ HO \end{matrix} \; + \; \begin{matrix} O—\overset{\overset{O}{\|}}{C}—(CH_2)_y—\overset{\overset{O}{\|}}{C}— \\ | \\ (CH_2)_x—O—\underset{\underset{O}{\|}}{C}— \end{matrix}$$

Neither the number of free functional groups, nor the number of molecules, nor the number-average degree of polymerization, is thereby altered. Thus, two equal-sized macromolecules could react with one another to give one very large and one very small macromolecule; conversely a very large and a very small macromolecule can be converted into two macromolecules of similar size. Independent of the initial distribution such exchange reactions will in each case lead to a state of dynamic equilibrium in which the rates of formation and consumption of molecules of a given degree of polymerization are equal. This results in an equilibrium distribution of molecular weights which is formally the same as that obtained by purely random polycondensation. Therefore, in normal polycondensations the exchange reactions will not affect the molecular weight distribution. On the other hand by mixing a low-molecular-weight and a high-molecular-weight polyester in the melt, especially if catalyst is still present, the molecular weight distribution very soon adjusts to an equilibrium distribution with a single maximum, instead of the two in the original mixture. Such exchange reactions between the end groups and the junction positions are known for polyamides, polysiloxanes, and polyanhydrides, as well as for polyesters.

Although the procedures for the preparation of condensation polymers are analogous in many respects to those for the condensation reactions of monofunctional compounds, some additional factors must be taken into account if one wishes to attain high molecular weights in polycondensations. These are essentially consequences of equations (1) to (5), (9) and (10). First the condensation reaction must be specific and proceed with the highest possible yield, otherwise only mixtures of oligomers will be obtained. Furthermore, for polycondensation reactions in which two or more components may participate, care must be taken to ensure strict equivalence in the proportions of reacting groups throughout the reaction. The equilibrium position of the reaction must be displaced as far as possible towards condensation. By analogy with the condensation of monofunctional compounds this may be achieved by removing the low-molecular-weight reaction product, e.g. water, as completely as possible from the reaction mixture. This can be done by distillation under high vacuum or by azeotropic distillation. It is advantageous to pass very dry inert gas through the well-stirred reaction mixture in order to facilitate the diffusion of the eliminated component from the viscous solution formed during polycondensation. High demands are placed on the purity of the starting materials. They must be especially free from monofunctional compounds since these block the end groups of the resulting macromolecules and so prevent further condensation. Only bifunctional compounds can be used for the preparation of linear condensation polymers, since polyfunctional compounds give rise to branching and crosslinking. Finally, polycondensation reactions should be carried out with the exclusion of oxygen, since oxidative decomposition can easily occur at the high temperatures of reaction that are frequently needed.

4.1.1. Polyesters

Polyesters are macromolecules whose monomeric units are joined through an ester group:

$$HO-(CH_2)_x-O\left[-\underset{\displaystyle O}{\overset{}{\underset{\|}{C}}}-(CH_2)_y-\underset{\displaystyle O}{\underset{\|}{C}}-O-(CH_2)_x-O-\right]_n\underset{\displaystyle O}{\underset{\|}{C}}-(CH_2)_yCOOH$$

Linear polyesters are generally soluble in chloroform, dichlorobenzene, and formic acid. Their properties depend markedly on their chemical composition: thus pure aliphatic polyesters generally have melting points below 100°C (with the melting point increasing with the number of methylene groups between the ester groups), and are easily hydrolyzed. In contrast, polyesters made from aromatic or cyclo-aliphatic dicarboxylic acids and diols are high-melting and resistant to hydrolysis, for example, polyesters from terephthalic acid and ethylene glycol or from terephthalic acid and 1,4-cyclohexylene dimethanol; these are suitable for the preparation of fibres and films.

High-molecular-weight polyesters are generally prepared by means of the following reactions[1,2], taking into account the factors discussed in Section 4.1:

— polycondensation of hydroxycarboxylic acids
— polycondensation of diols with dicarboxylic acids
— polycondensation of diols with derivatives of dicarboxylic acids (e.g. anhydrides or esters)
— ring-opening polymerization of lactones (see Section 3.2.4.3).

The establishment of the equilibrium is often accelerated by acidic or basic catalysts, as for example by strong acids (*p*-toluenesulfonic acid), metal oxides (lead dioxide, antimony trioxide), weak acid salts of the alkali metals or alkaline earths (acetates, benzoates), or by alcoholates.

4.1.1.1. Polyesters from hydroxycarboxylic acids[3]

Polycondensations of aliphatic hydroxycarboxylic acids, such as glycolic acid, lactic acid, or 12-hydroxy-9-octadecenoic acid (ricinoleic acid), are mostly carried out in the melt and yield high-molecular-weight polyesters. However, these reactions often result in the formation of cyclic esters (lactones). The tendency towards ring formation depends on the number of methylene groups between the hydroxyl and carboxylic acid groups; it is greatest for γ-hydroxybutyric acid, δ-hydroxyvaleric acid, and ε-hydroxycaproic acid. Aromatic carboxylic acids carrying an aliphatic hydroxyl group can also be condensed to high-molecular-weight polyesters.

4.1.1.2. Polyesters from diols and dicarboxylic acids

Polycondensation of diols with dicarboxylic acids is often performed in the melt. However, it does not always lead to high-molecular-weight polyesters. Thus the starting components or the resulting polyester are sometimes thermally unstable at the high temperature required for the condensation reaction; the elimination of water from diols and dicarboxylic acids frequently occurs rather slowly. In such cases suitable functional derivatives (anhydrides or esters) can be used with advantage in place of the acid.

---

[1] For detailed accounts see *R.E. Wilfong*, J. Polym. Sci. *54* (1961) 385; *E. Müller* in *Houben-Weyl 14/2* (1963) 1.
[2] *V.V. Korshak* and *S.V. Vinogradova*, "Polyesters", Pergamon Press, Oxford 1965.
[3] *E. Müller* in *Houben-Weyl 14/2* (1963) 6.

Very-high-molecular-weight polyesters can sometimes be obtained from diols and dicarboxylic acids by a solution condensation reaction at high temperature (see also Section 2.1.5.2).

Example 4–01:
Preparation of a low-molecular-weight, branched polyester from a diol and dicarboxylic acid by melt condensation

(a) Preparation of a slightly branched polyester
A 500 ml three-necked flask is fitted with a nitrogen inlet, a stirrer, a thermometer that reaches deep into the flask, a fractionating column of about 20 cm length packed with Raschig rings, at the upper end of which is another thermometer, and a condenser for distillation, with vacuum adaptor and graduated receiver. 146 g (1.0 mol) of adipic acid, 110.5 g (1.04 mol) of anhydrous bis(2-hydroxyethyl) ether, (diglycol), and 9.0 g (0.067 mol) of anhydrous 1,1,1-propanetriol are weighed in, some anti-bump granules added, and the air displaced by evacuation and filling with nitrogen. The flask is then heated slowly under a gentle stream of nitrogen; when the contents become fluid between 80 and 110°C, the stirrer is switched on. Polyesterification sets in at 130–140°C as evidenced by the formation of water. The internal temperature is now raised to 200°C at such a rate that the temperature at the head of the column does not exceed 100°C. During this period most of the water (36 g) is eliminated; it is important not to interrupt the stirring because of the danger of bumping.

As soon as the temperature at the head of the column drops much below 100°C, after the internal temperature has reached 200°C, the vacuum pump is attached and the pressure reduced to 12–14 torr. At this stage of the experiment, distillation of the diglycol must be completely avoided; if necessary one must evacuate more slowly. The reaction is now allowed to continue at 12–14 torr and 200°C, samples being taken at intervals of 5 h in order to determine the acid number; in order to take the samples, the apparatus is momentarily filled with nitrogen and then evacuated again. When the acid number is smaller than 2 (total time about 35–40 h) the experiment is terminated. The slightly branched polyester remains, after cooling under nitrogen, as a viscous, pale yellow mass. It has an OH number (see under (*d*)) of about 60 and can be directly processed to yield an elastic polyurethane foam (see Example 4–24).

(b) Preparation of a highly branched polyester
16.9 g (0.12 mol) of adipic acid, 42.5 g (0.15 mol) oleic acid, 44 g (0.30 mol) of phthalic anhydride and 105 g (0.78 mol) of anhydrous 1,1,1-propanetriol are weighed into the apparatus described under (*a*) and the air displaced by evacuation and filling with nitrogen. The mixture is slowly heated under a

stream of nitrogen; at 120°C the contents of the flask have melted and the stirrer can be started. The internal temperature is raised to 190°C over a period of 2 h; as soon as the temperature at the top of the column drops below 70°C, the pump is attached and the apparatus slowly evacuated to 30 torr over the course of 2 h. At this pressure and an internal temperature of 190°C, the mixture is stirred for a further 8 h. The pump is then switched off and the product allowed to cool under nitrogen. The highly branched polyester, which remains as a viscous liquid, has an acid number of 2 and an OH number of 350; it can be used directly for the preparation of a rigid polyurethane foam (see Example 4–25).

(c) Determination of the acid number

1–2 g of the polyester are dissolved by warming with 50 ml of acetone and, after cooling, are titrated as quickly as possible with 0.1 M alcoholic potassium hydroxide using phenolphthalein as indicator, until the red color remains for a second. The alkali requirement of the solvent is determined in a blank experiment. The acid number is given by the weight of KOH in mg, required to neutralize 1 g of substance:

$$\text{acid number} = 5.61 f(A - B)/E$$

where A = titre for the sample, in ml,
B = titre for the blank, in ml,
f = concentration of the alcoholic KOH, in mol l^{-1},
E = weight of sample, in g.

(d) Determination of the hydroxyl number[1]

To acetylate the free hydroxyl groups of a polyester a solution of 150 g of freshly distilled acetic anhydride (b.p. 140°C) in 350 g of freshly distilled dry pyridine is prepared. The solution, which turns slighly yellow with time, is stored in a dark bottle.

To determine the hydroxyl number, about 2.5 g of polyester are weighed to the nearest 10 mg into each of three conical flasks with ground joints. The samples are each dissolved with gentle warming in 10 ml of pyridine. After cooling, 10 ml of the acetylating solution are added with a pipette to each of the flasks. 10 ml of pyridine and 10 ml of acetylating solution are placed in each of two conical flasks with ground joints, for the determination of the blank value. All five flasks are closed with stoppers fastened with adhesive tape. They are then placed in an oven at 110°C for 70 min. (Caution when removing: use safety goggles and asbestos gloves!). After cooling, a few drops of 0.1% aqueous Nile blue chloride solution are

[1] *Houben-Weyl 14/2* (1963) 17.

added to each of the samples which are then titrated with 1 M NaOH solution until the colour changes suddenly to violet-red. The hydroxyl number indicates how many mg KOH are equivalent to the free hydroxyl groups present in 1 g of substance.

$$\text{OH number} = 56.1 f(B - A)/E$$

where B = titre for the blank, in ml,
A = titre for the sample, in ml,
f = concentration of the NaOH, in mol l^{-1},
E = weight of sample, in g.

The average value is taken of those values which do not differ by more than ± 2 from one another.

Example 4–02:
Preparation of a high-molecular-weight linear polyester from a diol and dicarboxylic acid by condensation in solution

The recycling apparatus shown in Fig. 2.9 is charged with 59.05 g (0.5 mol) of 1,6-hexanediol (purified by vacuum distillation), 59.09 g (0.5 mol) of recrystallized succinic acid, 200 ml of dry toluene and 1.5 g of pure *p*-toluenesulfonic acid; at the same time the siphon is also filled with toluene so that the circulation of the solvent can begin immediately. After adding some anti-bump granules the solution is heated on an oil- or air-bath to boiling; the toluene should flow briskly through the drying tube filled with soda-lime back into the flask. After some hours, when about three-quarters of the theoretical amount of water has collected in the separator, the soda-lime is renewed for the first time; it is renewed again after another 10 h.

The viscosity of the solution gradually increases and so does the temperature in the flask. In order to maintain a brisk rate of distillation (and therefore polycondensation), each time the internal temperature reaches 130°C about 50 ml of pure toluene is added. After about 25 h the flask is cooled to room temperature and the solution is added dropwise to a tenfold amount of methanol; the polymer is filtered off and dried to constant weight in vacuum at 40°C. Yield: about 90%.

The linear polyester so obtained is of higher molecular weight than that produced by melt polycondensation. It is soluble in benzene, toluene, chloroform, and formic acid, and readily hydrolyzable (see Example 5–19); the melting point is about 100°C.

The increase of molecular weight during polycondensation can be followed by taking samples from time to time (initially at short intervals) and determining the limiting viscosity numbers of the precipitated and dried samples; the carboxyl and hydroxyl end groups can also be quantitatively determined (see Section 2.3.2.2).

4.1.1.3. *Polyesters from diols and dicarboxylic acid derivatives*

The polycondensation of a diol and the diester of a dicarboxylic acid (say the dimethyl ester) can be carried out in the melt at a considerably lower temperature than for the corresponding reaction of the free acid. Under the influence of acidic or basic catalysts a transesterification then occurs with the elimination of the readily volatile alcohol. Instead of diesters of the carboxylic acids one can also use their anhydrides (see Example 4–06).

The reaction between dihydroxy compounds and dicarboxylic acid chlorides is especially useful for the preparation of polycarbonates.[1,2] The commercially very important polycarbonates are obtained from aromatic dihydroxy compounds (bisphenols) and phosgene. In contrast to the products of reaction of aliphatic diols and phosgene they possess high softening points and glass transition temperatures. They are soluble in a number of organic solvents (e.g. in chlorinated hydrocarbons and cyclic ethers) and show considerable resistance to water and mineral acids; but they are degraded by strong alkali, ammonia and amines. The reaction between dihydroxy compounds and phosgene proceeds even at room temperature, and can be accelerated by tertiary amines or quaternary ammonium salts.

$$HO-C_6H_4-C(CH_3)_2-C_6H_4-OH + COCl_2 \xrightarrow{-2HCl} \ldots-O-C_6H_4-C(CH_3)_2-C_6H_4-O-\underset{\underset{O}{\|}}{C}-\ldots$$

The reaction is generally carried out in anhydrous organic solvents (e.g. pyridine) or in mixtures, for example of methylene chloride or benzene with aqueous sodium hydroxide. In this way one obtains polycarbonates with molecular weights of 25 000 to above 75 000.

Example 4–03:
Preparation of a polyester from ethylene glycol and dimethyl terephthalate by melt condensation

Pure ethylene glycol is dried by refluxing for 1 h in the presence of 2 wt. % metallic sodium and is then distilled. Dimethyl terephthalate is recrystallized from methanol and carefully dried in vacuum (m.p. 141–142°C).

---

[1] *H. Schnell*, Angew. Chem. *68* (1956) 633; Polym. Rev. *9* (1964).
[2] *H. Krimm* in *Houben-Weyl 14/2* (1963) 48.

9.7 g (0.05 mol) of dimethyl terephthalate, 7.1 g (0.115 mol) of ethylene glycol, 0.015 g of pure anhydrous calcium acetate and 0.04 g of pure antimony trioxide are weighed into a 50 ml round-bottomed flask. The flask is then fitted with a Claisen still head, an air-cooled condenser, a vacuum adaptor and a graduated receiver (or a measuring cylinder). The air is removed by evacuating and filling with nitrogen, and the components melted on an oil or metal bath at 170°C. The Claisen head is now immediately fitted with a capillary tube (attached to a ground joint) that reaches to the bottom of the flask and a slow stream of nitrogen is passed through. The transesterification sets in almost at once; the progress of reaction is followed by the amount of methanol that has distilled over into the graduated receiver. When the methanol production slackens (after about an hour) the temperature is raised to 200°C for 2 h, whereby the remainder of the methanol distils over. Finally the excess ethylene glycol is distilled over during 15 min at 220°C, the temperature than being raised to 280°C. After 15 min the graduated rceeiver is replaced by a simple round-bottomed flask and the apparatus gradually evacuated while keeping the temperature steady. After another 3 h the polycondensation is complete. The flask is allowed to cool under nitrogen and then broken carefully with a hammer to remove the poly(ethylene terephthalate). The polyester is soluble in *m*-cresol and can be reprecipitated in ether or methanol. The limiting viscosity number is determined in *m*-cresol or in a mixture of equal parts phenol and tetrachloroethane (see Section 2.3.2.1); the softening range is also determined. Fibres can be spun from the melt or pulled out from the melt by means of a glass rod. The preparation of poly-(ethylene terephthalate) can also be carried out in the apparatus described in Example 4–08.

Example 4–04:
Preparation of a polycarbonate from 4,4'-isopropylidenediphenol (bisphenol A) and phosgene by condensation in solution

Under an efficient fume hood a phosgene cylinder is connected to three wash bottles in series: the first and third are reversed and empty while the second is filled with paraffin oil (not glycerol!). To the last wash bottle is connected a 500 ml three-necked flask, fitted with stirrer, thermometer, and gas inlet and outlet. The thermometer must dip into the reaction mixture and can be inserted through a T-piece that also serves as gas outlet. The inlet tube should be as wide as possible (diameter >5 mm) and dip into the mixture just above the paddles of the stirrer in order to prevent clogging by the precipitating pyridine hydrochloride. To the outlet tube is connected an empty wash bottle and bubbler, the tubing from which leads directly to the hood ducting.

22.5 g of pure 4,4′-isopropylidenediphenol (bisphenol A) (m.p. 155–157°C) and 230 ml pyridine are placed in the reaction flask. (Caution: bisphenol A powder is particularly irritating to the eyes!). Phosgene is now passed into the vigorously stirred mixture at such a rate that the bubbles can just be individually distinguished. (On opening the valve of the cylinder and during the later dismantling of the apparatus a gas mask with suitable filter must be worn).

The reaction begins at once and the contents of the flask gradually become opaque as a result of the separation of pyridine hydrochloride. The temperature of reaction should be about 25°C; if it rises above 30°C, the mixture should be cooled with a water bath. Towards the end of the polycondensation (after about 45–60 min) the reaction mixture becomes very viscous. The flow of phosgene is continued for another 5 min or so until it is present in slight excess as indicated by the yellow colour of the complex formed from phosgene and hydrogen chloride. The apparatus is then dismantled taking the precautions already mentioned.

After the reaction has finished the polymer is precipitated by replacing the gas inlet tube with a dropping funnel and running 250 ml of methanol dropwise into the vigorously stirred mixture over a period of about 10 min. The mixture is stirred for a few minutes longer and the polymer then filtered off and dried in vacuum at 50°C. The polymer is redissolved in a 20-fold amount of methylene chloride, precipitated again in methanol and redried at 50°C in vacuum. The polycarbonate is soluble in methylene chloride, chloroform, pyridine, 1,4-dioxane, and tetrahydrofuran; its melting range is between 225 and 250°C. The limiting viscosity number is determined in tetrahydrofuran at 20°C and the molecular weight derived (see Section 2.3.2.1).

4.1.1.4. Preparation and crosslinking (curing) of unsaturated polyesters

Polycondensation of a diol with a dicarboxylic acid, either of which may contain a double bond, results in an unsaturated polyester.[1,2,3,4] For this purpose suitable starting compounds are fumaric acid, maleic acid (or maleic anhydride) and 2-buten-1,4-diol. These can also be used mixed with saturated dicarboxylic acids or diols (copolycondensation) in order to vary the number of double bonds per macromolecule and thereby the properties of the polyester. Unsaturated polyesters are generally prepared by melt condensation. The resulting products are often viscous or waxy substances of relatively low molecular weight.

---

[1] *E. Müller* in *Houben-Weyl 14/2* (1963) 30.
[2] *H.V. Boenig*, "Unsaturated Polyesters: Structure and Properties", Elsevier Publishing Company, Amsterdam, London, New York 1964.
[3] *W. Funke*, Adv. Polym. Sci. *4* (1965) 157.
[4] *K. Demmler*, Angew. Makromol. Chem. *76/77* (1979) 209.

The incorporation of double bonds into polyesters allows the possibility of subsequent reactions. Of particular interest in this context are reactions that lead to crosslinking, which may be ionic or radical in nature.

One of the best examples of ionic crosslinking is the addition of polyhydroxy compounds to double bonds, which sometimes takes place during the actual preparation of the unsaturated polyester. However, radically induced crosslinking reactions are much more important. They may be brought about by the action of oxygen and light, especially in the presence of suitable catalysts such as cobalt (II) compounds ("air drying"). The possibilities are considerably widened if one carries out a crosslinking copolymerization: the unsaturated polyester is dissolved in a radically polymerizable monomer and the polymerization then initiated by addition of a radical-forming system. The temperature required depends on the initiator used. With peroxides such as dibenzoyl peroxide, cyclohexanone peroxide, or cumyl hydroperoxide the reaction is carried out at 70–100°C ("hot curing"); with redox systems it can be done at room temperature ("cold curing"). Suitable redox systems consist of a combination of peroxides with reducing agents that are soluble in organic media, for example metal salts (cobalt or copper naphthenates or octanoates) and tertiary amines (*N*,*N*-dimethylaniline). The most commonly used monomer for this crosslinking graft copolymerization is styrene; allyl compounds such as diallyl phthalate can also be employed. In all cases there are formed transparent, insoluble, three-dimensionally crosslinked products that exhibit good heat stability, and, especially when mixed with glass fibres, good mechanical properties. Their structure can be represented schematically as follows:

$$\ldots -OCO-CH-CH-COO-CH_2-CH_2-OCO-CH-CH-COO-CH_2-CH_2-O- \ldots$$

$$\left(CH_2-CH(C_6H_5)\right)_u \qquad \left(CH_2-CH(C_6H_5)\right)_v$$

$$\ldots -OCO-CH-CH-COO-CH_2-CH_2-OCO-CH-CH-COO-CH_2-CH_2-O- \ldots$$

$$\left(CH_2-CH(C_6H_5)\right)_w \qquad \left(CH_2-CH(C_6H_5)\right)_x$$

$$\ldots -OCO-CH-CH-COO-CH_2-CH_2-OCO-CH-CH-COO-CH_2-CH_2-O- \ldots$$

$$\left(CH_2-CH(C_6H_5)\right)_y \qquad \left(CH_2-CH(C_6H_5)\right)_z$$

The crosslinking graft copolymerization (curing) is generally carried out by first dissolving the unsaturated polyester (about 70 parts) by stirring in the monomer to be grafted (about 30 parts). A high temperature (up to 130°C) must often be used for this purpose, and to prevent premature polymerization it is best to add some inhibitor (0.1–0.5 wt. % hydroquinone or 4-*tert*-butylpyrocatechol). As soon as a homogeneous solution is obtained, it is cooled to the desired polymerization temperature and the initiator (about 1–3 wt.% with respect to the total solution) stirred in. The onset of polymerization is evidenced by gelation of the reaction mixture; the initially added inhibitor does not upset the catalyzed polymerization. The fully polymerized mass (allow up to 24 h according to the reaction conditions) often contains a low proportion of soluble material (homopolymer of the vinyl compound) that can be determined quantitatively by extraction of a well-ground sample using a suitable solvent (e.g. benzene when the monomer is styrene) for a period of 30 min. In this way one obtains an indication of the relative proportions of grafting and crosslinking.

Examples 4–05:
Preparation of an unsaturated polyester and its crosslinking (curing) with styrene

(a) Preparation of the unsaturated polyester
In a 500 ml three-necked flask, fitted with stirrer, nitrogen inlet, thermometer, and condenser for distillation with vacuum adaptor, are placed 80 g (1.05 mol) of propane-1,2-diol, 49 g (0.5 mol) of maleic anhydride, 74 g (0.5 mol) of phthalic anhydride, and 37 mg (0.02%) of hydroquinone (as polymerization inhibitor). The air is removed from the apparatus by evacuating and filling with nitrogen. The mixture is then heated under a slow stream of nitrogen; melting occurs at 80–90°C so that the stirrer can be set in motion. Esterification begins at 180–190°C as indicated by distillation of water. This temperature is maintained until the acid number (see Example 4–01) of the polyester has fallen to 50; this takes about 5–6 h. The melt is now cooled to 140°C and 100 g of monomeric styrene (to which 0.03% hydroquinone has been added after distillation) are introduced with stirring over a period of 1 min. The mixture is cooled immediately to room temperature with a water bath in order to prevent premature polymerization. One thus obtains a viscous, colourless to pale yellow, solution of the unsaturated polyester in styrene.

(b) Crosslinking (curing) of the unsaturated polyester with styrene
To crosslink the polyester at room temperature (cold curing), 10 g of the polyester solution prepared in (*a*) are placed in a small beaker or flat tin

can and the components of a suitable redox system stirred in one at a time (on no account at the same time!), for example:

(1) 0.06 ml of a 10% solution of cobalt (II) naphthenates (or octanoates) in styrene, and
(2) 0.2 ml of cyclohexanone peroxide or butanone peroxide (e.g. as 50% solution in dibutyl phthalate);

or

(1) 200 mg of dibenzoyl peroxide and
(2) 0.05 ml of pure dimethylaniline.

After 30–40 min the mixture warms up, indicating that the crosslinking graft copolymerization has begun. The sample gels and after 1 h has become almost solid, most of the styrene having polymerized.

To carry out the crosslinking at higher temperatures (hot curing), 0.1 g of dibenzoyl peroxide is dissolved in 10 g of the polyester solution prepared in (*a*) and heated to 80°C. The polymerization sets in after a few minutes (gelation) and is essentially complete after 15 min. The sample so obtained is not yet fully polymerized and to obtain optimum rigidity an after-curing treatment is necessary. A 1–2 h period at 70–100°C is suitable.

1 g each of the well-ground samples (see Section 2.4.1) obtained by cold-, hot-, and after-curing are boiled with 10 ml of benzene for 30 min, filtered after cooling and washed with benzene. The samples are thoroughly dried in vacuum at 60°C and the weight loss determined. The benzene solutions are dropped into methanol and any extracted polystyrene is thereby precipitated. The swellability of all samples is determined in an organic solvent.

Glass-fibre-reinforced plates can also be made in the above manner. For this purpose a matte of glass fibres is impregnated with the polyester solution already containing the radical initiator and cured at the appropriate temperature.

4.1.1.5 Preparation and crosslinking (curing) of alkyd resins

Alkyd resins[1] are defined as branched or crosslinked polyesters obtained, for example, by polycondensation of a dicarboxylic acid with a polyfunctional alcohol. Branching and crosslinking occur consecutively in a controllable manner. Thus in the polycondensation of glycerol with phthalic acid or phthalic anhydride, there is first formed a branched polyester that remains soluble and fusible so long as the polycondensation is interrupted

[1] *Houben-Weyl 14/2* (1963) 40.

before more than about 75 mole % of the hydroxyl or carboxyl groups have reacted. If this degree of condensation is exceeded the branched polyester transforms by further polycondensation (self-crosslinking) into completely insoluble products. Since one is here dealing with crosslinking by polycondensation, a temperature of about 200°C is generally required (baking varnishes), in contrast to crosslinking graft copolymerization of unsaturated polyesters. Of course, if carboxylic acids containing double bonds are incorporated into the alkyd resins they can be crosslinked at lower temperatures (e.g. "oil-modified alkyd resins", see Example 4–07). However, the crosslinking then proceeds by another mechanism.

The dicarboxylic acids normally employed are phthalic acid or its anhydride, or mixtures of them with, for example, adipic acid or unsaturated acids. The polyhydroxy compounds generally used are glycerol, hexanetriol or pentaerythritol. In the preparation and crosslinking of alkyd resins (with the exception of some unsaturated alkyd resins) the reaction is practically always carried out in the melt, taking the precautions discussed in Section 4.1.

Example 4–06:
Preparation and crosslinking of an alkyd resin from glycerol and phthalic anhydride

In a 250 ml three-necked flask, fitted with Claisen head, stirrer, condenser and nitrogen inlet, 74 g (0.5 mol) of phthalic anhydride and 30.5 g (0.33 mol) of anhydrous glycerol are heated to 200°C with stirring under a stream of nitrogen. At 10 min intervals samples of about 3 g are withdrawn from the melt by means of a wide tube fitted with a rubber squeeze-bulb, and poured quickly on to an aluminium foil. The acid number is determined on the cooled resin samples (see Example 4–01). The solubility is also tested in a 1:1 toluene/butyl acetate mixture; for this purpose 1 g resin is stirred with 5 ml solvent and the solubility checked after 1 h at room temperature. It is to be noted whether the polymer is completely soluble, partially soluble, or insoluble. The polycondensation is continued (about 1.5 h) until an acid number of 127–132 is attained, the branched polyester then still being soluble and fusible.

Further condensation, leading finally to crosslinking, is monitored by heating about 1 g of resin in each of ten test tubes at 200°C and determining the solubility and swellability of samples cooled at intervals of a few minutes. The onset of the crosslinking reaction is signified by gelation of the hot samples. The same investigation can also be made with films obtained by pouring the samples on to aluminium foil. Reaction time, acid number and solubility of the samples are tabulated.

Example 4–07:
Preparation of an unsaturated ("air-drying") alkyd resin

Air-drying lacquers, which are crosslinked by the action of atmospheric oxygen at room temperature, and which find application for example in oil paints, often consist of so-called oil-modified alkyd resins. The double bonds required for crosslinking are here incorporated into the chains by including unsaturated monocarboxylic acids in the reactants during the polycondensation.

A 500 ml multi-necked flask, fitted with stirrer, thermometer, dropping funnel, reflux condenser and nitrogen inlet, is flushed with nitrogen, charged with 118 g of linseed oil, and heated to 235°C under nitrogen. 0.5 g of finely powdered lead oxide (PbO) is added next, followed by 26 g (0.28 mol) of glycerol added slowly through the dropping funnel over a period of about 20 min with vigorous stirring. A homogeneous mixture is obtained within 30 min, to which is now added 54 g (0.36 mol) of phthalic anhydride in one shot. The anhydride is melted before its introduction in order to minimize ingress of trapped air. The dropping funnel is replaced by a water separator before polycondensation commences.

The mixture is heated with stirring at 250°C until the acid number is less than 5 (see Example 4–01). During this time, in order to prevent the ingress of atmospheric oxygen, the nitrogen flow should never be completely stopped. After 6 h the polycondensation is terminated, the resulting unsaturated alkyd resin being a viscous liquid which is soluble in aromatic hydrocarbons, butyl acetate and acetone.

A 50% solution of the unsaturated alkyd resin in benzene, to which 1.5% (with respect to resin) cobalt naphthenates has been added as autoxidation accelerator ("drier"), gives a smooth transparent film of lacquer when the solution is spread out thinly on a metal surface and exposed to the action of atmospheric oxygen at room temperature.

A white lacquer can be made by throughly mixing in a mortar 50 g of the resin solution as obtained above, together with 75 g of Lithopone ($ZnS/BaSO_4$ pigment), 25 g of zinc oxide, 27 g of turpentine oil and 1 g of cobalt naphthenates; this mixture forms a film of lacquer under the same conditions as the unpigmented resin.

4.1.2. Polyamides

Polyamides are macromoleculcs whose CRU's are joined together by amide groups, for example:

$$H_2N{-}(CH_2)_x{-}\underset{\displaystyle H}{\underset{|}{N}}{-}\left[\underset{\displaystyle O}{\underset{\|}{C}}{-}(CH_2)_y{-}\underset{\displaystyle O}{\underset{\|}{C}}{-}\underset{\displaystyle H}{\underset{|}{N}}{-}(CH_2)_x{-}\underset{\displaystyle H}{\underset{|}{N}}\right]_n{-}\underset{\displaystyle O}{\underset{\|}{C}}{-}(CH_2)_y{-}COOH$$

It has become the custom to name polyamides according to the number of carbon atoms of the diamine component (first named) and of the dicarboxylic acid. Thus the condensation polymer from hexamethylenediamine and adipic acid is called polyamide-6,6 (or Nylon-6,6), while the corresponding polymer from hexamethylenediamine and sebacoic acid is called polyamide-6,10 (Nylon-6,10). Polyamides resulting from the polycondensation of an aminocarboxylic acid or from ring-opening polymerization of lactams are indicated by a single number; thus polyamide-6 (Nylon-6) is the polymer from ε-aminocaproic acid or from ε-caprolactam.

Many properties of polyamides are attributable to the formation of hydrogen bonds between the —NH— and —CO— groups of neighbouring macromolecules. This is evidenced by their solubility in special solvents (sulfuric acid, formic acid, *m*-cresol), their high melting points (even when made from aliphatic components) and their resistance to hydrolysis. In addition, polyamides with regular CRU's crystallize very readily.

The melting points of aliphatic polyamides depend on the chain length of the starting materials: the melting point falls with increasing distance between the amide groups in the macromolecule, but polyamides made from components with an even number of carbon atoms melt at a higher temperature than the neighbouring odd-numbered members of the series. Alkyl side chains (on carbon or nitrogen) lower the melting point, but at the same time improve the solubility. Aliphatic polyamides are used, above all, for the manufacture of fibres and of components that have to withstand high mechanical stress. Aromatic polyamides, for example poly(iminocarbonyl-1,4-phenylene) [poly(1,4-benzamide)], have recently claimed attention as fibres with unusually high elastic modulus.[1]

The following are the main reactions called upon for the preparation of polyamides,[2,3,4] taking into account the factors discussed in Sections 2.1.5.1, 2.1.5.2 and 4.1:

- polycondensation of ω-aminocarboxylic acids,
- polycondensation of diamines with dicarboxylic acids,
- polycondensation of diamines with derivatives of dicarboxylic acids (e.g. acid chlorides),
- ring-opening polymerization of lactams (see Section 3.2.4.4).

In principle the attainment of chemical equilibrium can be accelerated by catalysts; however, in contrast to polyester formation, catalysts are not

---

[1] *P.W. Morgan*, "Macromolecules" *10* (1977) 1381, 1390, 1396.

[2] *H. Rinke* and *E. Istel* in *Houben-Weyl 14/2* (1963) 99.

[3] *V.V. Korshak* and *T.M. Frunze*, "Synthetic Hetero-chain Polyamides", Oldenbourne Press, London 1964.

[4] Kunststoffe-Handbuch, Vol. VI (1966), Polyamide.

absolutely essential in the above-mentioned polycondensations. The first two types of reaction are generally carried out in the melt; solution polycondensations at higher temperature, e.g. in xylenol or 4-*tert*-butylphenol are of significance only in a few cases on account of the poor solubility of polyamides. On the other hand polycondensation of diamines with dicarboxylic acid chlorides can be carried out either in solution at low temperature or as interfacial condensations.

4.1.2.1. Polyamides from ω-aminocarboxylic acids

The formation of polyamides by elimination of water from aminocarboxylic acids at high temperature (see Example 4–08) is generally only possible with acids having more than four methylene groups between the amino- and carboxyl-groups; under these conditions α- to δ- (2- to 5-) amino-carboxylic acids undergo preferential ring closure (e.g. glycine→2,5-dioxopiperazine; γ-aminobutyric acid→4-butanelactam [γ-butyro-lactam]).

Example 4–08:
Polycondensation of ε-aminocaproic acid in the melt

The polycondensation can be performed in the apparatus described in Example 4–03. For the preparation of small amounts of Nylon-6 the following procedure is especially suitable, and at the same time permits a quantitative determination of the water eliminated.

15 g of ε-aminocaproic acid are weighed into a thick-walled tube of about 50 ml capacity, carrying a ground joint. The tube is fitted with as small a distillation adaptor as possible and connected via a short tube to a weighed U-tube filled with anhydrous granular calcium chloride. The adaptor and attached tube are wrapped with aluminium foil and asbestos string. A slow stream of nitrogen is passed through the distillation head which serves both to carry away the eliminated water and to provide a buffer against atmospheric oxygen, the presence of which leads to strongly coloured products.

The tube is now immersed nearly up to its neck in an oil or metal bath and the ε-aminocaproic acid melted at a bath temperature of 220°C. Then it is heated quickly to 260°C and held at this temperature under a continuous stream of nitrogen for 15 min. Should any water condense on the wall of the connecting tube leading to the calcium chloride tube, it can be removed by warming with a hot-air blower before the hot bath is taken away and the melt allowed to cool under nitrogen. Finally the tube is broken and the solid polyamide removed. The calcium chloride tube is weighed again to determine the amount of water eliminated. The experiment is repeated twice more, extending the reaction times to 30 and 60 min respectively. The limiting viscosity numbers of the three samples are determined in concentrated sulfuric acid at 30°C ($c = 10\,g\,l^{-1}$) using an Ostwald viscometer (capillary diameter 0.6 mm). The increase of the η_{sp}/c values with reaction time is a measure of the progress of the polycondensation.

The resulting Nylon-6 has a melting point of 215°C; fibres can be drawn from the melt. The product still contains small amounts of cyclic and linear polyamides that can be extracted from the finely ground product with methanol in a Soxhlet apparatus (12 h). The extract contains ε-caprolactam as well as cyclic and linear oligomers up to pentamer; these can be quantitatively determined by evaporating off the methanol in vacuum. The ε-caprolactam is then dissolved out by digesting with anhydrous ether. The residue is taken up in methanol (1% solution) and passed through a column filled with a strongly acid cation exchange resin[1] (elute with the same volume of methanol). The linear oligomers are retained on the column while the cyclic oligomers are quantitatively determined by evaporating down the eluate. They can be separated by paper chromatography[2] using tetrahydrofuran/cyclohexane/water mixtures (volume ratio 186 : 14 : 10); the R_f values increase with decreasing ring size.

4.1.2.2. Polyamides from diamines and dicarboxylic acids

The polycondensation of diamines with dicarboxylic acids can be carried out simply by melting together the highly purified components under nitrogen at 180–300°C. However, considerable amounts of diamine can be carried over with the water which distils off, especially towards the end of the reaction when vacuum is applied; the equivalence of the reaction partners, which is a prerequisite for the attainment of high molecular weight, is thereby disturbed. Therefore, it is advantageous to start with an excess of diamine; the optimum excess must be determined for each individual case.

An elegant variation of this procedure is to carry out the polycondensation of the salt of a diamine and a dicarboxylic acid (see Example 4–09). Here the formation of salt (which is also the first step in the direct polycondensation of diamine and dicarboxylic acid) and the

---

[1] *P.F. van Velden, G.M. van der Want, D. Heikens, C.A. Kruissink, P.H. Hermans* and *A.J. Staverman*, Rec. Trav. Chim. Pays-Bas *74* (1955) 1376.

[2] *M. Rothe*, J. Polym. Sci. *30* (1958) 227.

polycondensation are carried out as two separate steps. The salts can be obtained in good crystalline form most simply by mixing equimolar amounts of diamine and dicarboxylic acid in a solvent in which the salt formed is insoluble (e.g. ethanol). In order to attain high molecular weights by such polycondensations the salts should be as neutral as possible (exactly equivalent amounts of amine and acid) and very pure (recrystallize, for example, from mixtures of ethanol/water).

Example 4–09:
Preparation of Nylon-6,6 from hexamethylene-diammonium adipate (AH-salt) by condensation in the melt

(a) Preparation of the AH-salt
36 g (0.245 mol) of pure adipic acid (m.p. 152°C) are dissolved in 300 ml of 95% ethanol and 29 g (0.25 mol) of pure hexamethylenediamine are dissolved in a mixture of 80 ml of ethanol and 30 ml of water. Both solutions are filtered if they are not perfectly clear.

The adipic acid solution is now placed in a 500 ml beaker and the diamine solution added dropwise with stirring over a period of 8 min, during which time the solution warms up to 40–45°C. After stirring for a further 30 min and allowing to cool, the crystallized AH-salt is filtered off with suction, washed twice with 95% ethanol and dried in vacuum. Yield: 90–95%; m.p. 183°C (with loss of water); pH value of 9.5% aqueous solution: 7.62. Impure AH-salt can be recrystallized from ethanol/water mixtures (volume ratio 3:1).

(b) Polycondensation of the AH-salt
A 50 ml pear-shaped flask, fitted with distillation head, air condenser, vacuum adaptor, and receiver, is filled three-quarters full with AH-salt and the air removed by evacuation and filling with nitrogen. It is then heated under nitrogen for 1 h on a metal bath at 220°C, and for three further hours at 260–270°. After cooling, the flask is broken carefully with a hammer. The polyamide from adipic acid and hexamethylenediamine melts at 265°C. It can be spun from the melt into threads which can be cold drawn. The viscosity number is determined in *m*-cresol, concentrated sulfuric acid or in 2 M KCl in 90% formic acid (see Section 2.3.2.1).

4.1.2.3. *Polyamides from diamines and dicarboxylic acid derivatives*

As in the preparation of polyesters, so in the preparation of polyamides, the reaction temperature can be considerably reduced by using derivatives of dicarboxylic acids instead of the free acids. Especially advantageous in this connection are the dicarboxylic acid chlorides which react with diamines at room temperature by the Schotten-Baumann reaction (also see Section 2.1.5.2); this polycondensation can be carried out in solution as well as by a special procedure known as interfacial polycondensation.

Example 4–10:
Preparation of Nylon-6,10 from hexamethylenediamine and sebacoyl dichloride

(a) By polycondensation in solution at low temperature (precipitation polycondensation)

The alcohol-free chloroform required for this experiment is first prepared by running 200 ml of chloroform through a column filled with basic aluminium oxide.

Sebacoyl dichloride is obtained as follows. 60 g of sebacoic acid and 150 g of thionyl chloride are refluxed on a water bath for 2 h. The excess thionyl chloride is then distilled off and the sebacoyl dichloride fractionated in vacuum (b.p. 142°C/2 torr).

5.3 g (50 mmol) of hexamethylene diamine and 15.3 ml of triethylamine are dissolved in 100 ml of alcohol-free chloroform in a 500 ml three-necked flask fitted with stirrer and reflux condenser. With vigorous stirring, 10.7 ml (50 mmol) of sebacoyl dichloride in 40 ml of alcohol-free chloroform are then added as quickly as possible (in about 10 s) at room temperature, stirring being continued for another 5 min. The polycondensation occurs instantaneously with considerable heat evolution and precipitation of the condensation polymer. The reaction mixture is cooled and filtered, the product being washed successively with chloroform, petroleum ether, 1 M hydrochloric acid, water and 50% acetone, and is finally dried in vacuum at 50°C. Yield: about 70%; a further low-molecular-weight fraction can be isolated from the filtrate by shaking with petroleum ether.

(b) By interfacial polycondensation

A solution of 3 ml (14 mmol) of freshly distilled sebacoyl dichloride (for preparation see (*a*)) in 100 ml of carbon tetrachloride are placed in a 250 ml beaker. A solution of 4.4 g (38 mmol) of hexamethylenediamine in 50 ml of water is carefully run on to the top of this solution, using a pipette. (The aqueous solution can be made more readily visible by colouring it with a few drops of phenolphthalein solution). A polyamide film is immediately formed at the interface and can be pulled out from the centre with forceps and laid over some glass rods; it can now be pulled out continuously in the form of a hollow strand and wound up on to a spool driven by a slow-running motor. The polycondensation comes rapidly to a standstill if the motor is stopped, but immediately recommences, even after some hours, when the motor is restarted.

The drawn-off polyamide strand is thoroughly washed in 50% ethanol or in acetone, then with water and dried in vacuum at 30°C. The Nylon-6,10 so obtained has the same physical properties as that obtained by condensation in solution (see under *(a)*) or by melt condensation. It melts at 228°C and can be spun from the melt by pulling with a glass rod; the threads can be cold drawn.

Example 4–11:
Microencapsulation of a dyestuff by interfacial polycondensation

By microencapsulation[1] is meant the envelopment of liquid droplets or solid particles with natural or synthetic polymers. The encapsulation of a substance with a polymer membrane is undertaken for various reasons, for example as protection against moisture, or to obtain delayed dissolution of fertilizers, herbicides or drugs by microencapsulation with semi-permeable membranes.

Various techniques have been developed for the preparation of microcapsules with diameters of 1–5000 μm; one of these involves the method of interfacial polymerization. The following example describes the microencapsulation of a dyestuff, which has practical application in the manufacture of carbon-free copy paper.

(a) Preparation of dye-containing microcapsules
The following ingredients are prepared in parallel:

(1) Dispersion system:
1 g of low-molecular weight poly(vinyl alcohol), which serves both as dispersing agent and protective colloid, is dissolved in about 150 ml of distilled water in a narrow 600 ml beaker. It is stirred with a paddle stirrer at room temperature for about 2 hours until dissolved. The polymer used should be a partially (88%) hydrolyzed poly(vinyl acetate),[2] whose 4% aqueous solution has a viscosity of 0.4 Pa s (4 cP).

(2) Acid chloride solution:
7.7 g (0.04 mol) of terephthalyl dichloride are finely powdered in a mortar and dissolved in 40 g of dibutyl phthalate by stirring at 70°C, taking care to exclude moisture. Dissolution takes about 30 min; any insoluble material is filtered off.

---

[1] *N. Sliwka*, Angew. Chem. *87* (1975) 556.
[2] For example Mowiol 4–88 from Hoechst A.G., Frankfurt-am-Main, W. Germany.

(3) Dyestuff solution:
1 g of crystal violet lactone (3,3-bis(*p*-dimethylaminophenyl)-6-dimethylaminophthalide[1]) is dissolved in 10 g of dibutyl phthalate by stirring at 90°C for 15 min. The solution is brown/red.

(4) 3 g (0.03 mol) of *N*-(2-aminoethyl)ethylenediamine, (diethylenetriamine), and 3.2 g of NaOH pellets are dissolved in 20 ml of distilled water with cooling; time required: 15 min. (Caution: use hood and avoid all contact of the amine with the skin or clothes!)

The paddle stirrer in the dispersion vessel is now replaced by a high-speed disperser with a strong shearing action. To minimize the formation of foam the contents are stirred at about 2000 r.p.m. Solutions (2) and (3) are then mixed together with gentle stirring (magnetic stirrer). This water-insoluble mixture is run dropwise into the dispersion vessel over a period of 30 s, the stirrer speed having been raised to about 7000 r.p.m. as the first drops enter. After a further 30 s the emulsion drop size has adjusted to about 6–10 μm. The particle size can be checked by examination of a sample under a transmission microscope (magnification about 500) with a built-in measuring scale.

The amine solution (4) is now immediately run in dropwise over a period of 30 s and the stirrer speed reduced to 2000 r.p.m. After 2–3 min the high-speed stirrer is replaced by a normal paddle stirrer and the dispersion stirred (500 r.p.m.) for a further 30 min at room temperature to complete the interfacial polycondensation between terephthalyl dichloride and diethylenetriamine.

(b) Testing the microcapsules

The microcapsules can be seen under the microscope as individual spherical particles or as small agglomerates.

The use of the microcapsules to make carbon-free copy paper can be demonstrated in principle. The dispersion is painted on to a piece of typing paper with a fine paint brush. To prevent the paper from curling up it is fixed to a glass plate with adhesive tape. After drying the paper with a hot air drier, the coated side is laid on a silica gel plate. The other side is then written on with gentle pressure. This breaks down the microcapsules, causing the colourless colour-forming solution to flow out on to the underlying silica gel plate, developing a blue-coloured imprint; thus a blue copy of the writing appears on the silica gel plate.[2] One must take care that the painted paper is completely dry, otherwise no copy appears. The

[1] *H. Moriga* and *R. Oda*, Kogyo Kagaku Zasshi *67* (1964) 1050; Chem. Abstr. *62*, 2851f.

[2] With paper as underlay the copying effect can only be made visible by a complicated pretreatment; the silica gel plate serves for simple demonstration purposes.

resulting blue coloration on the silica gel plate can be removed reversibly by moistening.

4.1.3 Preparation of polyurethanes (polycarbamates) by polycondensation

Linear polyurethanes are generally prepared by the addition polymerization reaction of diols with diisocyanates (see Section 4.2.1.1). However there is also the possibility of synthesizing them by condensation polymerization of chloroformic acid diesters with diamines in the presence of compounds that react with the liberated hydrogen chloride:

$$\mathrm{Cl{-}\underset{\underset{O}{\|}}{C}{-}O(CH_2)_x{-}O{-}\underset{\underset{O}{\|}}{C}{-}Cl + H_2N{-}(CH_2)_y{-}NH_2 \xrightarrow{NaOH} \left[\underset{\underset{O}{\|}}{C}{-}O{-}(CH_2)_x{-}O{-}\underset{\underset{O}{\|}}{C}{-}NH{-}(CH_2)_y{-}NH\right]}$$

Since this reaction can proceed in aqueous medium, it can with advantage be carried out as an interfacial polycondensation (see Section 2.1.5.2); this procedure is especially valuable for the preparation of high-molecular-weight or thermally unstable polyurethanes. If secondary diamines are used, one then obtains *N*-alkylated polyurethanes which are remarkably resistant to hydrolysis and possess special properties on account of the absence of hydrogen bonds. They cannot be prepared by addition polymerization reactions of diisocyanates.

Example 4–12:
Preparation of a linear polyurethane from ethylene dichloroformate and hexamethylenediamine by interfacial polycondensation

(a) Preparation of ethylene dichloroformate
A phosgene cylinder is connected to three wash bottles under an efficient hood. The middle bottle contains paraffin oil and serves as bubble counter; the other two are left empty. The third wash bottle is connected to a cold trap immersed in a methanol/dry-ice bath. The cold trap is connected in turn to an empty wash bottle and another filled with concentrated ammonia solution. A gas mask with a suitable filter must be worn when opening the valve of the cylinder and later when dismantling the apparatus.

55–60 ml (about 80 g) of phosgene are condensed in the cold trap and then poured carefully into a 100 ml two-necked flask immersed in an ice/salt mixture. In one neck is placed an immersion thermometer dipping into the phosgene. The other neck carries a Claisen head with dropping funnel and thermometer; a condenser for distillation, with adaptor and round-bottomed flask as receiver, is attached. 10 g of anhydrous ethylene glycol are dropped into the phosgene, the temperature of the mixture being held below 5°C. The mixture is allowed to warm slowly to room temperature. The evaporating phosgene passes through an empty wash bottle into a cold trap immersed in a methanol/dry-ice bath and protected by an empty wash bottle and another filled with ammonia solution. (The phosgene condensed in the cold trap is destroyed in the open after pouring ammonia solution on the ground). The reaction product is evacuated on the filter pump at 40–50°C for 30 min after inserting a capillary to assist boiling. Finally the ethylene dichloroformate is distilled under vacuum (oil pump); b.p. 81°C/0.8 torr.

(b) Preparation of a polyurethane by interfacial polycondensation
A solution of 5.8 g (0.05 mol) of pure hexamethylenediamine, 10.6 g (0.1 mol) of sodium carbonate and 1.5 g of an emulsifier (e.g. sodium dodecyl sulfate) in 150 ml of water is cooled to 5°C. A solution of 9.35 g (0.05 mol) ethylene dichloroformate in 125 ml of benzene is cooled to 10°C and dropped quickly into the vigorously stirred amine solution (best using a high-speed mixing apparatus). After a few minutes the polyurethane is filtered off under suction, washed with water and dried in vacuum. The flaky polyurethane (yield 70%) has a melting point of about 180°C. The limiting viscosity number is determined in *m*-cresol.

4.1.4. Phenol-formaldehyde resins

The high-molecular-weight products formed by the condensation of phenols with carbonyl compounds (especially with formaldehyde) are known as phenolic resins.[1] They are mixtures of structurally non-uniform compounds that are initially soluble and fusible but which can become crosslinked (cured) by subsequent reactions. One distinguishes between acid- and base-catalyzed condensations, since they lead to different end products; the properties of the condensation polymer are also affected by the mole ratio of phenol to formaldehyde.

Condensation in acid medium gives soluble, fusible phenolic resins, with an average molecular weight between 600 and 1500, and a structure consisting essentially of phenol residues linked by methylene groups in the ortho- and para-positions; they are called Novolaks. No further condensation occurs on heating this product for longer periods; but it can be

[1] Detailed compilations: *R. Wegler* and *H. Herlinger* in *Houben-Weyl 14/2* (1963) 193; Kunststoff-Handbuch, Vol. X; *A. Knop* and *W. Scheib*, "Chemistry and Application of Phenolic Resin", Springer Verlag, Berlin, Heidelberg, New York 1978.

crosslinked by reaction with suitable polyfunctional components, e.g. with additional formaldehyde. On the other hand in basic medium one obtains soluble, fusible hydroxymethylphenols, with a molecular weight between 300 and 700, containing one or more benzene nuclei; they are called Resols. In contrast to the Novolaks, the products undergo crosslinking through their reactive groups on heating, giving insoluble, infusible products (Resites).

By far the most important phenolic resins are those made from phenol and formaldehyde. They exhibit high hardness, good electrical and mechanical properties and chemical stability.

4.1.4.1. *Acid-catalyzed phenol-formaldehyde condensation (Novolaks)*

The condensation of phenol with formaldehyde in acid medium is an exothermic reaction (97 kJ/mol) yielding soluble phenolic resins which melt between 100 and 140°C depending on the molecular weight (600–1500). The structure of the Novolaks can only be indicated schematically, since the coupling of the phenol nuclei through methylene groups can occur in both ortho- and para-positions; coupling also occurs through oxymethylene groups so that there are a multitude of possibilities[1]:

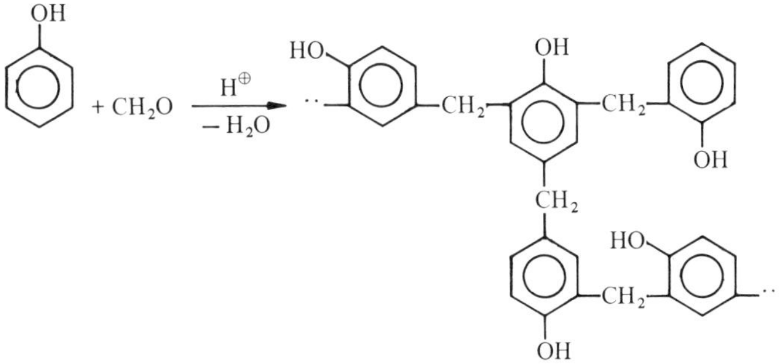

$$C_6H_5OH + CH_2O \xrightarrow[-H_2O]{H^{\oplus}} \cdots$$

OH HO OH CH$_2$ CH$_2$ OH CH$_2$ HO CH$_2$ OH

This polycondensation is practically always carried out in aqueous solution, either by dropping an approximately 30% solution of formaldehyde at 80–100°C slowly into an acidified phenol solution, or by mixing the components at room temperature and then heating. In order to avoid premature crosslinking, the mole ratio of phenol/formaldehyde should not be higher than 1:0.8. After all the formaldehyde has been added the mixture is allowed to react until the odour of formaldehyde has disappeared. The aqueous layer is separated off and the product washed with hot water to remove the acid as completely as possible. The residual water

[1] Monodisperse condensation polymers of phenol-formaldehyde have also been prepared: *H. Kämmerer* and *H. Schweikert*, Makromol. Chem. *60* (1963) 155; *83* (1965) 188; *H. Kämmerer*, Angew. Chem. *77* (1965) 965.

and the unconverted phenol are then removed under vacuum at higher temperature. The resins so obtained are generally soluble in alcohols, lower esters, ketones, and dilute alkali. As catalyst one should choose an acid that is easy to remove from the final product by washing or distillation. Hydrochloric acid, oxalic acid, or mixtures of the two are very suitable. Oxalic acid can be removed both by washing and by heating to 180°C.

Since Novolaks have no reactive groups that can lead to self-crosslinking, they must be subsequently crosslinked (cured) by addition of suitable di- or poly-functional compounds that react with phenols. Amongst these are formaldehyde, hydroxymethyl compounds, amino-benzyl alcohols, hydroxy- and amino-benzylamines, and bis(hydroxy-benzyl) ether. The most commonly used crosslinking agent is hexa-methylenetetramine (urotropin). Crosslinking is effected by mixing the components and heating for a short time (a few minutes) to 150–220°C. The structure of the crosslinked phenol-formaldehyde resin is very complex. If crosslinking is carried out with hexamethylenetet-ramine the crosslinks consist mainly of dibenzylamine and tertiary amine bridges:

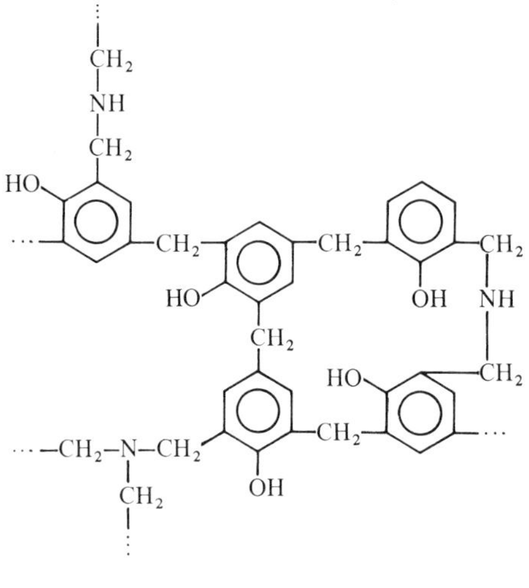

Example 4–13:
Acid-catalyzed phenol-formaldehyde condensation

130 g (1.38 mol) of phenol, 92 g of a 37% aqueous formaldehyde solution (1.13 mol), 15 ml of water and 2 g of oxalic acid dihydrate are inserted in a 500 ml three-necked flask, fitted with stirrer and reflux condenser, and heated under reflux on an oil bath for 1.5 h. 300 ml of water are then added, stirred briefly and allowed to cool, whereby the condensation

polymer settles out. The aqueous layer is separated and the residual water distilled off at 50–100 torr while slowly raising the temperature to 150°C. This temperature is maintained (at most for 1 h) until test samples solidify on cooling. The resin is poured out while still warm and solidifies to a colourless, brittle mass, soluble in alcohol.

The Novolak obtained is used to fabricate a moulding by mixing with sawdust and hexamethylenetetramine as follows: 50 g of finely ground Novolak, 50 g of dry sawdust, 7 g of hexamethylenetetramine, 2 g of magnesium oxide (to trap residual acid from the condensation), and 1 g of calcium stearate (as lubricant) are thoroughly mixed (best in a ball-mill or analytical mill) and then heated in a mould at 140 bar for 5 min at 160°C. The resulting moulding is infusible and insoluble.

4.1.4.2. Base-catalyzed phenol-formaldehyde condensation (Resols)

Condensation of phenols with an excess of formaldehyde in basic medium yields phenolic resins (Resols), with an average molecular weight of 300–700, which are generally soluble in water or alcohol. Like Novolaks they consist essentially of phenol nuclei linked to one another through methylene groups; they differ, however, especially in their content of hydroxymethyl groups. The latter make a number of reactions accessible, e.g. esterification and ether formation. The coupling of the phenol residues and the incorporation of hydroxymethyl groups again occurs in the ortho- and para-positions, with neither position preferred in alkaline medium. The structure of Resols may be indicated schematically as follows:

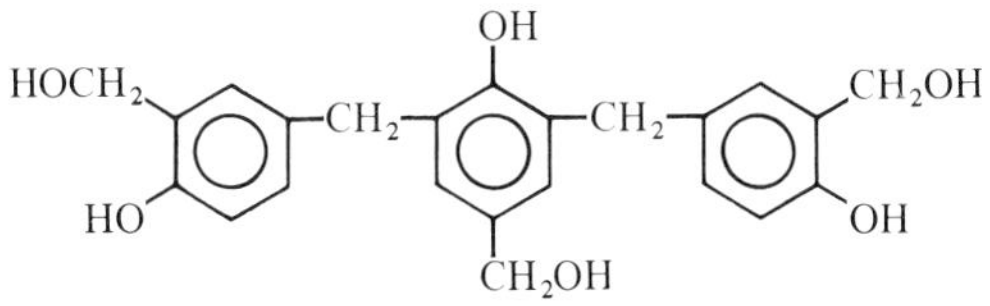

OH
$HOCH_2$ — CH_2 — CH_2 — CH_2OH
HO OH
CH_2OH

Resols are prepared by heating an aqueous phenol solution with a 10–50% molar excess of formaldehyde in basic medium. Suitable catalysts are the hydroxides of the alkali and alkaline earth metals, also primary and secondary amines, and especially ammonia; the use of amines can lead to their incorporation in the phenolic resin. The reaction temperature should be kept below 70°C if possible, otherwise the water solubility of the resulting Resol is reduced. Reaction times of 2–5 h are generally required. In contrast to the Novolaks, Resols can be transformed into insoluble, infusible products by self-crosslinking under mild conditions; water is eliminated from the hydroxymethyl groups, forming dimethyleneether bridges.[1] Crosslinking, accompanied by structural transformations (formaldehyde elimination) and rearrangements, occurs on heating to 150–200°C; depending on the chemical composition and structure of the Resol, this requires a few minutes to several hours to achieve complete hardening.

Example 4–14:
Base-catalyzed phenol-formaldehyde condensation

94 g (1.0 mol) of phenol, 125 g of a 37% aqueous formaldehyde solution (1.54 mol) and 4.7 g of $Ba(OH)_2{,}8H_2O$ are introduced into a 500 ml three-necked flask, fitted with reflux condenser, stirrer and thermometer, and heated for 2 h at 70°C. The pH value of the reaction mixture is then adjusted to 6–7 by addition of 10% sulfuric acid. Before testing the pH the stirrer is switched off and the aqueous phase allowed to separate. The reflux condenser is now replaced by a condenser for distillation and the water distilled off at 30–50 torr; the temperature must not exceed 70°C. The distillation is interrupted at intervals of 20 min in order to take a sample; if the sample sets to a hard, non-sticky mass, the pre-condensation is stopped. (When the resin becomes viscous, samples should be taken every 10 min; the polymer must be quite viscous at 70°C and still fluid when the condensation is stopped after about 20 h). The contents of the flask are then poured out and allowed to solidify to a soluble, fusible mass (Resol).

On further heating of the resin the condensation continues; one obtains insoluble but still swellable and fusible products that are called Resitols. Still further heating leads finally to the strongly crosslinked, insoluble and no longer fusible Resites. A sample of the polymer is heated in a test tube for a long time at 100°C until a hard block is formed; it is freed by breaking the tube. A comparison is made of the solubility of the Resite and Resol in ethanol. Mouldings can be made by mixing Resol with additives such as sawdust, chalk, pigments etc., followed by heating under pressure.

4.1.5. Urea- and 2,4,6-triamino-1,3,5-triazine (melamine)-formaldehyde condensation products[2]

4.1.5.1. *Urea-formaldehyde resins*

Condensation of urea with formaldehyde leads to products of various structures and properties according to the experimental conditions (pH,

---

[1] *H. Kämmerer*, *M. Grossmann* and *G. Umsonst*, Makromol. Chem. *39* (1960) 39.
[2] Detailed compilation: *R. Wegler* in *Houben-Weyl 14/2* (1963) 319.

temperature, reaction time, mole ratio of components). The condensation is generally carried out in basic medium, resulting essentially in the formation of hydroxymethyl compounds; some oxymethylene groups are also formed, particularly on heating. Mono-, di-, and trihydroxymethyl urea have been proved to be primary products of this condensation; whether tetrahydroxymethylurea is also formed is as yet uncertain:

$$NH_2{-}\underset{\underset{O}{\|}}{C}{-}NH_2 + CH_2O \xrightarrow{OH^{\ominus}} NH_2{-}\underset{\underset{O}{\|}}{C}{-}NHCH_2OH + HOCH_2NH{-}\underset{\underset{O}{\|}}{C}{-}NHCH_2OH$$

4-Oxo-perhydro-1,3,5-oxadiazine (urone) is also formed by intramolecular condensation:

$$\begin{matrix} HN{-}CO{-}NH \\ | \qquad\quad | \\ CH_2 \quad CH_2 \\ | \qquad\quad | \\ OH \quad\; OH \end{matrix} \xrightarrow{-H_2O} \begin{matrix} \quad CO \\ HN \quad NH \\ | \qquad\; | \\ H_2C \quad CH_2 \\ \quad O \end{matrix}$$

These "pre-condensates" are soluble in water and alcohol; they are transformed by further condensation with elimination of water, first into high-molecular-weight, poorly soluble materials and finally into crosslinked insoluble products. The structure of the crosslinked (hardened) urea-formaldehyde resins is not yet entirely understood.

The soluble hydroxymethyl compounds can be chemically modified, before crosslinking, by reaction with monofunctional compounds (e.g. by esterification or ether formation). The properties of the starting materials as well as the crosslinked end products can thereby be substantially altered. For example, by partial etherification with butanol the hydroxymethyl compounds, originally soluble only in polar solvents, become soluble also in non-polar solvents (toluene), without losing their ability to undergo self-crosslinking.

Urea-formaldehyde resins are generally prepared by condensation in aqueous basic medium. Depending on the intended application a 50–100% excess of formaldehyde is used. All bases are suitable as catalysts provided they are somewhat soluble in water. The most commonly used catalysts are the alkali hydroxides. The pH value of the alkaline solution should not exceed 8–9, on account of the possible Cannizzaro reaction of formaldehyde. Since the alkalinity of the solution drops in the course of the reaction, it is necessary either to use a buffer solution or to keep the pH constant by repeated additions of aqueous alkali hydroxide. Under these conditions the reaction time is about 10–20 min at 50–60°C. The course of the condensation can be monitored by titration of the unused formaldehyde with sodium hydrogensulfite[1] or hydroxylamine hydrochloride. These determinations must, however, be carried out quickly and at as low temperature as possible (10–15°C), otherwise elimination of formaldehyde from the hydroxymethyl compounds already formed can falsify the analysis. The isolation of the soluble condensation products is not possible without special precautions, on account of the facile back reaction; it can be done by pumping off the water in vacuum below 60°C under weakly alkaline conditions, or better by careful freeze-drying. However, the further condensation to crosslinked products is nearly always performed with the original aqueous solution. This can be done either by heating the neutral solution to 120–140°C (10–60 min) or catalytically in the presence of acids at low temperatures. The catalytic crosslinking (acid hardening) can be carried out not only with free acids (e.g. phosphoric acid), but also with compounds that become acidic on heating (latent hardeners). The latter include sodium salts of halogenated carboxylic acids, esters of phosphoric acid, ammonium chloride and pyridine hydrochloride. Addition of large amounts, for example, of phosphoric acid ($pH \approx 2$), causes crosslinking even at room temperature (cold glue). All crosslinking reactions of urea-formaldehyde resins occur by further condensation of hydroxymethyl compounds with expulsion of water. This water can, during the curing of large mouldings, lead to inhomogeneities and fissures; however, these difficulties can be overcome by addition of water-absorbing fillers such as cellulose or other polyhydric alcohols (Example 4–15).

Example 4–15:
Urea-formaldehyde condensation

(a) Preparation of a urea-formaldehyde resin
120 g (2 mol) of urea are dissolved in 243 g of a 37% aqueous formaldehyde solution (3 mol) heated to 50°C in a 500 ml three-necked flask fitted with thermometer, stirrer and reflux condenser. 10 ml of concentrated ammonia solution are then added and the temperature raised to 85°C. After about 20 min the solution becomes cloudy and the viscosity increases; at the same time the pH value falls to about 5. After a total period of 1 h the heating is removed and a small sample is tested to see if the condensation product is still soluble in water.

Cellulose powder is now stirred into the cooled solution until the mass can still just be stirred. The contents of the flask are then transferred to a large beaker and more cellulose powder is kneaded in by hand (wear rubber gloves!) until a total of 140 g has been added. The crumbled mixture is dried for 24 h at 50°C in the vacuum oven. Finally, the dried product

[1] *G. Lemme*, Chem.-Ztg. *27* (1903) 896.

is finely ground in a mortar, mixed with 1% of ammonium chloride and 1% of zinc stearate (as lubricant) and crosslinked (cured) by heating to about 160°C for 10 min in a simple press at 300–400 bar. A thin transparent plate is obtained that is no longer soluble in or attacked by water; this is more convincing if a small piece is allowed to stand overnight in a beaker of water.

If a heatable press is not available, one may proceed as follows. Two flat iron plates of about 2 cm thickness are heated in an oven to 240°C and then taken out. One of these plates is laid on the lower jaw (use asbestos underlay) of a horizontally mounted vice. It is allowed to cool until a crumb of the filled condensation product no longer decomposes (colours) when placed on the plate. A few grammes of the mixture are then quickly put on the plate, the second plate placed on top followed by a piece of asbestos and the whole tightened in the vice.

(b) Preparation of a urea-formaldehyde foam

178 g (2.2 mol) of a 37% aqueous formaldehyde solution, neutralized with 1 M NaOH, are heated to 95°C in a 250 ml three-necked flask fitted with thermometer, stirrer and reflux condenser. 120 g (2 mol) of urea are dissolved in this solution with stirring, the pH value being adjusted to 8 by addition of 1 M NaOH. The pH falls and after 15 min is adjusted to pH 6 by addition of NaOH. After a further 1.5 h at 95°C it is acidified with 20% formic acid (pH 5) and the condensation stopped after 15 min by cooling and neutralizing with NaOH.

For the preparation of the foam, a solution of 1 g of technical sodium diisobutylnaphthalene sulfonate in 50 ml of 3% phosphoric acid is first made up. 20 ml of this solution are placed in a 1 litre beaker and beaten with a fast-running household mixer until sufficient air has been taken up to give 300–400 ml of creamy dispersion. 20 ml of the prepared urea-formaldehyde resin are next stirred in so that the resin is uniformly dispersed. After 3–4 min, under the action of the acid catalyst, the resin sets to a form permeated with water/air pores. Crosslinking is complete in 24 h. After drying for 12 h at 40°C in a current of air one obtains a brittle plastic foam.

This material is hydrophobic and has a large internal surface. It can take up about 30 times its own weight of petroleum ether. This property can be used to separate volatile and low viscosity mineral oils from oil/water dispersions. To demonstrate this 20 ml of ligroin or petroleum ether is dispersed in 200 ml of water in a tall 400 ml beaker with the aid of a high-speed stirrer, and 5 g of the granulated urea-formaldehyde foam are then stirred in. After 5 min the mixture is filtered through a fluted filtered paper. The aqueous runnings are optically free from dispersed hydrocarbons. A crude-oil/water dispersion can be separated in the same way.

4.1.5.2. 2,4,6-Triamino-1,3,5-triazine (melamine)-formaldehyde resins

Formaldehyde resins with better water- and temperature-stability are obtained if the urea is partly (copolycondensation) or wholly replaced by melamine (aminoplasts). These condensations are likewise carried out mainly in alkaline medium, again yielding soluble "pre-condensates", consisting essentially of *N*-[tris- and hexakis-(hydroxymethyl)] compounds of melamine.

$$\text{melamine } (C_3N_3(NH_2)_3) \xrightarrow[OH^-]{+\,3\,CH_2O} C_3N_3(NH{-}CH_2OH)_3$$

$$C_3N_3(NH{-}CH_2OH)_3 \xrightarrow{+\,3\,CH_2O} C_3N_3[N(CH_2OH)_2]_3$$

$$\text{melamine } (C_3N_3(NH_2)_3) \xrightarrow[OH^-]{+\,6\,CH_2O} C_3N_3[N(CH_2OH)_2]_3$$

These pre-condensates are most stable at pH 8–9; they are transformed by further condensation (essentially by elimination of water from hydroxymethyl groups and free NH groups) into poorly soluble and finally insoluble, crosslinked products. Chemical modification of the soluble pre-condensate, for example by esterification or ether formation, is again possible.

The practical preparation of melamine-formaldehyde resins is done under the same conditions as for urea-formaldehyde resins. Melamine is at first insoluble in the aqueous reaction mixture but dissolves completely as the condensation proceeds. Because of the greater stability of the *N*-hydroxymethylmelamines compared with the corresponding urea compounds the reaction can easily be followed by titration of the unconverted formaldehyde with sodium hydrogensulfite (cf. Section 4.1.5.1).

Crosslinking (hardening) of these pre-condensates can be carried out exactly as for the urea-formaldehyde resins, best at a pH value of 3.5–5. Melamine-formaldehyde condensates crosslink most quickly if prepared using a 2.8–3 fold excess of formaldehyde.

Urea- and melamine-formaldehyde resins are used as mouldings, lacquers, and adhesives (for wood), also as textile additives (increased crease resistance) and paper additives (improved wet strength).

Example 4–16:
Melamine-formaldehyde condensation

(a) Preparation of the polymer
The experimental arrangement consists of a 250 ml three-necked flask equipped with stirrer, double-necked head with reflux condenser and thermometer, and a rubber stopper through which passes a glass rod. 63 g (0.5 mol) of melamine and 150 g of a 40% aqueous formaldehyde solution (2 mol) are placed in the flask; the aqueous suspension is stirred and adjusted to pH 8.5 by adding a few drops of 20% NaOH. The pH is checked using special pH paper (pH 5.5–9.5), taking care to make the colour comparison 2–3 s after moistening the paper. The mixture is heated on a water bath to 80°C within 5–10 min, with continuous stirring. Complete solution is reached at 70 to 80°C. During this warm-up period the fall of pH of the solution must be continually compensated by dropwise addition of 20% NaOH. The stirred solution is now heated at constant pH of 8.5 until the precipitation ratio (see below) reaches 2:2. Next the solution is cooled and filtered from small amounts of insoluble material.

Melamine dissolves in aqueous formaldehyde solution on warming, with the formation of *N*-hydroxymethyl compounds. The latter are crystalline substances that dissolve in hot water but are only slightly soluble in cold water. If a sample of the reaction mixture is cooled immediately after the melamine has been completely taken into solution, the poorly soluble *N*-hydroxymethyl compounds precipitate out.

With further heating, condensation polymers are obtained that at first are completely miscible with water. However they still contain a relatively large number of low-molecular-weight *N*-hydroxymethyl compounds so that the solutions quickly become cloudy on cooling. With longer heating times the aqueous solutions of the resins remain clear. At this stage of condensation however the solutions are clear only if a limited amount of water is present; they precipitate on dilution. With increasing condensation time the compatibility with water is diminished further until eventually the melamine-formaldehyde condensation polymer separates out from the reaction solution.

(b) Determination of the precipitation ratio
After about 50–60 min condensation time a small sample of the reaction mixture is dropped into iced water and the cloudiness observed. From this moment on, samples of exactly 2 ml are taken at regular intervals of 10 min and, after allowing to cool to 20°C, distilled water at 20°C is added dropwise with stirring. The condensation experiment is stopped when the sample becomes slightly cloudy after the addition of 2 ml of water.

(c) Impregnation of paper
The resin solution prepared above is transferred to a porcelain dish and in it are immersed about 10–20 circular filter papers (diameter 9 cm). After about 1–2 min the filter papers are lifted out with tweezers and the excess solution allowed to drip off. The impregnated filter papers are fastened with clips to a line and allowed to dry overnight.

(d) Fabrication of a laminated moulding
Ten of the resin-impregnated papers are stacked on top of each other. This stack is laid between two aluminium foils (15 cm × 15 cm) and pressed in a hydraulic press at 135°C and a pressure of 40–100 bar for 15 min. After releasing the pressure the sample is removed while it is still hot.

The fabrication of laminated plastics with good transparency requires pressures of the magnitude indicated. All heatable, hydraulic, laboratory or commercial presses are suitable for this work. It is also possible to use two nickel-plated iron plates, 2 cm thick, heated to about 140°C in an oven, and clamped in a horizontally mounted vice (see Section 2.4.2.1). The resistance of the hardened melamine-formaldehyde laminated plastic is tested against solvents and chemicals.

4.1.6. Poly(thioalkylene)s; [poly(alkylene sulfide)s]

The reaction of suitable aliphatic dihalogen compounds with alkali or alkali-earth polysulfides results in the formation of linear, rubbery or resinous, poly(thioalkylene)s[1]:

$$n\ Cl{-}R{-}Cl + n\ Na_2S_x \longrightarrow {-}[R{-}S_x]_n + 2n\ NaCl$$

The most widely used dihalides are 1,2-dichloroethane, bis-(2-chloroethyl) ether, and bis(2-chloroethyl) ketone. The use of polyhalides (e.g. 2% 1,2,3-trichloropropane) results in the formation of branched or crosslinked products. Sodium tetrasulfide (Na_2S_4) is generally used as polysulfide since it contains scarcely any of the monosulfide which reacts with dihalides to form cyclic by-products with unpleasant odour.

Sulfur can be removed from the poly(thioalkylene)s with the aid of sulfur-binding agents; for example, by treatment of an aqueous dispersion with Na_2S, NaOH, or Na_2SO_3 at 30–100°C, the sulfur content can be reduced to two atoms per CRU of the macromolecule:

$$\text{-}[R{-}S_4]\text{-}_n + 2n\ S^{2-} \rightarrow \text{-}[R - S_2]\text{-}_n + 2n\ S_2^{2-}$$

[1] Detailed review: *G. Spielberger* in *Houben-Weyl 14/2* (1963) 591.

Further desulfurization results in degradation of the poly(thioalkylene)s to low-molecular-weight products. Vulcanization of the linear poly-(thioalkylene)s yields crosslinked elastic materials which are commercially important because of their solvent and oil resistance. They are also less sensitive to oxygen and light than most synthetic rubbers. The technological properties can be modified especially by changing the sulfur content.

Poly(thioalkylene)s are prepared by allowing the dihalide to drop slowly under vigorous stirring into a moderately concentrated aqueous polysulfide solution (generally in 10–20% excess). Temperature and reaction time depend mainly on the dihalide being used: for 1,2-dichloroethane, temperatures between 50 and 80°C, and reaction times of about 5 h suffice; on the other hand bis(2-chloroethyl) ether requires 20–30 h at 100°C. Since the poly(thioalkylene)s are insoluble in water and very easily agglomerate into lumps, thereby making further reaction and subsequent washing very difficult, it is expedient to carry out the polycondensation in the presence of a dispersing agent, for example 2–5% magnesium hydroxide. Emulsifiers are not to be recommended since they make the work-up difficult. When the reaction has finished, the mixture is freed from sodium chloride and unreacted sodium polysulfide by slurrying several times with water. It is then acidified with concentrated hydrochloric acid in order to coagulate the poly(thioalkylene)s. Poly(thioalkylene)s can also be prepared by ring-opening polymerization of episulfides.[1]

Example 4–17:
Preparation of a poly(thioalkylene) from 1,2-dichloroethane and sodium tetrasulfide

(a) Preparation of a solution of sodium tetrasulfide
Sodium polysulfides can be prepared by dissolving sulfur in aqueous sodium sulfide or sodium hydroxide:

$$Na_2S + 3S \longrightarrow Na_2S_4$$

$$6NaOH + 10S \longrightarrow Na_2S_2O_3 + 2Na_2S_4 + 3H_2O$$

The sodium thiosulfate produced by the second reaction does not affect the reaction with dihalide.

(i) From sodium sulfide and sulfur: 240 g (1.0 mol) of technical, crystalline sodium sulfide (Na_2S, $9H_2O$) are heated with 96 g (3.0 g atom) of

[1] *W. Cooper*, Brit. Polym. J. *3* (1971) 28; *F. Lautenschlaeger*, J. Macromol. Sci., Chem. *6* (1972) 1089.

ground sulfur and 50 ml of water on the water bath, with stirring and exclusion of air, until the sulfur is dissolved (30–60 min). The dissolution of the sulfur can be accelerated by addition of a few ml of ethanol or a trace of a wetting agent. Heating is continued for 30 min at 90–95°C. The solution is then filtered through a Büchner funnel to remove traces of suspended iron (III) sulfide and diluted with 190 ml of water.

(ii) From sodium hydroxide and sulfur: 160 g (5.0 g atom) of ground sulfur are dissolved in a mixture of 300 g of 40% sodium hydroxide solution (3.0 mol) and 120 ml of water by stirring at 90–95°C.

(b) Preparation of the poly(thioalkylene)
290 g of a solution of sodium tetrasulfide (0.5 mol) are placed in a 500 ml three-necked flask, fitted with a powerful stirrer, thermometer, reflux condenser, and dropping funnel. To this are added with stirring 3 g of crystalline magnesium chloride in 10 ml of water and 7.5 ml of 2 M sodium hydroxide. 40 g (0.40 mol) of 1,2-dichloroethane are added dropwise to the stirred mixture at a steady rate over a period of 2 h, while the solution is maintained at 50–60°C. If the stirrer should stop, the addition of 1,2-dichloroethane must be interrupted, otherwise lumps will form. After the addition is complete, the mixture is heated for another 2 h at 70–80°C; the cold milky dispersion is then poured into about 1 l of water. The poly(thioalkylene) is separated by centrifugation, the upper aqueous layer decanted off, and the polymer washed free of sodium chloride and unreacted sodium tetrasulfide by slurrying several times with water. The washed dispersion is now acidified with 5 ml of concentrated hydrochloric acid, causing the poly(thioalkylene) to coagulate and settle as a spongy cake. After filtration the unpleasant smelling, yellow-white polymer is dried in a vacuum desiccator over P_2O_5. The dry product is partially soluble in carbon disulfide and swells in carbon tetrachloride and benzene. It softens on a Kofler hot block at about 130°C.

4.1.7. Polysiloxanes[1,2]

Polysiloxanes[3] are macromolecules in which silicon atoms are joined together via oxygen atoms, with the two other valencies of the silicon

[1] *E.G. Rochow*, "An Introduction to the Chemistry of Silicones", 2nd ed., Wiley, New York 1951.

[2] *W. Noll*, "Chemie und Technologie der Silicone", 2nd ed., Verlag Chemie, Weinheim 1968.

[3] Besides the scientific term, polysiloxane, these polymers are commonly referred to as silicones, especially in commercial practice (silicone oil, grease and rubber).

atoms linked to at least one organic residue. A linear polysiloxane accordingly has the following general structure:

$$-\underset{\underset{R}{|}}{\overset{\overset{R}{|}}{Si}}-O-\underset{\underset{R}{|}}{\overset{\overset{R}{|}}{Si}}-O-\underset{\underset{R}{|}}{\overset{\overset{R}{|}}{Si}}-O-\underset{\underset{R}{|}}{\overset{\overset{R}{|}}{Si}}-O- \qquad R = \text{alkyl or aryl}$$

In their properties polysiloxanes are intermediate between purely organic polymers and the inorganic silicates; the structure may be varied in numerous ways so as to shift the pattern of properties of silicones either in one direction or the other. Commercial polysiloxanes generally contain methyl substituents.

Various functional silanes (e.g. R_2SiCl_2 or $RSiCl_3$) can be used as starting materials for the preparation of polysiloxanes. The silanes are first hydrolyzed to the corresponding silanols which are very unstable and easily undergo polycondensation with the elimination of water and the formation of —Si—O—Si— linkages:

$$n\,Cl-\underset{\underset{R}{|}}{\overset{\overset{R}{|}}{Si}}-Cl \xrightarrow{H_2O} n\,(HO-\underset{\underset{R}{|}}{\overset{\overset{R}{|}}{Si}}-OH) \xrightarrow{-(n-1)H_2O} HO-\left[\underset{\underset{R}{|}}{\overset{\overset{R}{|}}{Si}}-O\right]_n-H$$

Linear polysiloxanes obtained by the hydrolysis of dichlorosilanes are of relatively low molecular weight; they can, however be condensed further through the terminal OH groups by thermal after-treatment.

The siloxanes have a great tendency towards ring formation so that in the hydrolysis products of dichlorosilanes one finds not only linear polysiloxanes but also cyclic oligosiloxanes with 3 to 9 Si—O units in the ring; under the right reaction conditions these ring compounds can even be the main product (see Example 4–18). Cyclic siloxanes can be polymerized by ring-opening both cationically (e.g. with Lewis acids) and anionically (e.g. with alkali hydroxides); trimers and tetramers of dimethylsiloxane are especially suitable as monomers (see Example 4–18):

$$\text{cyclo-}[(CH_3)_2SiO]_4 \xrightarrow{KOH} \left[\underset{\underset{CH_3}{|}}{\overset{\overset{CH_3}{|}}{Si}}-O-\underset{\underset{CH_3}{|}}{\overset{\overset{CH_3}{|}}{Si}}-O-\underset{\underset{CH_3}{|}}{\overset{\overset{CH_3}{|}}{Si}}-O-\underset{\underset{CH_3}{|}}{\overset{\overset{CH_3}{|}}{Si}}-O\right]$$

The polysiloxanes obtained by anionic initiation have considerably higher molecular weights than those obtained by cationic initiation.

Ring-opening polymerization of cyclic siloxanes with cationic initiators allows the possibility of introducing stable end groups by the use of suitable chain transfer agents. Thus, polysiloxanes with trimethylsilyl end groups are formed when the cationic polymerization of octamethylcyclotetrasiloxane is carried out in the presence of hexamethyldisiloxane as transfer agent:

$$n\,[(CH_3)_2SiO]_4 + CH_3-\underset{CH_3}{\overset{CH_3}{|}}{Si}-O-\underset{CH_3}{\overset{CH_3}{|}}{Si}-CH_3 \xrightarrow{H^{\oplus}} CH_3-\underset{CH_3}{\overset{CH_3}{|}}{Si}-O-\left[\underset{CH_3}{\overset{CH_3}{|}}{Si}-O\right]_{4n}-\underset{CH_3}{\overset{CH_3}{|}}{Si}-CH_3$$

The acid-catalyzed degradation of a high-molecular-weight polysiloxane, containing OH end groups, when carried out in the presence of hexamethyldisiloxane, also leads to low-molecular-weight polysiloxanes with trimethylsilyl end groups[1] (see Example 4–19). Polysiloxanes that do not possess OH end groups cannot condense further on heating; the molecular weight and therefore the viscosity of this product remains constant on heating, which is a very important property of silicone oils.

Low-molecular-weight polysiloxanes are oily or waxy substances (silicone oils). High-molecular-weight polysiloxanes on the other hand are elastomeric; they can be converted to silicone rubbers by crosslinking (vulcanization). This crosslinking is carried out industrially by means of peroxides (see Example 4–18) and probably occurs by abstraction of hydrogen atoms from methyl groups giving carbon radicals on the polysiloxane molecules. Combination of carbon radicals of different macromolecules then causes crosslinking by the formation of C—C bonds.

Example 4–18:

Ring-opening polymerization of a cyclic oligosiloxane to a linear, high-molecular-weight polysiloxane with hydroxyl end groups; curing of the polymer

(a) Preparation of octamethylcyclotetrasiloxane

200 g of dimethyldichlorosilane are dropped slowly from a dropping funnel (protected from moisture by a $CaCl_2$ tube) into 600 ml of water with vigorous stirring at room temperature. The organic phase is then taken up in about 200 ml of ether, separated from water and washed twice with

[1] This reaction involve an adjustment towards equilibrium and is also described as an "equilibration".

distilled water; the ethereal solution is dried over magnesium sulfate. The ether is taken off on a rotary evaporator or by distillation; an oil is left behind, consisting essentially of cyclic oligosiloxanes that can be separated by distillation. A small amount (about 0.5%) of hexamethylcyclotrisiloxane comes over first (b.p. 134°C/760 torr, m.p. 64°C), followed by the main product octamethylcyclotetrasiloxane (b.p. 175°C/760 torr, b.p. 74°C/20 torr, m.p. 17.5°C); yield: about 40%. Finally a few per cent of pentamer (b.p. 101°C/20 torr) and hexamer (b.p. 128°C/20 torr) distil over at temperatures around and above 200°C.

The residue from distillation is a viscous oil consisting of high-molecular-weight compounds. Further cyclic oligomers, especially trimer and tetramer, can be obtained by heating this oil to about 400°C. To do this, a new receiver is fitted to the distillation apparatus and the flask heated to 400–450°C. The more volatile products are carried over in a slow stream of nitrogen and the distillate is then fractionated as before.

(b) Polymerization of an oligosiloxane

60 g of distilled octamethylcyclotetrasiloxane are mixed with 30 g of very finely powdered potassium hydroxide in a 250 ml conical flask which is then placed in an oil bath at 140°C. The increase in viscosity of the mixture can easily be observed by occasional swirling. After 20–30 min the liquid has reached the consistency of thin honey. Half the product is taken out and cooled (further work-up under (*c*) and (*d*)). The residue is heated again until, after 2–3 h, a plastic, putty-like mass is produced. It is allowed to cool, whereby a rubbery polymer is obtained. Yield: 90–95%.

(c) Hot curing of the rubbery polymer

The polysiloxane from experiment (*b*) is soluble in benzene and toluene. It can be converted by hot vulcanization into an insoluble silicone rubber. Using a small blender 10 g of the polymer are kneaded with 10 g of quartz powder or 7.5 g of ground kieselguhr, and 0.6 g of 2,4-dichlorodibenzoyl peroxide paste (50% in silicone oil). To work the additives into the silicone rubber without a mechanical blender is very tedious and difficult to achieve completely. The mixture is then heated for 10 min in the oven at 110°C, to bring about crosslinking. The solubilities of the mixture and the resulting silicone rubber are tested in benzene.

(d) Cold curing of the syrupy polymer at room temperature

10 g of the honey-like polysiloxane from experiment (*b*) are mixed with 5 g of quartz powder or ground kieselguhr in a mortar. The viscous syrup is poured into a beaker and 0.3 g of tetraethylsilicate[1] (crosslinking agent) stirred in, together with 0.3 g of dibutylzinc didodecanoate as vulcanization

[1] Preparation: *A.W. Dearing* and *E.E. Reid*, J. Am. Chem. Soc. *50* (1928) 3058.

accelerator. The mass solidifies in 1–2 h to an elastic silicone rubber. Such a silicone rubber retains its elasticity over an unusually large temperature range (−90 to 300°C); it is also very resistant towards atmospheric effects.

Example 4–19:
Conversion of a silicone elastomer to a silicone oil with trimethylsilyl end groups (equilibration)

10 g of the rubbery silicone with hydroxyl end groups, made in Example 4–18 (*b*) are taken up in 20 g of toluene. 0.5 g of 96% sulfuric acid and 0.2 g of hexamethyldisiloxane[1] are added to the very viscous solution, with stirring or shaking until the high-molecular-weight material has disappeared. 0.3 ml of water are then added and stirring continued for 2 h. The solution is next washed with water in a separating funnel until the washings have a neutral reaction. The toluene is distilled off leaving a clear, mobile silicone oil (viscosity 200–500 Pa s). Yield: 70%.

4.1.8. Cyclopolycondensation

By cyclopolycondensation will here be understood a procedure in which two separate reaction steps lead to macromolecules with carbocyclic or heterocyclic main chains. This structure confers especially high chemical and physical stability at high temperature. The synthetic principle consists of reacting two tetrafunctional monomers so that in the first reaction step two of their functional groups are used to form a linear macromolecule. In the next step the two other functional groups (or newly formed functional groups) cyclize by condensation to yield a macromolecule with a ladder-like structure. The first reaction step may be either an addition or a condensation polymerization.

(a) Addition polymerization of one monomer, followed by cyclopolycondensation
An example of this type is the radical polymerization of methyl vinyl ketone followed by the cyclic condensation of poly(methyl vinyl ketone) to a ladder polymer with a carbocyclic main chain. The first reaction, namely radical polymerization, is carried out according to the usual procedures at temperatures below 80°C; the second reaction, the cyclic condensation, occurs with the elimination of water at 300°C:

[1] Hexamethyldisiloxane is prepared as follows, according to *R. O. Sauer*, J. Am. Chem. Soc. 66 (1944) 1707: 5 g of trimethylchlorosilane are hydrolyzed in boiling water. After a few minutes the organic layer is separated off and distilled over P_2O_5 (b.p. 100°C). Yield: about 80%.

1st reaction (addition polymerization):

$$CH_2{=}CH{-}\underset{\underset{O}{\|}}{C}{-}CH_3 \longrightarrow \cdots$$

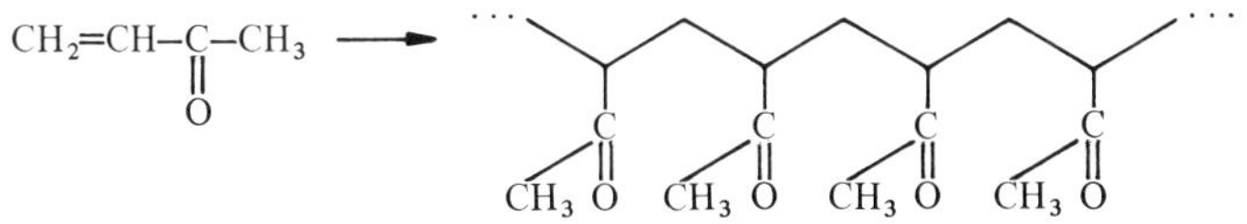

2nd reaction (cyclopolycondensation):

$$\xrightarrow[-H_2O]{300°C} \cdots$$

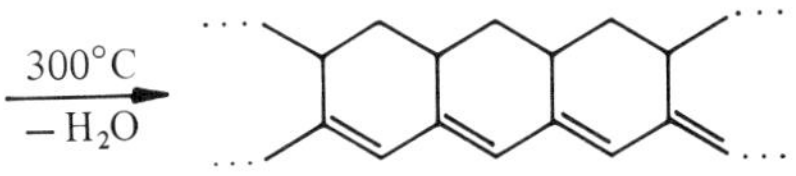

The resulting polymer is red, insoluble in all solvents, but still fusible.

(b) Stepwise addition polymerization of two monomers, followed by cyclopolycondensation

The formation of polyimides[1] is a typical example of this type. The first reaction consists of the addition of an aromatic diamine to a dianhydride of a tetracarboxylic acid, and occurs very rapidly at 20–40°C in a strongly polar solvent (dimethylformamide, *N*,*N*-dimethylacetamide, *N*-methylpyrrolidone). The highly viscous solution of the poly(acid-amide) thereby produced, is cast in a thin layer and heated to 150–300°C; the solvent evaporates and the second reaction occurs at the same time, resulting in ring closure with elimination of water to give the polyimide (see Example 4–20):

1st reaction (stepwise addition polymerization involving two monomers)

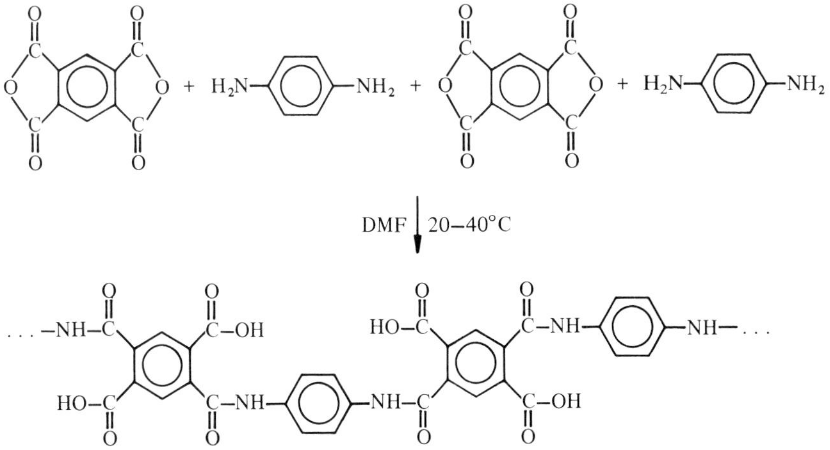

$$+\ H_2N{-}C_6H_4{-}NH_2\ +\ \ldots\ +\ H_2N{-}C_6H_4{-}NH_2$$

DMF 20–40°C

$$\ldots{-}NH{-}C(=O)\ \ldots\ C(=O){-}OH \quad HO{-}C(=O)\ \ldots\ C(=O){-}NH{-}C_6H_4{-}NH{-}\ldots$$

$$HO{-}C(=O)\ \ldots\ C(=O){-}NH{-}C_6H_4{-}NH{-}C(=O)\ \ldots\ C(=O){-}OH$$

poly(acid-amide)

[1] *C.E. Stroog*, J. Polym. Sci. (Polymer Rev.) *11* (1976) 161.

2nd reaction (cyclopolycondensation):

150–250°C ↓ $-H_2O$

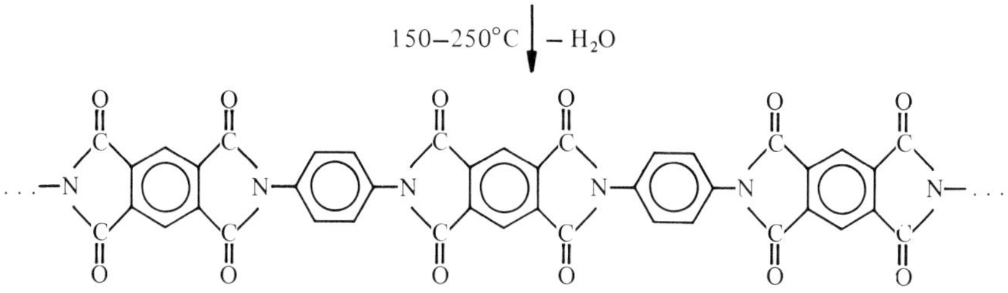

polyimide

The polyimide made from pyromellitic dianhydride and 4,4′-oxydianiline exhibits long-term stability in air above 200°C.

(c) Polycondensation followed by cyclopolycondensation

To this group belong especially the polybenzimidazoles prepared by Marvel et al.[1] These syntheses are carried out as follows. In the first step an aromatic tetramine is condensed with the diphenyl ester of an aromatic dicarboxylic acid at 220–260°C, giving a poly(amino-amide) with elimination of phenol. Ring closure with elimination of water occurs in the second step, conducted at 400°C and yielding the polyimidazolobenzimidazole:

1st reaction (polycondensation):

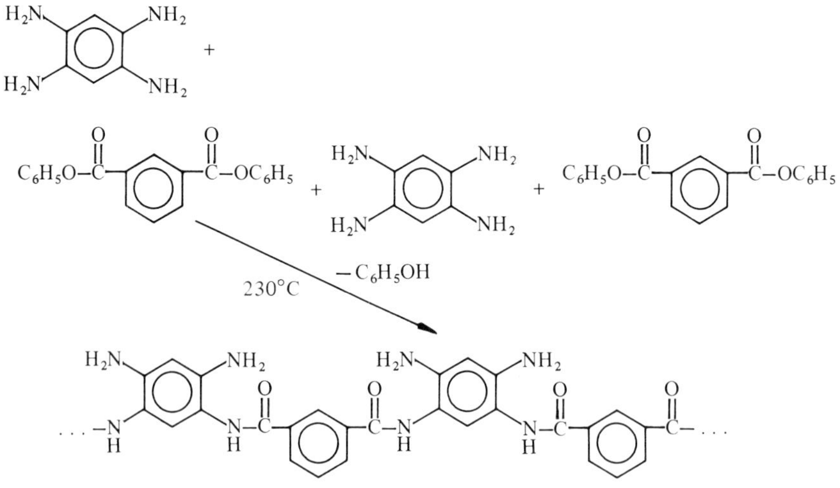

230°C $-C_6H_5OH$

poly(amino-amide)

[1] *C.S. Marvel*, J. Macromol. Sci., Macromol. Rev. *13* (1975) 219; also see *L.R. Belohlav*, Angew. Makromol. Chem. 40/41 (1974) 465.

2nd reaction (cyclopolycondensation):

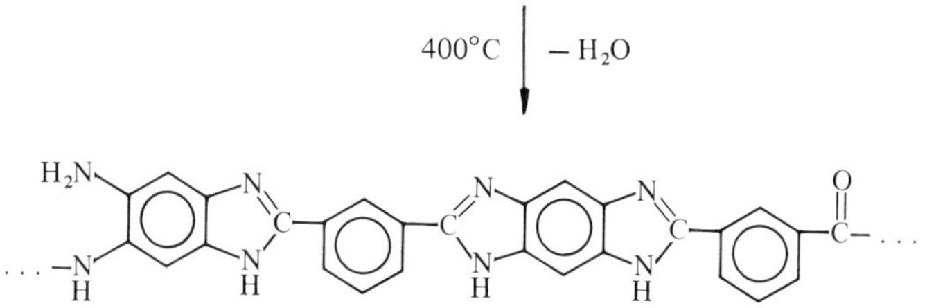

400°C ↓ $-H_2O$

polyimidazolobenzimidazole

Polyimidazolobenzimidazoles are stable under nitrogen to 600°C, and even at 900°C lose only 30% of their weight per hour.

The advantage of these synthetic procedures, using two separate polyreactions, is that useful objects can be made from ladder polymers in spite of their insolubility and virtual infusibility. Thus, films and fibres can be made from the soluble and fusible linear macromolecules formed by the first reaction step; the second step of cyclopolycondensation, leading to the ladder polymer, is then carried out on the fabricated article.

Ladder polymers are amongst the class of polymers that are resistant to high temperature, and frequently decompose only above 400°C. Their mechanical properties are likewise maintained over a wide temperature range.

Example 4–20:
Preparation of a polyimide from pyromellitic dianhydride and 4,4'-oxydianiline by cyclopolycondensation

(a) Preparation of the polyamide
4.0 g (0.02 mol) of 4,4'-oxydianiline (recrystallized from benzene and dried in vacuum for 8 h at 50°C) is dissolved in 30 ml of pure dimethylformamide in a dry 100 ml round-bottomed flask, fitted with a drying tube and magnetic stirrer. 4.36 g (0.02 mol) of pyromellitic dianhydride (sublimed in high vacuum) are added in portions over a period of a few minutes, with continuous stirring. The addition polymerization begins at once and is completed by stirring the viscous solution for a further hour at 15°C.

About 20 ml of the poly(acid-amide) solution are then dropped into 300 ml of water with vigorous stirring. The precipitated white polymer is filtered with suction, washed several times with water and dried in vacuum at 100°C for 20 min. The limiting viscosity number is determined in dimethylformamide at 25°C (Ostwald viscometer, capillary diameter 0.4 mm).

(b) Preparation of the polyimide
The polyimide is formed by the thermal cyclopolycondensation of the poly(acid-amide). For this purpose 5 ml of poly(acid-amide) solution are placed on a watch-glass (diameter 10 cm) and kept in a vacuum oven at 50°C for 24 h. The solvent evaporates and at the same time cyclization to the polyimide takes place; the resulting film is insoluble in dimethylformamide. The formation of the polyimide can also be followed by infrared spectroscopy: the NH band at 3.08 μm disappears, while the imide bands make their appearance at 5.63 and 13.85 μm.

The polyimide formation can be accelerated by heating the poly(acid-amide) film to 300°C in the vacuum oven for about 45 min, once the initial drying process has raised the solid content to 65–75%.

4.1.9. Dehydrogenation of aromatic compounds

Dehydrogenations of aromatic compounds, in which low-molecular-weight compounds are joined together to form macromolecules by loss of hydrogen, are formally to be classified as polycondensation reactions; here too, polymers are formed from monomers by the continuous elimination of a low-molecular-weight compound (hydrogen), while the polymer retains its activity towards the growth reaction. However, there is a difference from the usual polycondensation reactions in that a catalyst is always necessary in the dehydrogenation of aromatic compounds. These catalysts interact with the monomer to form a complex which reacts in the desired manner.

The synthesis of poly(1,4-phenylene) from benzene, catalyzed by Lewis acids, is of this type:[1,2]

$$n\ C_6H_6 \xrightarrow[-H_2]{AlCl_3/O_2} \cdots\!-\!C_6H_4\!-\!C_6H_4\!-\!C_6H_4\!-\!\cdots$$

likewise the preparation of poly(oxy-1,4-phenylene)s from substituted phenols[3], for example poly(oxy-2,6-dimethyl-1,4-phenylene) from 2,6-

[1] *P. Kovacic, M.B. Feldman, J.P. Kovacic* and *J.B. Lando*, J. Appl. Polym. Sci. *12* (1968) 1735.

[2] For summary on polyphenylenes see *J.G. Speight*, J. Macromol. Sci. *5* (1971) 295; *G.K. Noren* and *J.K. Stille*, J. Polym. Sci., Macromol. Rev., *5* (1972) 385.

[3] *A.S. Hay*, J. Polym. Sci. *58* (1962) 581; Adv. Polym. Sci. *4* (1967) 496; Polym. Eng. Sci. *16* (1976) 1.

dimethylphenol in the presence of a copper/pyridine complex:

$$n\ \text{H}-\text{C}_6\text{H}_2(\text{CH}_3)_2-\text{OH} \xrightarrow[\text{Cl(HO)Cu(Py)}_2]{+n/2\ \text{O}_2,\ -n\ \text{H}_2\text{O}} \cdots-[\text{C}_6\text{H}_2(\text{CH}_3)_2-\text{O}]_n-\cdots$$

For the synthesis of poly(oxy-1,4-phenylene)s by dehydrogenation of phenols, the monomer must conform to certain requirements. To obtain linear, essentially uncrosslinked, products, the phenolic nucleus must be substituted in the 2- and 6-positions. Halogen atoms do not inhibit the reaction but may be eliminated during the polymerization with the formation of branch points. The phenol must be relatively easily oxidizable; substituents that raise the oxidation potential lead to a retardation or inhibition of the dehydrogenation reaction. Thus, 2,6-dichlorophenol gives only a low-molecular-weight poly(oxy-1,4-phenylene), and 2,6-dinitrophenol does not react at all. The substituents in the 2- and 6-positions must not exceed a certain geometrical size, since otherwise, instead of regular head-tail coupling leading to the poly(oxy-1,4-phenylene)s, there is simply a tail-tail coupling of the monomers, two at a time, forming diphenoquinones [4,4′-dioxo-1,1′-bi(2,5-cyclohexadiene ylidine)s]

$$\text{HO}-\text{C}_6\text{H}_3\text{R}_2 + \text{R}_2\text{C}_6\text{H}_3-\text{OH} \xrightarrow{-\text{H}_2} \text{O}{=}\text{C}_6\text{H}_2\text{R}_2{=}\text{C}_6\text{H}_2\text{R}_2{=}\text{O}$$

In the dehydrogenation of phenols with relatively small substituents this quinone formation occurs only to a minor extent, but its formation increases considerably with increasing temperature. The pale yellow coloration of poly(oxy-2,6-dimethyl-1,4-phenylene) may likewise be caused by the presence of quinones.

Suitable catalysts are copper(I) salts (e.g. chloride, bromide, and sulfate) in combination with amines. However, high-molecular-weight poly(oxy-1,4-phenylene)s are only obtained with a few amines, such as pyridine. The amine/copper salt ratio must be made as large as possible, to minimize the formation of diphenylquinone and to give a high molecular weight.

Example 4–21:
Preparation of poly(oxy-2,6-dimethyl-1,4-phenylene)

Poly(oxy-2,6-dimethyl-1,4-phenylene) can be prepared by dehydrogenation of 2,6-dimethylphenol with oxygen in the presence of copper(I) chloride/pyridine as catalyst at normal temperature. It is known that the mechanism involves a stepwise reaction, probably proceeding via a copper-phenolate complex that is then dehydrogenated.[1]

0.4 g of copper(I) chloride are slurried in a mixture of 100 ml of chloroform and 20 ml of pyridine contained in a 500 ml three-necked flask fitted with stirrer, gas inlet and thermometer. Oxygen is passed in with vigorous stirring. After 10–20 min a clear, dark green solution is formed. 10 g (0.082 mol) of pure 2,6-dimethylphenol is added to this solution and more oxygen passed in with vigorous stirring. The temperature of the solution rises slowly to about 40°C and the colour of the solution becomes yellow-brown. After the temperature has reached its maximum (about 30 min) the viscous solution is poured into 1 l of methanol containing 10 ml of concentrated hydrochloric acid to destroy the copper complex. The polymer is washed with methanol and precipitated again from chloroform (100 ml) into 1 l methanol containing 10 ml of HCl. Yield: quantitative.

With larger quantities the temperature can be kept constant by external cooling. A temperature of 30°C is advantageous for the formation of a high-molecular-weight product. The higher the 2,6-dimethylphenol concentration, the higher is the molecular weight of the polymer formed.

Poly(oxy-2,6-dimethyl-1,4-phenylene) has a softening point of about 250°C. It is soluble in chlorinated hydrocarbons such as chloroform, carbon tetrachloride, and tetrachloroethane, also in nitrobenzene and toluene. The limiting viscosity number is determined in chloroform at 25°C.

To make a film, 2 g of the polymer is pressed for 5 min between two metal plates heated to 320°C (see Section 2.4.2.1). After chilling the metal plates with water, the film can be peeled off.

[1] *G.F. Endres*, *A.S. Hay* and *J.W. Eustance*, J. Org. Chem. *28* (1963) 1300; *A.S. Hay*, Adv. Polym. Sci. *4* (1967) 496.

4.2. STEPWISE ADDITION POLYMERIZATIONS INVOLVING TWO MONOMERS

The addition polymerizations described here involve a stepwise reaction of at least two bifunctional compounds, leading to the formation of macromolecules. In contrast to condensation polymerization, no low-molecular-weight compounds are eliminated; the coupling of the monomer units is, instead, a consequence of the migration of a hydrogen atom (cf. the equation for the formation of a polycarbamate (polyurethane) from a diol and a diisocyanate, Section 4.2.1). Like condensation polymerization, this kind of addition polymerization is also a stepwise reaction, consisting of a sequence of independent individual reactions, so that here too the average molecular weight of the resulting polymer steadily increases during the course of the reaction. The oligomeric and polymeric products formed in the individual steps possess the same functional end groups and the same reactivity as the starting materials; they can be isolated without losing their reactivity, in contrast to the products of addition polymerization. As stepwise reactions these addition polymerizations are governed by kinetic laws similar to those for condensation polymerization (see equations (1) to (4) in Section 4.1); the experimental procedure is also similar.

4.2.1. Polyurethanes (polycarbamates)[1]

Polyurethanes are macromolecules in which the CRU's are coupled with one another through oxycarbonylamino (urethane or carbamate) groups. They are prepared almost exclusively by addition polymerization reactions of di- or poly-functional hydroxy compounds with di- or poly-functional isocyanates:[2]

$$n\ HO{-}(CH_2)_x{-}OH + n\ O{=}C{=}N{-}(CH_2)_y{-}N{=}C{-}O$$

$$\downarrow$$

$$HO{-}(CH_2)_x{-}O{-}\left[\underset{\substack{\| \\ O}}{C}{-}\underset{\substack{| \\ H}}{N}{-}(CH_2)_y{-}\underset{\substack{| \\ H}}{N}{-}\underset{\substack{\| \\ O}}{C}{-}O{-}(CH_2)_x{-}O\right]_{n-1}\underset{\substack{\| \\ O}}{C}{-}\underset{\substack{| \\ H}}{N}{-}(CH_2)_y{-}N{=}C{=}O$$

This addition reaction proceeds readily and quantitatively. Side reactions[3] can give amide, urea, allophanamide (biuret), allophanate, and isocyanurate groupings, so that the structure of the product can deviate

[1] *O. Bayer*, Angew. Chem. *59* (1947) 257; *E. Müller* in *Houben-Weyl 14/2* (1963) 57; *J.H. Saunders* and *K.C. Frisch*, "Polyurethanes", High Polymers *16/1* and *2*, 1962–1964; *E. Windemuth*, Kunststoffe *57* (1967) 377; Kunststoff-Handbuch, Vol VII, Hanser-Verlag.

[2] Preparation of polyurethanes by polycondensation: see Example 4–12.

[3] *R. Merten, D. Lauerer, G. Braun* and *M. Dahm*, Makromol. Chem. *101* (1967) 337.

from that above; such side reactions are sometimes desired (see Section 4.2.1.2).

Linear polyurethanes made from short chain diols and diisocyanates are high-melting, crystalline, thermoplastic substances whose properties are comparable with those of the polyamides because of the similarity in chain structure. However, they generally melt at somewhat lower temperatures and have better solubility, for example in chlorinated hydrocarbons. The thermal stability is lower than for polyamides: depending on the structure of the polymer the reverse reaction of the urethane groups begins at temperatures as low as 150–200°C with regeneration of the initial functional groups; the cleavage of the allophanate groups begins at the still lower temperature of 100°C. Polyurethanes are used for the fabrication of fibres, while crosslinked polyurethanes are employed as lacquers and adhesives, as coatings for textiles and paper, and as elastomers and foam plastics.

A key factor in the preparation of polyurethanes is the reactivity of the isocyanates.[1] Aromatic diisocyanates are more reactive than aliphatic diisocyanates, and primary isocyanates react faster than secondary or tertiary isocyanates. The most important and commercially the most readily accessible diisocyanates are hexamethylene diisocyanate, 4,4′-methylenedi(phenyl isocyanate), 1,5-naphthalene diisocyanate, and a 4:1 mixture of 2,4- and 2,6-toluene diisocyanates.

As already mentioned, polyurethanes, on heating, decompose into isocyanates and hydroxy compounds, the decomposition temperature depending on the structure of the urethane. Use is made of this fact when reacting polyhydroxy compounds with so-called "capped isocyanates" or "isocyanate splitters". One may use, for example, diurethanes of aromatic diisocyanates and phenols that revert to the original components at 150–180°C. Such diurethanes can be mixed with polyhydroxy compounds at room temperature and stored without change; on heating, the diisocyanate is split off and reacts with the free hydroxyl groups forming urethane groups of greater stability, while the phenol distils away.

The addition of isocyanates to hydroxy compounds is inhibited by acid compounds (e.g. hydrogen chloride or *p*-toluenesulfonic acid); on the other hand it can be accelerated by basic compounds (e.g. tertiary amines such as pyridine, *N*,*N*-dimethylbenzylamine, and especially 1,4-diazabicyclo[2.2.2]octane[2]) and also by certain metal salts or organometallic compounds (e.g. bismuth nitrate, iron or zinc acetylacetonate,

[1] For preparation see *W. Siefken*, Liebigs Ann. Chem. *562* (1949) 75.
[2] Trade name: DABCO

dibutylzinc didodecanoate). These catalysts are often effective in amounts of much less than 1 wt.%.

4.2.1.1. Preparation of linear polyurethanes

The addition polymerization reaction of dihydroxy compounds with diisocyanates sets in on mixing the two components and gently warming. Under proper conditions (see Section 4.2.1) polyurethanes with molecular weight up to about 15000 can thereby be obtained. As in the case of polyamides and polyesters, the softening point of the aliphatic polyurethanes depends on the number of carbon atoms between the urethane groups. The polyurethane obtained from 1,4-butanediol and hexamethylene diisocyanate has achieved commercial significance because of its favorable properties, e.g. high melting point (about 184°C) and greater resistance to hydrolysis than Nylon-6,6. The polymerization can be conducted both in bulk (melt) and in solution.

To carry out the polymerization in the melt, the diisocyanate is dropped into the carefully purified, anhydrous dihydroxy compound with thorough mixing. For smooth reaction it is important to keep the temperature as steady as possible; this can be achieved by controlled addition of the diisocyanate and, if necessary, by cooling (heat liberated = 224 kJ per mol urethane groups). The polymerization in bulk requires relatively high temperatures, and, in addition, the polyurethane formed is exposed to the action of the diisocyanate throughout the duration of the reaction, so that secondary reactions can easily take place (see Section 4.2.1). For the preparation of polyurethanes with high molecular weight and with as linear a structure as possible, polymerization in solution is, therefore to be preferred. Suitable inert solvents are toluene, xylene, chlorobenzene, and 1,2-dichlorobenzene. The diisocyanate is normally dropped into the solution of dihydroxy compound at the desired temperature, which may conveniently be the boiling point of the solvent. The resulting polyurethane often separates out from the reaction mixture and so is much less vulnerable to secondary reactions than when the polymerization is carried out in bulk.

Linear polyurethanes can also be prepared by polycondensation of diamines with dichloroformates (see Section 4.1.3).

Example 4–22:
Preparation of a linear polyurethane from 1,4-butanediol and hexamethylene diisocyanate in the melt

22.5 g (0.25 mol) of pure 1,4-butanediol (m.p. 19.7°C, b.p. 120°C/10 torr) are weighed into a dry 25 ml three-necked flask, fitted with a strong glass or metal stirrer, dropping funnel with pressure equalizer and calcium chloride tube, and nitrogen inlet; the air is removed by evacuation and the flask filled with nitrogen. The flask is then heated to 55°C under a steady stream of nitrogen and 42 g (0.25 mol) of carefully distilled hexamethylene diisocyanate (b.p. 122–125°C/10 torr) dropped in, with vigorous stirring, over a period of 60 min according to the following programme, using an oil bath to adjust the temperature as appropriate:

Weight of hexamethylene diisocyanate added in g	Total dropping time in min	Temperature in °C
5	12	95
10	24	125
15	35	145
20	45	165
25	50	175
30	55	190
42	60	200

As soon as all the hexamethylene diisocyanate has been added the internal temperature falls somewhat; reaction is allowed to continue at 195°C for 15 min and the polyurethane melt is then poured into a beaker or porcelain dish, where it solidifies to a hard block. The polyurethane from 1,4-butanediol and hexamethylene diisocyanate melts at 181–183°C and is soluble in *m*-cresol and formamide. The limiting viscosity number and the colour of this product is compared with that prepared by precipitation addition polymerization (see Example 4–23).

Example 4–23:
Preparation of a linear polyurethane from 1,4-butanediol and hexamethylene diisocyanate in solution (precipitation addition polymerization)

22.5 g (0.25 mol) of pure 1,4-butanediol, 42 g (0.25 mol) of pure hexamethylene diisocyanate, and 125 ml of anhydrous chlorobenzene are placed in a dry 500 ml three-necked flask fitted with stirrer, thermometer, reflux condenser with attached drying tube, and nitrogen inlet; the air is removed by evacuation and the flask filled with nitrogen. The mixture is then heated carefully in an oil bath under a slow stream of nitrogen. At about 95°C the initially cloudy reaction mixture suddenly becomes clear; the temperature now climbs rather rapidly and under some circumstances may exceed the prevailing oil bath temperature. About 15 min after the solution has come to the boil (132°C), a faint cloudiness appears (recognizable as a blue rim at the vessel wall) which visibly strengthens. Finally the high-molecular-weight polyurethane settles out in the form of a sand. The reaction is completed by heating for a further 15 min. After cooling, the polyurethane powder is filtered off with suction. (If the mixture is filtered hot (100°C), about 1-3% of the low-molecular-weight portion remains in solution and can be isolated by precipitation with methanol). The residual absorbed chlorobenzene is best removed by steam distillation. In this way one obtains 62 g (96%) of a white powder that melts at 181–183°C. The limiting viscosity number and colour are compared with that of the product obtained by polymerization in the melt (Example 4–22).

4.2.1.2. Preparation of branched and crosslinked polyurethanes

Essentially two methods can be used for the preparation of branched and crosslinked polyurethanes.

(a) The reactions of diisocyanates with compounds that possess more than two hydroxyl groups per molecule.

The degree of crosslinking here depends essentially on the structure and functionality of the polyhydroxy compound so that the properties of the polyurethane can be altered by variation of this component. This procedure is applied mainly to the preparation of lacquers (reactions with diisocyanates at low temperature in anhydrous solvents such as butyl acetate) or mouldings (usually with "capped" diisocyanates at higher temperatures).

(b) The reaction of linear polyurethanes, which possess either hydroxyl or isocyanate end groups, with suitable reactive compounds.

Following O. Bayer[1], this procedure, which is used especially for the preparation of elastomeric polyurethanes, is carried out in two separate stages. First a carefully dried, relatively low-molecular-weight, aliphatic polyester or polyether with hydroxyl end groups is reacted with an excess of diisocyanate. A "chain extension" reaction occurs in which two to three linear diol molecules are coupled with diisocyanate, so as to yield a linear polymer with some in-chain urethane groups and with isocyanate end groups. Suitable starting compounds are polyesters from ethylene glycol and adipic acid, also poly(propene oxide) or poly(oxytetramethylene) with molecular weight around 2000, whose hydroxyl end groups can be reacted with very reactive diisocyanates such as 1,5-naphthalene diisocyanate, 1,4-phenylene diisocyanate, and 4,4′-methylenedi(phenyl isocyanate).

The chain-extended, linear poly(ester-urethanes) so obtained can now be crosslinked in a second stage, involving reaction with water, a glycol or a diamine.

---

[1] *O. Bayer* and *E. Müller*, Angew. Chem. *72* (1960) 934.

In crosslinking with water, pairs of isocyanate end groups in the chain-extended polymer OCN~~~NCO first react with a molecule of water; this results in a linear coupling through urea groupings, with simultaneous elimination of CO_2:

$$\mathrm{OCN{\sim\sim\sim}NCO} \quad \mathrm{OCN{\sim\sim\sim}NCO} \quad \mathrm{OCN{\sim\sim\sim}NCO} \quad (+\ 2\,H_2O) \xrightarrow{-2\,CO_2} \mathrm{OCN{\sim\sim\sim}N(H){-}C({=}O){-}N(H){\sim\sim\sim}N(H){-}C({=}O){-}N(H){\sim\sim\sim}NCO} \tag{1}$$

The subsequent crosslinking probably occurs by reaction of the hydrogen atoms of the resulting urea groups with isocyanate groups still present in the starting polymer or the chain-extended polymer, with the formation of allophanamide (biuret) groups:

$$\begin{array}{c} \mathrm{{\sim\sim}N(H){-}C({=}O){-}N(H){\sim\sim}} \\ \mathrm{OCN{\sim\sim\sim}NCO} \\ \mathrm{{\sim\sim}N(H){-}C({=}O){-}N(H){\sim\sim}} \end{array} \longrightarrow \begin{array}{c} \mathrm{{\sim\sim}N(C{=}O){-}C({=}O){-}N(H){\sim\sim}} \\ \mathrm{|\ NH{\sim\sim\sim}NH\ |} \\ \mathrm{{\sim\sim}N(C{=}O){-}C({=}O){-}N(H){\sim\sim}} \end{array} \tag{2}$$

Crosslinking with glycols and diamines proceeds according to a similar scheme (but without elimination of CO_2). The first reaction is again a linear chain-extension reaction. With glycols this occurs with the formation of urethane groups, which can then react with residual isocyanate end groups to give crosslinking with formation of allophanate groups. With diamines the linear coupling occurs through urea groups and the crosslinking reaction then proceeds as formulated in equation (2).

Crosslinking with glycols and diamines plays a major role in the preparation of polyurethane elastomers. The properties of the resulting products can be widely varied by choice of starting components and the number of crosslinks ("mesh width").

Crosslinking with water is mainly used in the preparation of polyurethane foams, the liberated carbon dioxide (equation (1)) serving to expand the foam. The properties of polyurethanes crosslinked in this way depend on the starting materials and on the reaction conditions. Flexible foams are obtained if the crosslink density is kept low; in this case it is appropriate to use high-molecular-weight linear polyesters or polyethers as diols with a relatively low OH number[1] of 40–60. Rigid foams result if the crosslink density is raised by use of short-chain or branched polyols. Suitable for this purpose are adducts of propylene oxide with polyfunctional alcohols, having an OH number of 300–600. As diisocyanate, the commercially available mixture of 80% 2,4-toluene- and 20% 2,6-toluenediisocyanates is especially suitable. If foam formation is to take place at room temperature, and especially when hydroxy compounds with secondary hydroxyl groups are used [poly(propylene glycol)s,], the presence of a catalyst is generally required (see Section 4.2.1).

The homogeneity of foamed materials and the proportion of open and closed pores can be influenced by additives such as emulsifiers and stabilizers. Emulsifiers (e.g. sodium, potassium or zinc salts of long chain fatty acids) bring about a uniform distribution of water in the reaction mixture, ensuring homogeneous foaming, while stabilizers (certain silicone oils) prevent a breakdown of the cell structure at the beginning of the reaction and also act as pore regulators.

For laboratory experiments it is sufficient to mix the individual components with vigorous stirring. For the preparation of uniform foams with optimum properties, however, the proportions of polyhydroxy compound, diisocyanate and additives must be very carefully balanced. For practical application, blending must be carried out with suitable dosing arrangements.

Example 4–24:
Preparation of a flexible polyurethane foam

100 g of a polyester with an OH number of 60 (see Example 4–01a) and 35 g of toluene diisocyanate are well mixed in a 600 ml beaker for 1 min using a wooden stick. A mixture of 1 g of *N*,*N*-dimethylbenzylamine, 2 g of a 50% aqueous solution of a non-ionic emulsifier, 1 g of a 50% aqueous solution of sodium dodecyl sulfate, 0.25 g of a poly (dimethylsiloxane), and 1 g of water are then added with intensive stirring. After 20–30 s the mixture begins to foam and is then poured quickly into a paper-lined container of at least 2 l capacity. After 1 min the expansion of the foam is practically

[1] *Houben-Weyl 14/2* (1963) 17; also see Example 4–01.

finished; after a further 20–30 min (depending on the prevailing atmospheric conditions) the surface of the polyurethane foam is no longer tacky. The flexible foam (total volume 1500 cm^3) can now be removed from the paper casing and mechanically stressed.

Example 4–25:
Preparation of a rigid polyurethane foam

110 g of a strongly branched polyester with an OH number of about 350 (see Example 4–01b) and 85 g of toluene diisocyanate are intimately mixed in a 600 ml beaker using a wooden stick. A mixture of 3 g of a non-ionic emulsifier and 2 g of sodium dodecyl sulfate in 2 ml of water are added with stirring until the mixture begins to foam. It is then immediately poured into a paper bag with a flat bottom (minimum capacity 1 l), which the resulting foam will fill almost completely. After standing for an hour the rigid, brittle foam can be removed from the paper casing (total volume about 1000 cm^3).

4.2.2. Epoxy resins[1,2]

Epoxy resins are usually understood to mean the products of reaction of polyfunctional hydroxy compounds with 1-chloro-2,3-epoxypropane (epichlorohydrin) in basic medium. In the simplest case 2 mol of epichlorohydrin react, for example, with 1 mol of 4,4′-isopropylidenediphenol (bisphenol A), according to the following scheme:

$$ClCH_2\text{–}\underbrace{CH\text{–}CH_2}_{O} + HO\text{–}C_6H_4\text{–}C(CH_3)_2\text{–}C_6H_4\text{–}OH + \underbrace{CH_2\text{–}CH}_{O}\text{–}CH_2Cl$$

$$\downarrow \qquad (1)$$

$$\underset{Cl}{CH_2}\text{–}\underset{OH}{CH}\text{–}CH_2\text{–}O\text{–}C_6H_4\text{–}C(CH_3)_2\text{–}C_6H_4\text{–}O\text{–}CH_2\text{–}\underset{OH}{CH}\text{–}\underset{Cl}{CH_2}$$

$$NaOH \downarrow -HCl \qquad (2)$$

$$\underbrace{CH_2\text{–}CH}_{O}\text{–}CH_2\text{–}O\text{–}C_6H_4\text{–}C(CH_3)_2\text{–}C_6H_4\text{–}O\text{–}CH_2\text{–}\underbrace{CH\text{–}CH_2}_{O} \qquad \mathbf{1}$$

---

[1] Also see *R. Wegler* and *R. Schmitz-Josten* in *Houben-Weyl 14/2* (1963) 462.
[2] Kunststoff-Handbuch, Vol. XI.

Higher-molecular-weight products **2** result from coupling of epoxide **1** with further bisphenol:

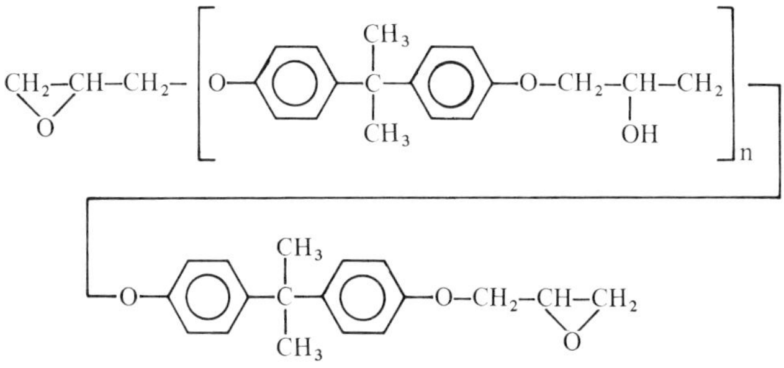

$$\underset{\backslash O /}{CH_2-CH}-CH_2-\left[O-C_6H_4-\underset{CH_3}{\overset{CH_3}{C}}-C_6H_4-O-CH_2-\underset{OH}{CH}-CH_2\right]_n-O-C_6H_4-\underset{CH_3}{\overset{CH_3}{C}}-C_6H_4-O-CH_2-\underset{\backslash O /}{CH-CH_2}$$

2

However, side reactions can also occur. Thus, bisphenol may add only one epoxy group and some of the epoxy groups may be hydrolyzed during the preparation so that the number of hydroxyl groups will deviate from that shown in formula **2**.

Depending on the conditions, reactions (1) and (2) can be carried out either concurrently or consecutively. If one works from the outset in alkaline medium, for example by dropping the desired amount of epichlorohydrin into the mixture of hydroxy compound and the equivalent amount of aqueous alkali hydroxide at 50–100°C, then the addition reaction (1) and the HCl-elimination (2) occur side by side (Example 4–26). On the other hand if the hydroxy compound and epichlorohydrin are allowed to react in non-aqueous medium in the presence of acid catalysts, the corresponding chlorohydrin is first formed; this can then be transformed into the epoxy compound in a second step by reaction with an equivalent amount of alkali hydroxide (Example 4–27).

The structure and molecular weight of the resulting epoxy resin are strongly influenced by the reaction conditions: a large excess of epichlorohydrin (about 5 mol per mol phenolic hydroxyl groups) favours the formation of terminal epoxy groups; however, the molecular weight (and hence the softening point) of the product decreases with increasing amount of epichlorohydrin. The reaction temperature is also important: high temperatures promote secondary reactions such as hydrolytic cleavage of epoxy groups, leading to the formation of additional hydroxyl groups.

The commercially most important epoxy resins are those prepared from 4,4′-isopropylidenediphenol (bisphenol A) and epichlorohydrin. They have molecular weights between 450 and 4000 (n in formula **2** between 1 and 12) and softening points between 30 and 155°C. Such epoxy resins are still soluble but become insoluble and infusible through subsequent crosslinking reactions.

Crosslinking (curing) can in principle be brought about with any di- or poly-functional compound that adds to epoxy groups.[1] Moreover, self-crosslinking can also be achieved by addition of catalytic amounts of a tertiary amine or acid compound, such as a sulfonic acid or a Friedel-Crafts catalyst (generally in the form of their adducts with ether or alcohols); this reaction often occurs even at low temperature but does not proceed uniformly. Crosslinking of epoxy resins is an exothermic reaction, liberating 92–109 kJ per mol epoxy groups. In most cases the amount of crosslinking agent used is equivalent to the analytically determined content of epoxy groups. The crosslinking is generally carried out in the bulk, but sometimes in solution (lacquers). A variant of crosslinking in solution is to use reactive thinners: these are substances which on the one hand lower the viscosity of the epoxy resin, but at the same time act as crosslinking agents (e.g. low viscosity mono-and di-epoxides or allyl 2,3-epoxypropyl ether). Most crosslinking reactions only set in at higher temperatures so that the epoxy resin and crosslinker can be mixed and stored at room temperature. In practice, polybasic carboxylic acids, acid anhydrides, and amines are generally used as curing agents. In contrast to carboxylic acids, which only react sufficiently fast at high temperatures (above 180°C), carboxylic acid anhydrides (e.g. phthalic anhydride) allow crosslinking to be achieved at 100°C, especially in the presence of catalytic amounts of a tertiary amine. Optimum crosslinking is obtained with about 0.5 mol of anhydride per mol epoxy groups.

Crosslinking with amines can be carried out either catalytically with tertiary amines (e.g. with *N,N*-dimethylbenzylamine), or especially by equimolar conversion with primary or secondary oligoamines at elevated temperatures. This reaction is catalyzed by compounds that are capable of forming hydrogen bonds (water, alcohols, phenols, carboxylic acids, etc.). The most favorable ratio of amine/epoxide is not necessarily the stoichiometric ratio and must, therefore, be determined empirically in each case.

In the crosslinked state, epoxy resins are highly resistant to chemicals and solvents and are also endowed with good electrical properties. They are therefore employed, for example, as resistant lacquers and coatings. Moreover, they possess excellent adhesive power for many plastics, wood and metals ("two-component adhesives").

---

[1] In many cases the hydroxyl groups of the epoxy resin are also involved in the crosslinking reaction.

Example 4–26:
Preparation of epoxy resins from 4,4'-isopropylidenediphenol (bisphenol A) and 1-chloro-2,3-epoxypropane (epichlorohydrin) in one step

(a) Preparation of an epoxy resin with a molecular weight of 900
45.6 g (0.2 mol) of pure bisphenol A (recrystallized from dilute acetic acid) are mixed, under a hood,[1] with a solution of 15 g (0.375 mol) of NaOH in 150 ml water contained in a 250 ml three-necked flask, fitted with thermometer, reflux condenser, and a powerful stirrer (good stirring is essential since the reaction mixture becomes heterogeneous). The contents of the flask are vigorously stirred and heated on an oil bath to 50°C within 10 min; 29.0 g (0.314 mol) of freshly distilled epichlorohydrin are then added in one batch. (The mole ratio epichlorohydrin/bisphenol A is thus 1.57). The temperature is now raised to 95°C within 20 min and held steady for 40 min. The temperature should not be allowed to exceed 95°C since high temperature favours side reactions. The reflux condenser is now removed and the stirrer switched off so that the resin formed can settle out. The clear aqueous upper layer is carefully siphoned off and the resin washed by vigorously stirring with hot distilled water at 80–95°C. It is allowed to settle again and the wash water removed. This washing procedure is repeated several times until 100 ml of the wash water is equivalent to less than 0.15 ml of 0.1 M HCl (indicator: methyl red). In order to remove the trapped water the washed resin is heated to 150°C with moderate stirring for 30 min, until it becomes clear. Finally it is poured into a porcelain dish where it solidifies on cooling. The solid epoxy resin has a softening point of about 70°C and a molecular weight of approximately 900 ($n = 2$ in formula **2**). The epoxy value (see below) is about 0.2, corresponding to 1.8 epoxy groups per molecule of resin (equivalent weight 500). It is soluble in aromatic hydrocarbons, tetrahydrofuran and chloroform.

(b) Preparation of an epoxy resin with a molecular weight of 1400
Under similar conditions to those used in (*a*), 45.6 g (0.2 mol) of bisphenol A, 11.1 g (0.28 mol) of NaOH (dissolved in 112 ml water) and 22.6 g (0.245 mol) of epichlorohydrin are allowed to react with one another. Because of the smaller mole ratio of epichlorohydrin to bisphenol A (1.22) the resulting epoxy resin is of higher molecular weight than that produced in (*a*). The epoxy value is approximately 0.1, corresponding to 1.44 epoxy

[1] Bisphenol A powder and epichlorohydrin irritate the eyes and the respiratory system. Moreover, by prolonged action on the skin, epichlorohydrin is absorbed by the body and causes poisoning. All contact with these substances is, therefore, to be avoided; if necessary, wash off with plenty of water. Wear safety goggles and rubber gloves!

groups per molecule of resin (equivalent weight approximately 970). The molecular weight of the resin, softening at 97–103°C, is found to be about 1 400 (n = 3.7 in formula **2**).

Determination of the epoxy value: This method of determination depends on the addition of hydrogen halide to epoxy groups (1 mol of hydrogen halide is equivalent to 1 mol of epoxy groups) and is carried out as follows. 0.5–1.0 g of epoxy resin are refluxed with an excess (50 ml) of pyridine hydrochloride solution (16 ml pure concentrated hydrochloric acid are made up to 1 l with pure pyridine) for 20 min and, after cooling, are back-titrated with 0.1 M NaOH, using phenolphthalein as indicator. The epoxy number represents the gramme-equivalents of epoxide-oxygen per 100 g resin:

$$\text{epoxy value} = \text{epoxide equivalents/100 g resin}$$

$$= \frac{(B - A)M}{10E}$$

where B = titre of the pyridine hydrochloride solution, in ml,
A = titre for the sample (back titration), in ml,
M = concentration of NaOH used for titration, in mol l^{-1},
E = weight of resin, in g.

The equivalent weight is the amount of resin that contains 1 equivalent of epoxide; this is equal to 100/(epoxy value).

(c) Crosslinking (curing) of epoxy resins

With an amine: After determining the epoxy value and equivalent weight of the resin prepared according to (*b*), a small sample is melted in the oven at 150°C with the equivalent amount (0.25 mol per mol epoxy groups) of finely powdered 4,4′-methylenedianiline and the mixture well stirred for 30 s. After heating for 1 h at 150°C the sample is taken out; it has now become insoluble and infusible.

With a carboxylic acid anhydride: 5 g of the resin prepared according to (*a*) are melted in a beaker at 120°C and 1.5 g of phthalic anhydride (0.6–0.8 equivalents per equivalent of epoxy groups) are stirred into the melt. The mixture is held at 120°C for 1 h (after this time the resin is still soluble in acetone or chloroform) and then cured at 170–180°C for 1–2 h.

Crosslinking of epoxy resins with carboxylic acid anhydrides is catalyzed by tertiary amines; thus, if 50 mg *N,N*-dimethylaniline are added to the initial mixture in the above example, the curing process is already complete after 1 h at 120°C.

Example 4–27:
Preparation of an epoxy resin from glycerol and 1-chloro-2,3-epoxypropane (epichlorohydrin) in two steps

(a) Preparation of the chlorohydrin
46 g (0.5 mol) of distilled glycerol and 0.25 ml of BF_3-etherate are placed in a 1 l three-necked flask, fitted with stirrer, thermometer and reflux condenser, and warmed on a water bath at 40°C. With good stirring 140 g (1.5 mol) of freshly distilled epichlorohydrin are now added dropwise over a period of 45 min. A spontaneous exothermic reaction takes place. During the addition of epichlorohydrin the temperature is held at 44–45°C by external cooling; the stirring is then continued for another 2 h without cooling. The epoxy value (see Example 4–26b) must now be zero.

$$\begin{array}{l} CH_2{-}OH \\ | \\ CH{-}OH \\ | \\ CH_2{-}OH \end{array} + 3\ \underset{\diagdown O \diagup}{CH_2{-}CH}{-}CH_2Cl \xrightarrow{BF_3} \begin{array}{l} CH_2{-}O{-}CH_2{-}CHOH{-}CH_2Cl \\ | \\ CH{-}O{-}CH_2{-}CHOH{-}CH_2Cl \\ | \\ CH_2{-}O{-}CH_2{-}CHOH{-}CH_2Cl \end{array}$$

$$\xrightarrow{+\ 3\ NaOH} \begin{array}{l} CH_2{-}O{-}CH_2{-}\overset{\diagup O \diagdown}{CH{-}CH_2} \\ | \\ CH{-}O{-}CH_2{-}\overset{\diagup O \diagdown}{CH{-}CH_2} \\ | \\ CH_2{-}O{-}CH_2{-}\overset{\diagup O \diagdown}{CH{-}CH_2} \end{array} + 3\ NaCl + 3\ H_2O$$

(b) Preparation of the epoxy resin
The chlorohydrin, thus obtained, is dissolved by addition of 100 ml of a mixture of butanol and toluene (volume ratio 1:1) and then cooled to −5°C. 135 ml of 45% NaOH are slowly added dropwise over a period of 30 min, keeping the temperature at −5 to 0°C; reaction is allowed to continue at the same temperature for another 3 h. In order to remove the sodium chloride completely, another 60 ml of water are now added and the aqueous layer (pH 7–8) drawn off using a separating funnel. The solvent mixture is distilled off cautiously in vacuum. The residual highly viscous epoxy resin is mixed with some kieselguhr (to facilitate filtration) and filtered off using a sintered glass filter. The clear, highly viscous liquid has an epoxy value of about 0.6 and molecular weight of about 300. Yield: 80%. It can be crosslinked as described in Example 4–26 (*c*) to yield a hard insoluble product.

5 Reactions of Macromolecular Substances

5.1. CHEMICAL CONVERSION OF MACROMOLECULAR SUBSTANCES[1–5]

The preparation of macromolecular substances by addition and condensation polymerization relies upon the reactivity of low-molecular-weight, polyfunctional compounds. Numerous chemical reactions of macromolecules[6] are also known, some of which are carried out on a commercial scale. Such reactions of high polymers can occur at the functional groups of the constitutional repeating units (CRU's), with retention of the macromolecular skeleton and average degree of polymerization, or they may involve degradation of the chains. In many cases both occur at the same time. There are also reactions involving enlargement of the macromolecules; they lead either to chain extension, chain branching, or crosslinking; some reactions can, therefore, be used for the formation of block and graft copolymers (cf. Section 3.3.2). The chemical reactions of macromolecules are also suited to the investigation of their structure.[5]

In chemical reactions between low-molecular-weight compounds, new compounds are formed that, in principle, can be separated from the unconverted reactants and by-products, albeit sometimes with difficulty; deductions concerning the structure of the reactants can be made from the nature of the products. Such reactions can, therefore, be used to elucidate the constitution of low-molecular-weight compounds.

---

[1] *W. Kern* and *R.C. Schulz*, Angew. Chem. *69* (1957) 153.
[2] *W. Kern*, *R.C. Schulz* and *D. Braun*, Chem.-Ztg. *84* (1960) 385.
[3] *P. Schneider* in *Houben-Weyl 14/2* (1963) 661.
[4] *E.M. Fettes* (Ed.), "Chemical Reactions of Polymers", Interscience Publishers, New York, London, Sydney 1964.
[5] *D. Braun*, J. Polym. Sci., Polym. Symp. *50* (1975) 149.
[6] General discussion, see Section 2.1.6.

With chemical reactions of macromolecular substances the situation is more complicated in that the main reaction and side reactions take place on the same molecular framework. If, for example, only 80 out of 100 CRU's in a polymer chain react in the desired sense, while the rest either do not react at all or react in some other way, the last 20 units cannot be separated from the others since they all belong to the same macromolecule; consequently one cannot obtain a chemically uniform reaction product (see Section 2.1.6).

So long as all CRU's react in the same way during the reaction of a macromolecular compound, without any chain fission or significant side reactions, the molecular weight is changed but not the degree of polymerization, i.e. the average number of CRU's per macromolecule remains constant. For example in the quantitative hydrolysis of unbranched poly(vinyl acetate) to poly(vinyl alcohol), the molecular weight of the CRU, and hence the average molecular weight of the polymer, changes, but the average degree of polymerization remains the same. Such conversions, in which the macromolecular skeleton remains intact, are called "polymer-analogous" reactions. They play an important part in the elucidation of the structure of macromolecular compounds. In the development of macromolecular chemistry, they are also of fundamental significance in leading to the recognition that polymers are composed of macromolecules and do not consist of some form of associate of low-molecular-weight compounds. The molecular weights of the CRU's corresponding to reactant, main product and by-product in the reaction of a macromolecule are generally different, so that it is necessary to define the yield or conversion in a special way (see Section 2.1.6).

In the above-mentioned example of the polymer-analogous hydrolysis of poly(vinyl acetate) the reactant and product differ in their properties, for example in their solubility; however, both compounds have the same average degree of polymerization. The poly(vinyl alcohol) obtained by hydrolysis can in principle be esterified back to poly(vinyl acetate) with the original molecular weight; the reacetylated polymer then has the same properties as the original material. The viscosity number may be used to check whether in fact any chain splitting has occurred during the reaction sequence of hydrolysis and reacetylation (cf. Example 5–01).

Besides reactions in which, ideally, each individual CRU of the polymer chain is involved, there are also those in which two neighbouring units take part. A commercially important example is the acetalation of poly(vinyl alcohol), in which hydroxyl groups of neighbouring CRU's undergo ring closure by reaction with the carbonyl group of an aldehyde. The statistically maximum possible conversion of functional groups in such reactions is 86.5% (see Section 2.1.6.1).

We cannot here embark upon a discussion of the large number of known chemical transformation of macromolecular compounds.[1] Sometimes such reactions allow the production of polymers whose monomers are unknown, for example poly(vinyl alcohol), or are difficult to prepare, or will either not polymerize or do so only with great difficulty. An example of the last kind is vinylhydroquinone, which, like hydroquinone itself, is an inhibitor of radical polymerization. However, the "masking" of the phenolic OH groups by acetylation yields a polymerizable derivative; after its polymerization the protecting groups can be removed. Likewise it is possible to prepared stable polyradicals, i.e. macromolecules with unpaired electrons in the CRU's, which up till now have been obtained only by reactions of polymers, not by polymerization reactions of stable low-molecular-weight radicals.[2] Likewise, in the preparation of many ion-exchange resins, suitable functional groups are introduced by secondary reactions of macromolecular substances (that are generally crosslinked).

Amongst the important chemical conversions of macromolecular substances are the various reactions of cellulose.[3,4] The three hydroxyl groups per CRU can be partially or completely esterified or etherified. The number of hydroxyl groups acetylated per CRU are indicated by the names, cellulose triacetate, cellulose 2.5-acetate etc. Another commercially important reaction of cellulose is its conversion to dithiocarboxylic acid derivatives (xanthates). Aqueous solutions of the sodium salt are known as "viscose"; they are spun into baths containing mineral acid, thereby regenerating the cellulose in the form of an insoluble fibre known as viscose rayon.

Reactions that take place only at the end groups, with preservation of the chain structure, also belong to the class of polymer-analogous conversions. Examples are the reactions of hydroxyl end groups of polyesters with phthalic anhydride, or of end carboxylic acid groups with diazomethane. Such conversions are especially important for the capping of labile end groups. Thus the poly(oxymethylene)s obtained from formaldehyde or 1,3,5-trioxane are thermally unstable because of the semi-acetal end groups; but they can be stabilized by acetylation, which, under suitable conditions, can be carried out without causing significant degradation (Example 5–09).[5] Another reaction of this type is that of the amino end

[1] See *P. Schneider* in *Houben-Weyl 14/2* (1963) 661.
[2] *D. Braun*, Pure Appl. Chem. *30* (1972) 41.
[3] *E. Husemann* and *R. Werner* in *Houben-Weyl 14/2* (1963) 862.
[4] *L.S. Gal'braikh* and *Z.A. Rogovin*, Adv. Polym. Sci. *14* (1974) 87.
[5] *W. Kern*, *H. Cherdron* and *V. Jaacks*, Angew. Chem. *73* (1961) 177.

groups in the terminal aminoacid units of polypeptides and proteins with, for example, 2,4-dinitrofluorobenzene; the end unit can then be removed by hydrolysis of the peptide link[1] and identified as the dinitrophenyl-aminoacid. For the determination of end groups, see Section 2.3.2.2.

Crosslinked polymers with functional groups have recently been used ever more frequently as reagents for the synthesis of low-molecular-weight organic compounds, since they are easily separated after conversion and sometimes can easily be regenerated.[2,3] The immobilization of enzymes by attachment to crosslinked polymers should also be mentioned. This technique has already found industrial application.[4,5]

Example 5–01:
Poly(vinyl alcohol) by transesterification of poly(vinyl acetate); reacetylation of poly(vinyl alcohol)

(a) Preparation of poly(vinyl alcohol)

50 ml of 1% methanolic NaOH are placed in a 500 ml three-necked flask, fitted with stirrer, reflux condenser and dropping funnel, and heated to 50°C on a water bath. A solution of 15 g of poly(vinyl acetate) (see Example 3–06) in 100 ml of methanol are added dropwise with vigorous stirring over a period of 30 min. The transesterification sets in immediately as indicated by the precipitation of poly(vinyl alcohol). After the addition is complete, stirring is continued for a further 30 min; the powdery precipitate is then filtered off, washed with methanol and finally dried in vacuum at 30–40°C. Poly(vinyl alcohol) is soluble or swellable in only a few organic solvents (e.g. in warm dimethylformamide), but dissolves easily in warm water.

(b) Reacetylation of poly(vinyl alcohol)

5 g of the poly(vinyl alcohol) obtained above are placed in a 100 ml round-bottomed flask fitted with a reflux condenser, together with 75 ml of acetylating reagent (consisting of a mixture of pyridine, acetic anhydride and acetic acid in the volume ratio 1:10:10) and heated at 100°C for 24 h. The excess reagent is then removed in a rotary evaporator in vacuum at room temperature and the polymer precipitated from methanol into water.

---

[1] *G. Braunitzer*, Angew. Chem. *69* (1957) 189.
[2] *W. Heitz*, Adv. Polym. Sci. *23* (1977) 1.
[3] *G. Manecke* and *P. Reuter*, J. Polym. Sci., Polym. Symp. *62* (1978) 227.
[4] *L. Goldstein* and *G. Manecke*, "The Chemistry of Enzyme Immobilization" in Appl. Biochem. Bioeng. Vol. 1 (*L.B. Wingard* and *E. Katchalski* (Eds.), Academic Press, New York 1976).
[5] *I. Chibata* (Ed.) "Immobilized Enzymes: Research and Development", Kodansha Scientific Ltd., J. Wiley & Sons Ltd., 1979.

The acetyl group content[1] of the polymer is compared with that of the original poly(vinyl acetate), and the solubility behaviour is also examined. The limiting viscosity number is determined, both for the original poly-(vinyl acetate) and the reacetylated sample, in acetone at 30°C. It will be found that the hydrolysis and reacetylation has caused a fall in η_{sp}/c, indicating partial degradation of the macromolecules. This may be due to cleavage of side chains formed as a result of the branching reactions which occur when the polymerization of vinyl acetate is taken to high conversion.[2]

The change of solubility of poly(vinyl alcohol) with the degree of acetylation can easily be followed by taking samples at various times and testing their solubility in water, acetone and methanol. When the degree of acetylation is more than 20 mol% the solubility in water is lost, but the solubility in organic solvents increases.

Example 5–02:
Preparation of polymeric butyraldehyde divinyl acetal

3.3 g of distilled butanal (butyraldehyde) is placed in a 250 ml three-necked flask fitted with stirrer, reflux condenser and dropping funnel; into this is dropped a solution of 5 g of poly(vinyl alcohol) in 50 ml of water which has been warmed to 65°C and to which 0.3 g of concentrated sulfuric acid has been added. The addition should take about 2 min and during this time the stirring rate should be adjusted to give good mixing, at the same time avoiding splashing the walls of the flask. The polymeric butyraldehyde divinyl acetal precipitates immediately. 1 g of 50% sulfuric acid is then added and the mixture allowed to react for another hour at 50–55°C. After cooling to room temperature the polymer is filtered off and washed with water until neutral. The polymer is reprecipitated from methanol solution into water, filtered and dried in vacuum at 40°C. Yield: about 6 g. The solubility of the polymer is determined and compared with that of poly(vinyl acetate); it varies markedly with the degree of acetal formation.

Example 5–03:
Hydrolysis of a copolymer of styrene and maleic anhydride

2 g of a copolymer of styrene and maleic anhydride (see Example 3–52) are heated to boiling with 50 ml of 2 M sodium hydroxide in a 100 ml round-bottomed flask fitted with a reflux condenser. The polymer goes into

[1] See, for example, *Houben-Weyl 2* (1953) 344; suitable instructions for poly(vinyl acetate) are given by *A. Beresniewicz*, J. Polym. Sci. *39* (1959) 63; for acetyl group determination by the *p*-toluenesulfonic acid-induced conversion to acetic acid, see *O. Aydin*, *B.U. Kaczmar* and *R. C. Schulz*, Angew. Makromol. Chem. *24* (1972) 21.

[2] Cf. *H. Staudinger* and *H. Warth*, J. Prakt. Chem. *155* (1940) 261.

solution in a few minutes. After 1 h the solution is cooled and the polymeric acid precipitated by running about 500 ml of 2 M hydrochloric acid into the alkaline solution; it is allowed to settle and then filtered off. If after some time the polymer has not settled out, the acid dispersion should be shaken in a separating funnel with about 100 ml of diethyl ether; the polymer then separates into sticky rubbery lumps at the interface of the two liquids. The polymer is filtered and washed with a small amount of water (it is water soluble!). It is now pressed well and allowed to dry in air. The polymeric acid is purified by dissolving it in 50 ml tetrahydrofuran or 1,4-dioxane and precipitating in 500 ml of benzene; the polymer settles after some hours and can then be filtered off and dried in vacuum at 50°C.

The resulting styrene/maleic acid copolymer is soluble in hot water, in contrast to the starting material; the aqueous solution of the product gives a distinctly acid reaction.

Example 5–04:
Esterification of poly(methacrylic acid) with diazomethane

1.2 g of poly(methacrylic acid) (for preparation see Example 3–10) are dissolved in 10 ml of methanol and cooled in an ice/water bath. To the cold solution is added dropwise with shaking an ethereal solution of diazomethane[1] until the yellow colour of the diazomethane persists. During this operation the completely esterified poly(methacrylic acid) is precipitated. The polymer is filtered off and dried to constant weight in vacuum at 50°C. The limiting viscosity number of the esterified poly(methacrylic acid) is determined in acetone at 25°C and the molecular weight derived (see Section 2.3.2.1).

Example 5–05:
Preparation of poly[1-(4-acetylphenyl)ethylene]; [poly(4-vinylacetophenone)]

100 ml of carbon disulfide and 27 g of anhydrous, powdered aluminium trichloride are placed in a 500 ml three-necked flask, fitted with stirrer, reflux condenser and dropping funnel. 12 g (0.15 mol) of freshly distilled acetyl chloride are added to this heterogeneous mixture and a solution of 10.4 g (0.1 base-mol) of polystyrene[2] in 100 ml of carbon disulfide is slowly

[1] Preparation of diazomethane, see *T.J. De Boer* and *H.J. Backer*, Org. Synth. Coll. Volume IV, p. 250 (1963). Caution required in the preparation of diazomethane!

[2] It is best to use a relatively low-molecular-weight polystyrene (Example 3–02 or 3–11), otherwise the amount of solvent needed to provide a solution of manageable viscosity becomes too large.

added dropwise; a brisk reaction sets in with evolution of HCl. The mixture is heated to boiling for another hour, the reflux condenser is then replaced with an adaptor and condenser for distillation, and the carbon disulfide is distilled off. The residue is powdered and treated thoroughly with ice-cold, dilute hydrochloric acid in order to destroy the aluminium chloride; the liquid is decanted off, the residue washed again first with cold hydrochloric acid, then several times with water, and finally with methanol. The polymer is filtered off with suction and dried in vacuum at 50°C. The partially converted polystyrene thus obtained is soluble in freshly distilled acetophenone, in hot 1,4-dioxane, acetone and glacial acetic acid; it is insoluble in methanol. It can be depolymerized to monomeric 4-vinylacetophenone (see Example 5–16).

Example 5–06:
Acetylation of cellulose

10 g of cotton wool or shredded filter paper[1] are covered with a solution of 0.5 g of concentrated sulfuric acid in 50 ml of glacial acetic acid in a 250 ml wide-necked bottle with a ground glass stopper. Uniform wetting of the cellulose is ensured by stirring with a glass rod, the closed bottle then being allowed to stand for 1 h at room temperature. After this pretreatment, a mixture of 50 ml of 95% acetic anhydride and 20 ml of glacial acetic acid are added, and the bottle again closed and placed in a water bath at 50°C. The cellulose goes into solution after about 15 min, the reaction being complete after another 15 min. This so-called "primary solution" is divided into two equal parts which are used for the preparation of cellulose triacetate and cellulose 2.5-acetate respectively.

(a) Preparation of cellulose triacetate
25 ml of 80% acetic acid at 60°C are carefully stirred into one half of the primary solution in order to destroy the excess acetic anhydride; care must be taken that there is no precipitation of cellulose acetate during this addition. The solution is held at 60°C for another 15 min, then poured into a 600 ml beaker; 25 ml of water are carefully stirred in. After the addition of another 200 ml of water, the cellulose triacetate precipitates as a white, crumbly and readily washed powder. The product is filtered from the dilute acetic acid, slurried with 300 ml of distilled water, and the supernatant

[1] The type of starting material has a marked effect on the rate of reaction; filter papers of different kinds react at different rates.

liquid decanted after 15 min; this procedure is repeated until the washings give a neutral reaction. The polymer is dried as far as possible by suction or centrifugation and then in the oven at 105°C. Yield: about 7 g. The product is soluble in methylene chloride/methanol (volume ratio 9:1), but practically insoluble in acetone or in boiling benzene/methanol mixture (volume ratio 1:1).

(b) Preparation of cellulose 2.5-acetate
Into the other half of the primary solution, 50 ml of 70% acetic acid at 60°C and 0.14 ml of concentrated sulfuric acid are slowly added with stirring, in order to bring about partial hydrolysis of the cellulose triacetate. The stoppered bottle is held at 80°C for 3 h and then worked up as described for cellulose triacetate. Yield: 6–6.5 g. Cellulose 2.5-acetate (acetyl group content 40%) is soluble in acetone and methylene chloride/methanol (volume ratio 9:1), as well as in boiling benzene/methanol (volume ratio 1:1).

The progress of the partial hydrolysis can be checked by a simple solubility test. About 1 ml of the solution is withdrawn and the cellulose acetate precipitated with water. The small sample is quickly washed free of acid and dried as much as possible by pressing between two filter papers. Some fibres of the still damp material are placed in a test tube with 15–20 ml of benzene/ethanol (volume ratio 1:1) and heated to boiling in a water bath. If the fibres go into solution, then after about 15 min the whole charge can be worked up as described above.

Example 5–07:
Preparation of trimethylcellulose

10 g of cellulose 2.5-acetate (40% content of acetyl groups, see Example 5–06b) are dissolved in 200 ml of acetone on a water bath in a 1 l three-necked flask, fitted with stirrer, reflux condenser and two dropping funnels. The solution is warmed to 55°C under a hood. Over a period of 1.5 h, 120 ml of dimethyl sulfate (Caution!) and 320 ml of 30% sodium hydroxide are run in simultaneously from the two dropping funnels in ten equal portions, while stirring continuously. The acetone is now distilled off, and the residue filtered while still hot and thoroughly washed with boiling water in order to remove the alkali. The trimethylcellulose is purified by extraction with acetone, then with diethyl ether, finally being dried in vacuum at room temperature.

Trimethylcellulose is soluble in chloroform, benzene, and hot cyclohexanone, but insoluble in ethanol, diethyl ether and petroleum ether.

Example 5–08:
Preparation of sodium carboxymethylcellulose

15 g of finely divided cellulose[1] and 400 ml of 2-propanol are placed in a 1 l three-necked flask fitted with stirrer, reflux condenser, dropping funnel and nitrogen inlet; the air is displaced with a stream of nitrogen. 50 g of a 30% solution of sodium hydroxide are added with vigorous stirring over a period of 15 min, the stirring then being continued for another half hour. A solution of 17.5 g of monochloroacetic acid in 50 ml of 2-propanol are added dropwise over a period of 30 min. Stirring is continued for 4 h on a water bath at 60°C. The solution is then neutralized by addition of a few drops of glacial acetic acid using phenolphthalein as indicator, and is filtered while still hot. The raw fibrous sodium carboxymethylcellulose is dispersed in 400 ml 80% aqueous methanol at 60°C, filtered, and washed with a little aqueous methanol; this washing process is repeated two or three times until the product is free of sodium chloride. Finally it is washed with pure methanol and dried at 80°C. Yield: 22–25 g. The solubility of the dried sodium carboxymethylcellulose is tested in water.

Example 5–09:
Acetylation of the semi-acetal end groups of poly(oxymethylene) with acetic anhydride

Poly(oxymethylene) is prepared by the polymerization of anhydrous formaldehyde (see Example 3–36) or of 1,3,5-trioxane (see Example 3–40); paraformaldehyde, made by polycondensation of formaldehyde hydrate, cannot be acetylated in heterogeneous medium. Acetic anhydride and *N*,*N*-dimethylcyclohexylamine are carefully fractionated and sodium acetate is dehydrated by heating.

(a) Acetylation in heterogeneous medium
In a 100 ml flask fitted with air condenser and calcium chloride tube, 3 g of finely powdered poly(oxymethylene) are refluxed (139°C) with 30 ml of acetic anhydride and 30 mg of anhydrous sodium acetate for 2 h with continuous stirring. The polymer is then filtered off with suction and thoroughly washed five times with warm (50°C) distilled water to which some methanol has been added. It is then boiled with acetone for 1 h while stirring, and again filtered. The polymer is stored in a desiccator over

[1] Filter paper, soaked in water, is worked into an aqueous pulp by kneading and shredding; it is then filtered and dried.

calcium chloride .and sodium hydroxide pellets. Yield: 92 wt.% of the original polymer. Properties of the acetylated polymer: melting range 174–177°C (determined in a sealed tube); thermally stable portion: 88% (after 10 h at 190°C under nitrogen; see Example 5–15).

(b) Acetylation in the melt

3 g of poly(oxymethylene) together with 6 ml of acetic anhydride and 2 ml of *N*,*N*-dimethylcyclohexylamine (to prevent hydrolysis) are sealed in a glass ampoule and heated to 200°C for 30 min. The polymer is then filtered off at a sintered glass crucible and thoroughly washed twice with ethanol. It is now boiled with acetone for 1 h while stirring, and is filtered off under suction until the polymer is odourless; it is stored in a desiccator over calcium chloride and sodium hydroxide pellets. Yield: 90 wt.%. Properties of the acetylated polymer: melting range 170–174°C (in a sealed tube); thermally stable portion: 98% (after 10 h at 190°C under nitrogen; see Example 5–15).

Example 5–10:
Chlorination of Nylon-6,6

Polyamides can be chlorinated under nitrogen and, like low-molecular-weight *N*-halocarboxamides, then behave as strong oxidizing agents. Although they are quite stable at room temperature they lose chlorine at higher temperatures or by u.v.-irradiation, with the regeneration of the carboxamide group, amongst other reactions.

$$-R'-\underset{\displaystyle Cl}{\underset{|}{N}}-\underset{\displaystyle O}{\underset{\|}{C}}-R- \xrightarrow{h\nu} -R'-\underset{\displaystyle H}{\underset{|}{N}}-\underset{\displaystyle O}{\underset{\|}{C}}-R-$$

The change in properties of the polymer associated with the photochemical reaction provide a simple example of photochemical production of pictures through the use of light-sensitive polymers.[1]

Nylon-6,6 is here chlorinated with *tert*-butyl hypochlorite and a glass plate is coated with the product. Part of the light-sensitive layer is shielded and the rest exposed to u.v.-radiation. In the exposed part chlorine is split off by the photochemical reaction. In this area the polymer becomes insoluble in chloroform and loses its oxidative properties. If the irradiated plate is dipped into chloroform the unirradiated part of the polymer is

[1] *A. Banihashemi* and *R.C. Schulz*, Makromol. Chem. *179* (1978) 855.

dissolved away, while the irradiated insoluble part can be coloured with a dyestuff, yielding a negative image. If on the other hand the plate is dipped, after irradiation, into aqueous potassium iodide solution, iodide ions are oxidized to iodine which is bound to the polymer by adsorption, giving a brown-coloured positive image.

(a) Reprecipitation of Nylon-6,6

40 g of ground commercial Nylon-6,6 or the polymer made in Example 4–09 are dissolved in 200 ml of concentrated (98–100%) formic acid. The solution, which may be coloured pale brown, is added dropwise to 1 l of methanol. After 1 h the viscous suspension is treated with 1 l of distilled water; after another 1.5 h the precipitated polymer is filtered at a Büchner funnel and washed several times with hot water. The damp powder is suspended in benzene in a 500 ml two-necked flask fitted with a stirrer, and the retained water distilled off azeotropically with the aid of a water separator. After further filtration, washing with diethyl ether, and drying over phosphorus pentoxide, the polymer is obtained in finely divided form. Yield: 36 g (90%).

(b) Preparation of tert-butyl hypochlorite

500 ml of bleaching liquor (containing about 12% of active chlorine) are cooled to about 10°C; to this are added, with vigorous stirring and protection from light, 37 ml of *tert*-butanol and 24.5 ml of glacial acetic acid. After 3 min the raw product is separated in a separating funnel, washed first with 50 ml of 10% sodium hydrogencarbonate solution and then with 50 ml of water, and finally dried with sodium sulfate or calcium chloride. It can be purified by distillation under reduced pressure (water pump) with the condenser and receiver cooled to at least 0°C; however, for the following experiment the distillation is not absolutely necessary.

(c) Chlorination of Nylon-6,6

1.13 g (0.01 base-mol) of reprecipitated Nylon-6,6 is suspended in 20 ml of 1,1,2,2-tetrachloroethane. After about 30 min, 0.1 ml of concentrated formic acid and 2.1 ml (0.02 mol) of *tert*-butyl hypochlorite are added at about 10°C. The reaction mixture is stirred until a clear solution is obtained (about 3 h). This solution is dropped into 200 ml of pure diethyl ether with stirring. After stirring for 2 h the solvent is decanted and the polymer then filtered off, washed several times with diethyl ether and dried in vacuum. One obtains an *N*-chloropolyamide-6,6, with an *N*-chloro group content of 87–97%, which can be kept in the cold and dark for extended periods. The content of active chlorine can be determined by iodimetric titration in concentrated formic acid to which sodium acetate has been added.

(d) Coating the glass plates

1 g of freshly prepared *N*-chloropolyamide-6,6 is dissolved in 100 ml of chloroform and filtered, if necessary, from insoluble swollen constituents. This solution is now run on to well-cleaned, exactly level glass plates by means of a pipette, using 1 ml of solution for every 25 cm^2 of surface. The solvent is allowed to evaporate in indirect light or in the dark. After about 30 min the plates are ready for use. They should be stored in the refrigerator in the dark until required.

(e) Irradiation and development

Part of the coated plate is covered with light-proof material, e.g. by means of an aluminium foil, and the plate is then irradiated with a u.v.-lamp (e.g. Heraeus Q700, distance 11 cm; 5 min). During the irradiation the plate is cooled by means of a fan. For development one makes use either of the difference in solubility of the irradiated and unirradiated regions or of their difference in reactivity.

(i) The irradiated plate is covered with chloroform in a flat dish and gently rocked for about 30 s, causing the unirradiated parts to dissolve away. The plate is allowed to drain and then immersed in a 0.1% methanolic solution of amido-black; this causes the irradiated, chloroform-insoluble parts to colour blue-black, giving a negative image of the masking foil. Instead of amido-black one can also use benzoazurin G (0.1% in water), azoeosin 6 (0.8% in water) or azoblue (0.6% in methanol).

(ii) The irradiated plate is immersed for about 5 min in an aqueous 1 M potassium iodide solution acidified with acetic acid, and then carefully rinsed with water. In the unirradiated regions the *N*-chlorocarboxamide groups oxidize iodide to iodine which is adsorbed by the polymer giving it a brown coloration. This results in a positive image of the masking foil.

5.2. EXPERIMENTS WITH ION EXCHANGERS

Ion exchangers are polyelectrolytes that generally consist of solid, crosslinked and hence insoluble macromolecular compounds carrying acidic or basic groups on the macromolecular framework. The long-known inorganic, naturally occurring or synthetically prepared materials (e.g. zeolites and permutites respectively) today play only a minor role as ion exchangers. Nowadays the bulk of ion exchangers are made either by secondary introduction of ionic groups into crosslinked addition or condensation polymers or by direct synthesis from appropriate low-molecular-weight starting materials.[1–5]

The usual ion exchangers are, thus, macromolecular, insoluble polyvalent acids or bases; because of their insolubility they are well suited for the exchange of $H^{\oplus}$ or $OH^{\ominus}$ ions. Thus if an insoluble polyacid is suspended in water with a low-valent salt, the cations of the salt are exchanged with hydrogen ions (cation exchanger); correspondingly the anions of a low-valent salt can be exchanged with hydroxyl ions by using a basic anion exchanger. This principle is applied, for example, in the preparation of pure water from water which contains salts. The processes which occur in these exchanges can be represented schematically as follows:

$$\mathrm{Pol}^{\ominus} H^{\oplus} + Na^{\oplus} \rightleftharpoons \mathrm{Pol}^{\ominus} Na^{\oplus} + H^{\oplus}$$

$$\mathrm{Pol}^{\oplus} OH^{\ominus} + Cl^{\ominus} \rightleftharpoons \mathrm{Pol}^{\oplus} Cl^{\ominus} + OH^{\ominus}$$

$$\mathrm{Pol}^{\ominus} H^{\oplus} + \mathrm{Pol}^{\oplus} OH^{\ominus} + Na^{\oplus} Cl^{\ominus} \rightleftharpoons \mathrm{Pol}^{\ominus} Na^{\oplus} + \mathrm{Pol}^{\oplus} Cl^{\ominus} + H_2O$$

where $\mathrm{Pol}^{\ominus}$ and $\mathrm{Pol}^{\oplus}$ denote the ionic sites in the insoluble crosslinked resin, to which various exchangeable ions are bound. Depending on the type of exchanger used, the acidic and basic groups may be contained in the same polymer, or the anionic and cationic exchangers can be mixed, or used in tandem.

Exchangers that have been loaded with cations or anions can be regenerated by treatment with acid or alkali, respectively, since one is always dealing with an equilibrium reaction. Commercially available ion exchangers are frequently delivered in the form of salts so that before use they must be converted into the free acids or bases. Metal cations can also be directly exchanged with one another.

The most suitable acidic groups for synthetic ion exchangers are sulfonic acid and carboxylic acid groups; phosphoric acid groups are less common. For anion exchangers, primary, secondary, and tertiary amino groups are often used, also polymeric quaternary ammonium bases. For the preparation of such ion exchangers it is usual to start from a polymer; especially

[1] *R. Griessbach*, "Austauschadsorption in Theorie und Praxis", Akademie-Verlag, Berlin 1957.

[2] *F. Helfferich*, "Ionenaustauscher", Vol. 1, Verlag Chemie, Weinheim/Bergstr. 1959.

[3] *K. Dorfner*, "Ionenaustauscher", 3rd Edn., Walter de Gruyter and Co., Berlin 1970.

[4] *J.X. Khym*, "Analytical Ion-Exchange Procedures in Chemistry and Biology", Prentice Hall Inc., Englewood 1974.

[5] *J. Inczêdy*, "Analytische Anwendungen von Ionenaustauschern", Verlag der Ungarischen Akademie der Wissenschaften, Budapest 1964.

useful for this purpose are the copolymers of styrene and 1,4-divinylbenzene (cf. Example 3–50), which can be conveniently used in the form of bead polymers with a particle diameter of about 0.1–2 mm. The content of 1,4-divinylbenzene in the monomer mixture used for polymerization determines the degree of crosslinking; it is generally indicated as wt. % 1,4-divinylbenzene. The higher the degree of crosslinking, the lower is the swellability of the polymer; however, the degree of swelling also depends on the nature of the counter ion and on some other factors. Under certain conditions, for example in the presence of organic solvents in bead polymerization, so-called macroreticular or macroporous networks are formed. In this case the crosslinks are distributed irregularly over the whole volume of the material, unlike the normal networks. This results in a porous structure combining the properties of high permeability for solvent and comparatively low swellability; such polymers sometimes have advantageous properties as starting materials for making ion exchangers. The ionic functional groups are then introduced by chemical reactions. One can also make ion exchangers by polymerizing monomers that already contain functional groups; thus, methacrylic acid can be polymerized to give a weak cation exchanger.

Another possibility, which is used commercially, is to prepare insoluble condensation polymers, for example from phenol and formaldehyde, into which ionic groups are subsequently introduced, as with addition polymers (Example 5–12); the ionic groups may also be present in the monomer before condensation, e.g. phenolsulfonic acid.

Ion exchangers can also be made from cellulose, especially for scientific applications. They are prepared from alkali cellulose by reaction, for example, with chloroacetic acid (for preparation of sodium carboxymethylcellulose, see Example 5–08). By conversion with β-chloroethyldiethylamine one obtains so-called DEAE-cellulose, an anion exchanger carrying 2-diethylaminoethyl groups, $(C_2H_5)_2NC_2H_4$.

Besides grain size, degree of crosslinking and swellability, an important characteristic of an ion exchanger is its capacity. This denotes the number of equivalents of exchangeable counter ions on a high polymer network, with respect either to the weight or volume of dry or swollen exchanger. For laboratory use the capacity is usually expressed in milliequivalents g^{-1}; for the softening of hard water the usable exchange capacity is often given in g of CaO per litre of exchanger.

Various procedures can be applied to effect ion exchange. The simplest method is to work batchwise, whereby the exchanger is left in contact with a solution of the ions to be exchanged until equilibrium is reached. This method is applicable to those exchange reactions where the equilibrium is in favour of the desired product; this can of course always be achieved by employing a sufficient excess of the ion exchanger. However, ion exchange is more usually carried out using columns. As in column chromatography, the solution to be exchanged is allowed to run through the column of ion exchanger from top to bottom. The ion exchanger used to fill the column must already be in the swollen state before it is washed into the column, otherwise the pressure caused by swelling can lead to bursting of the glass tube.

Ion exchangers are not only used commercially (e.g. for water softening, recovery of metals from waste water, refining of raw sugar) but also to an increasing extent in the laboratory. Thus, ion exchangers provide a convenient and clean way of preparing free acids or bases (e.g. free thiocyanic acid by exchange of ammonium thiocyanate with an acid cation exchanger), or of purifying aqueous solutions (e.g. removal of formic acid from solutions of formaldehyde). Ion exchangers can also be used to catalyze chemical conversions, being readily removed by filtration after the required reaction has occurred; examples of this type are esterifications or protein hydrolyses catalyzed by acid ion exchangers. By introducing complex-forming groups into crosslinked polymers, numerous exchangers have also been prepared which allow selective extraction of certain metals from solution or provide a means of enrichment of elements that are present in low concentration.[1,2] The synthetic routes for this purpose are generally very adaptable; for example crosslinked poly (aminostyrene) can be diazotized and then coupled with suitable phenol derivatives. Numerous other applications and working instructions can be found in the literature.

Example 5–11:
Preparation of a cation exchanger by sulfonation of crosslinked polystyrene

(a) Sulfonation of crosslinked polystyrene
Insoluble polystyrene crosslinked with divinylbenzene can easily be converted by sulfonation to a usable ion exchanger. For this purpose a mixture of 0.2 g of silver sulfate and 150 ml of concentrated sulfuric acid are heated to 80–90°C in a 500 ml three-necked flask fitted with stirrer, reflux condenser and thermometer. 20 g of a bead polymer of styrene and divinylbenzene (see Example 3–49) are then introduced with stirring; the temperature climbs spontaneously to 100–105°C. The mixture is maintained at 100°C for 3 h, then cooled to room temperature and allowed to stand for some hours. Next the contents of the flask are poured into a 1 l conical flask that contains about 500 ml of 50% sulfuric acid. After cooling,

[1] *R. Hering*, "Chelatbildende Ionenaustauscher", Akademie-Verlag, Berlin 1967.
[2] *E. Blasius* and *K.P. Janzen*, Chem.-Ing.-Tech. *47* (1975) 594.

the mixture is diluted with distilled water, and the gold-brown coloured beads filtered off at a sintered glass filter and washed copiously with water.

(b) Determination of the ion-exchange capacity

To determine the ion-exchange capacity the sulfonated polymer is washed into a glass tube closed at one end with a stopcock (chromatographic column) above which there is a plug of glass wool. 100 ml of 2 M NaCl solution are allowed to run through the column, followed by 100 ml of 2 M HCl. Finally, the column is washed with distilled water, the runnings being collected in 10 ml portions and titrated with 0.1 M NaOH using phenolphthalein as indicator. When the concentration has fallen below 0.001 M the washing is stopped. The water remaining in the column is allowed to run off and the damp ion exchanger is poured into a beaker. Three samples, each of 2 g are weighed into tared 100 ml conical flasks as quickly as possible. One is heated to constant weight in an oven at 110°C in order to determine the water content of the sample. 50 ml of 0.1 M NaOH are added to each of the other two flasks and vigorously shaken. After 30 min the mixtures are filtered, the resin washed with a little water, and the filtrate back-titrated with 0.1 M HCl. The ion-exchange capacity is expressed in milliequivalents per g of dry exchanger.

Example 5–12:

Preparation of a cation exchanger by sulfonation of a phenol-formaldehyde condensation polymer

40 g of an uncrosslinked phenol-formaldehyde condensation polymer (see Example 4–12) are gradually warmed to 140°C in 120 g of 95% sulfuric acid in a 250 ml round-bottomed flask fitted with a reflux condenser. As soon as the resin has completely dissolved, the solution is cooled to room temperature, poured into an iron dish and 30 ml of 37% aqueous formaldehyde solution is stirred in with a spatula as quickly as possible. The metal dish is then placed in an oil bath at 110°C and the resin allowed to harden for 2 h. The cooled product is washed with water, broken up with a hammer and ground down to pieces of 1–3 mm size in a mortar. The particles are then washed with water until the washings are clear. The exchange capacity is determined as described in Example 5–11.

Example 5–13:

Preparation of an anion exchanger from crosslinked polystyrene by chloromethylation and amination

(a) Chloromethylation of crosslinked polystyrene

20 g of a crosslinked styrene bead polymer (e.g. the styrene/divinylbenzene copolymer from Example 3–50) and a solution of 50 g

chloromethyl methyl ether[1] in 40 ml of tetrachloroethylene are placed in a 250 ml three-necked flask fitted with stirrer and reflux condenser (with drying tube attached), the third neck being closed with a ground glass stopper. This mixture is stirred for 30 min at room temperature, causing the beads to swell somewhat. 10 g of anhydrous zinc chloride are added at 40–60°C over a period of 60 min with continuous stirring; stirring is continued at this temperature for another 2 h. The unconverted chloromethyl methyl ether is now destroyed by careful addition of water; the beads are washed several times with water and dried in vacuum at 50°C. The chlorine content of the beads should be around 15 wt.%.

(b) Amination of the chloromethylated polystyrene

20 g of the chloromethylated polystyrene obtained in (*a*) are refluxed with 50 ml of benzene for 30 min in a 100 ml three-necked flask fitted with stirrer, reflux condenser, thermometer and gas inlet. The mixture is cooled to 30–35°C and gaseous anhydrous trimethylamine is passed in with stirring, while the temperature is raised steadily to 50–55°C. (The gaseous trimethylamine is prepared from an aqueous solution by dropping it into concentrated sodium hydroxide in a separate vessel and passing the gas through a drying tube filled with NaOH pellets into the three-necked flask). The flow of trimethylamine is stopped after 4 h and the mixture allowed to stand for another 3 h at room temperature. The beads are filtered off, washed a few times with benzene and dried in vacuum at 50°C. The dry beads are then treated for 2 h with 100 ml of 5% hydrochloric acid and finally washed thoroughly with water until free of acid (test with methyl red).

(c) Determination of the ion-exchange capacity

Three samples of the moist beads obtained in (*b*), each of about 2 g, are weighed into tared 100 ml conical flasks. One flask is heated to constant weight in the oven at 110°C in order to determine the water content of the beads. 10 ml of 15% sodium hydroxide are added to each of the other two flasks and the mixture stirred magnetically for 30 min to form the quaternary ammonium base. Next the excess alkali is removed by washing with water (test with phenolphthalein) and filtering. The moist beads are transferred quantitatively back to the conical flask, and shaken back and forth with 50 ml of 0.1 M hydrochloric acid. After 30 min the beads are filtered off, washed with a little water and the filtrate back-titrated with 0.1 M sodium hydroxide. The ion-exchange capacity is given in milliequivalents per g of dry exchanger (in the form of the chloride).

---

[1] Chloromethyl methyl ether is very poisonous; it is essential to work under a hood.

5.3. DEGRADATION AND CROSSLINKING OF MACROMOLECULAR SUBSTANCES[1-3]

Macromolecules can be cleaved by physical as well as by chemical action. The most frequent initial result is the formation of macromolecules with the same chain structure but with lower average degree of polymerization. The polymer homologous series can gradually degrade further until eventually low molecular weight fragments are produced. The analysis and characterization of the oligomeric and monomeric species formed by degradation can provide valuable evidence concerning the structure of macromolecules. However, with some polymers degradation proceeds by an "unzipping" mechanism; in this case monomer molecules are broken off continuously either from the chain end or from a cleavage point within the chain, while the molecular weight of the remaining polymer chains is unaffected.

Thermally and chemically initiated degradation of polymers is particularly important; chain cleavage by mechanical forces, for example in blenders and extruders, or by light or high energy radiation is also of considerable practical significance. Ultrasonic degradation of macromolecules can also occur.[4]

The generally undesired changes in the chemical and physical properties of polymers during use, under the action of air, light and heat, is termed aging. Such processes can be retarded or entirely eliminated by the addition of protecting agents (stabilizers, anti-oxidants, u.v.-absorbers etc.).[5-8]

The thermal degradation of high polymers is often a radical process. Chain cleavage can take place either at random within the chain, or preferentially at weak links, for example in the neighbourhood of branches

---

[1] *C.H. Bamford* and *C.F.H. Tipper* (Eds.), "Comprehensive Chemical Kinetics", Vol. 14 "Degradation of Polymers", Elsevier Scientific Publishing Co., Amsterdam, Oxford, New York 1975.

[2] *G. Geuskens* (Ed.), "Degradation and Stabilization of Polymers", Applied Science Publishers, London 1975.

[3] *H.H.G. Jellinek* (Ed.), "Aspects of Degradation and Stabilization of Polymers", Elsevier Scientific Publishing Co., Amsterdam, Oxford, New York 1978.

[4] *A.M. Basedow* and *K.H. Ebert*, Adv. Polym. Sci. *22* (1977) 83.

[5] *J. Voigt*, "Die Stabilisierung der Kunststoffe gegen Licht und Wärme", Springer-Verlag, Berlin 1967.

[6] *K. Thinius*, "Stabilisierung und Alterung von Plastwerkstoffen", 2 Vols., Verlag Chemie, Weinheim 1969.

[7] *B. Dolezel*, "Die Beständigkeit von Kunststoffen und Gummi" (edited by *C.M. von Meysenbug*), Carl Hanser Verlag, München, Wien 1978.

[8] *W.L. Hawkins* (Ed.), "Polymer Stabilization", Wiley Interscience, New York 1972.

or structural irregularities, or from unstable chain ends. With some polymers thermal degradation gives either none or very little of the monomers used in the preparation of the polymer (e.g. polyethylene, polypropene, poly(acrylic esters), polyacrylonitrile, polybutadiene); here one speaks only of degradation. On the other hand with other polymers degradation results in relatively large amounts of monomer (e.g. polystyrene, poly(α-methylstyrene), polyisoprene, poly(methyl methacrylate), poly(oxymethylene)) in which case the degradation may also be termed depolymerization. If the thermal degradation proceeds from labile end groups, the stability of such polymers can be substantially improved by blocking these end groups (cf. Example 5–15, thermal depolymerization of poly(oxymethylene)).

The kind of fragments formed on degradation depends mainly on the structure of the polymer and on the decomposition temperature. Thus, under the same conditions, the thermal decomposition of polymers of acrylic esters yields practically no monomer while that of polymers of methacrylic esters gives monomer almost exclusively. Degradation of polystyrene at 250°C gives mainly oligomers of styrene such as the dimer, trimer, and higher homologues, while at 350°C monomeric styrene is also to be found in the decomposition products. As the temperature is raised further, ethylbenzene, toluene, and benzene are formed in increasing amounts.

Finally, there are those polymers which undergo an elimination reaction on heating but without any initial breakdown of the molecular chains. Poly(vinyl chloride) belongs to this group, eliminating hydrogen chloride on heating. Such degradation is very undesirable both on account of the corrosive action of hydrogen chloride vapour and of the concomitant darkening of the polymer. This decomposition process can be suppressed to a certain extent, and in practice for considerable lengths of time, by addition of stabilizers.[1] Similarly poly(vinyl acetate) eliminates acetic acid on heating. Poly(*tert*-butyl methacrylate) yields isobutene, water and poly(methacrylic anhydride) at about 250°C.

In the presence of oxygen the thermal degradation of polymers is complicated by oxidation reactions, making the course of the reaction rather obscure. Chemical degradation reactions can be caused by oxidation, autoxidation or hydrolysis, also by the action of light. Under suitable conditions, especially at elevated temperatures, macromolecular substances are autoxidizable, like low-molecular-weight compounds.[2] The

---

[1] *D. Braun*, Gummi, Asbest, Kunstst. *24* (1971) 902, 1116.

[2] *A. Rieche*, Kunststoffe *54* (1964) 428; *L. Dulog*, *E. Radlmann* and *W. Kern*, Makromol. Chem. *60* (1963) 1; *80* (1964) 67.

hydroperoxides that are formed in the process are generally unstable at these temperatures and decompose into radicals, rendering the reaction autocatalytic; the secondary products formed can initiate further reactions. Both chain degradation and crosslinking are observed. In commercial practice such processes are suppressed by the addition of anti-oxidants such as certain phenols or amines; some macromolecular substances are, therefore, "stabilized" immediately after their preparation.[1] This problem is accentuated in polymeric dienes, the double bonds of which give rise to autoxidation particularly easily. In addition to oxidation of polymers by molecular oxygen, other oxidizing agents can also cause degradation. For example, the number of head-head linkages in poly(vinyl alcohol) can be determined by oxidation of the 1,2-diol groups with periodic acid; at each of these positions the carbon chain is broken during the oxidation process so that the degradation can easily be followed viscometrically (Example 5–18). Conclusions can then be drawn concerning the number of irregularly bound CRU's formed during the preparation of the precursor, poly(vinyl acetate), by the radical polymerization of vinyl acetate. Such irregularities can arise either by reverse addition of monomer in the propagation step or by combination of polymer radicals in the termination step. A further example of chemical degradation of polymers is the ozonolysis of natural rubber, resulting in 4-oxovaleraldehyde and 4-oxovaleric acid. Photochemical degradation of polymers in the presence of atmospheric oxygen is generally accompanied by oxidation and is one of the most important causes of aging of synthetic polymers.[2]

Hydrolytic degradation is especially important in polymers with hydrolyzable links between the CRU's. Thus, polyesters can be hydrolyzed to yield the starting materials from which they were formed. Acetal links in synthetic polymers such as poly(oxymethylene), or in natural polymers such as cellulose, can be hydrolyzed with acids. However, the resistance to hydrolysis depends very much on the structure of the polymer; for example polyesters of terephthalic acid are very difficult to hydrolyze while aliphatic polyesters are generally easily hydrolyzed. Polyamides are normally much more resistant to hydrolysis than polyesters; they may be cleaved by the methods usually employed for polypeptides and proteins.

Grinding or milling causes degradation of many polymers.[3] The process of mastication of natural rubber involves a mechanically initiated, aut-

---

[1] *G. Scott*, "Atmospheric Oxidation and Antioxidants", Elsevier, Amsterdam 1965.

[2] *B. Ranby* and *J.F. Rabek*, "Photodegradation, Photo-oxidation and Photostabilization of Polymers", John Wiley & Sons, London, New York, Sydney, Toronto 1975.

[3] *K. Murakami*, Mechanical Degradation in "Aspects of Degradation and Stabilization of Polymers" (Ed. H.H.G. Jellinek), Elsevier Scientific Publishing Co., Amsterdam, Oxford, New York 1978, p. 295.

oxidative degradation which lowers the molecular weight to a level where the material is easier to process on a commercial scale.[1] The radical chain fragments resulting from mechanical working can initiate the formation of block and graft copolymers in the presence of polymerizable monomers.[2]

Example 5–14:
Thermal depolymerization of poly(α-methylstyrene) and of poly(methyl methacrylate)

Exactly 5 g of polymer (from Examples 3–26 or 3–27, and 3–05 respectively) are weighed into a 100 ml round-bottomed flask. The flask is then connected to two cold traps maintained at −78°C in a methanol/dry-ice bath. The apparatus is evacuated to about 0.1 torr and the flask immersed in a metal bath whose temperature is regulated to ± 3°C by means of a bunsen burner and gas relay. The metal is first liquefied by heating to 100°C, the flask[3] inserted and then quickly heated to the appropriate depolymerization temperature (within 6–8 min). For poly(α-methylstyrene) a temperature of 280°C is suitable; for poly(methyl methacrylate), 330°C. Depending on the rate of decomposition, the temperature is maintained for 1–2 h while evacuating the apparatus with an oil pump. The experiment is stopped by removing the heating bath, and also the cold baths round the traps; as soon as the solid monomer, collected in the traps, has melted, the vacuum is released. The yield is determined by weighing the residue and the trapped monomer. The monomer is identified by measuring the refractive index (α-methylstyrene $n_D^{20} = 1.5386$; methyl methacrylate $n_D^{20} = 1.4140$). The decrease of molecular weight of the polymer during depolymerization is determined from the limiting viscosity numbers of the starting polymer and the residue left in the flask (measured in benzene at 20°C). For comparison one can also decompose polystyrene which, under these conditions, yields very little monomer.

Example 5–15:
Thermal depolymerization of poly(oxymethylene)[4]

The decomposition of poly(oxymethylene) can be conveniently performed in the vessel described in Section 2.3.8.1. 100 mg of each of the following samples are weighed into small test tubes with as constant an internal

---

[1] *W.F. Watson*, Makromol. Chem. *34* (1959) 240.
[2] *A. Casale* and *R.S. Porter*, Adv. Polym. Sci. *17* (1975) 1.
[3] It is recommended that the bottom of the flask be first coated with soot with the aid of a luminous flame so that the metal does not stick to the glass on removal from the bath.
[4] *W. Kern* and *H. Cherdron*, Makromol. Chem. *40* (1960) 101.

diameter as possible:

(a) poly(oxymethylene) with OH end groups (from Example 3–36; the product from Example 3–40 is not suitable),
(b) poly(oxymethylene) with acetyl end groups (from Examples 5–09a and 5–09b).

The tubes are now placed in the decomposition vessel which is then evacuated and filled with pure nitrogen three times. A slow stream of nitrogen is passed through the vessel which is heated in an oil bath or air thermostat to 190°C. At intervals of 1 h the decomposition vessel is taken out of the hot bath, the tubes allowed to cool under nitrogen for 15 min and then individually weighed. The weight per cent residue is plotted against time.

The thermal depolymerization of poly(oxymethylene) at 190°C starts from the hydroxyl end groups, but the oxidative and acid-catalyzed hydrolytic degradation takes place within the main chain. Hence if poly(oxymethylene) is heated in air or in the presence of strong acids (e.g. *p*-toluenesulfonic acid or poly(phosphoric acid)) samples with blocked end groups will also degrade.

Example 5–16:
Thermal depolymerization of poly[1-(4-acetylphenyl)ethylene]; [poly-(4-vinylacetophenone)]

A 100 ml round-bottomed flask, connected to two cold traps in series, is used as decomposition vessel; the second cold trap is cooled to −50°C in a methanol/dry-ice bath. 5 g of poly(4-vinylacetophenone) (see Example 5–05) are weighed into the flask which is then evacuated with an oil pump to about 0.1 torr. It is now heated in a metal bath to at least 300°C, whereby the depolymerization sets in and a yellow oil collects in the first cold trap. The heating is continued until no more product appears (about 30 min) and only a carbonized residue is left in the flask. The pyrolysis products are then taken up in dry diethyl ether and dried over calcium chloride for 1 h. The drying agent is filtered off, 1% of hydroquinone (with respect to product) is added to the solution and the ether is removed under reduced pressure (water pump). Finally the oil is distilled under reduced pressure (0.1–0.5 torr), the middle fraction (b.p. ≈75°C) being a clear oil that solidifies on cooling to give colourless crystals (m.p. 33°C). Yield of 4-vinylacetophenone: ≈3 g.

The oxime is prepared for identification purposes. 0.5 g of freshly distilled 4-vinylacetophenone, 0.5 g of hydroxylamine hydrochloride, 10 ml of ethanol, 5 ml of water and 6 sodium hydroxide pellets (the mixture must be alkaline) are refluxed for 15 min. 20 ml of 50% ethanol are then added and the mixture acidified with dilute sulfuric acid, causing it to go milky. It is allowed to crystallize in the refrigerator. After recrystallization from aqueous ethanol the 4-vinylacetophenone oxime has a m.p. of 116–118°C.

A small sample of the recovered 4-vinylacetophenone is mixed with 0.1 mol % 2,2′-azoisobutyronitrile and polymerized at 60°C under nitrogen in a tube fitted with a ground joint and tap. Since side reactions can cause crosslinking to occur during the polymerization of this monomer, leading to insoluble products, the experiment is stopped at relatively low conversion (after about 6 h). The contents of the tube are dissolved in a 10-fold amount of acetone, the solution decanted from any insoluble material, and dropped into a 10-fold amount of methanol. The polymer is filtered with suction and dried in vacuum at 50°C. Its solubility is then tested in the solvents indicated in Example 5–05.

Example 5–17:
Thermal dehydrochlorination of poly(vinyl chloride)

Two test tubes are each charged with 1 g of poly(vinyl chloride) (see Example 3–49), 100 mg of lead stearate having been previously added as stabilizer to one of the samples (well mixed in a mortar!). The tubes are then loosely stoppered with corks which carry, on the lower side (pinched into a slit), a 4 cm long strip of moistened universal indicator paper. Both tubes are now heated for 10 min in a beaker containing colourless silicone oil at 170–175°C. The indicator paper above the sample stabilized with lead stearate shows scarcely any change, but that over the unstabilized sample shows very clearly by its colour that hydrogen chloride has been evolved. The coloration of the polymer during heating is also strongly indicative: the unstabilized polymer becomes red to brown, while the stabilized sample only darkens a little. The evolution of hydrogen chloride can also be proved by passing the vapour over a beaker containing ammonia, or by passing it into acidified silver nitrate solution.

Example 5–18:
Oxidative degradation of poly(vinyl alcohol) with periodic acid

2.0 g of poly(vinyl alcohol) are placed in a 250 ml beaker containing 70 ml of distilled water (see Example 5–01); solution is hastened by warming somewhat and stirring with a glass rod. Care must be taken not to splash the solution. As soon as a homogeneous solution is obtained it is cooled to room temperature and filtered through a sintered glass disc in order to remove dust particles, the filtrate being collected in a 100 ml graduated flask. The beaker is washed several times with a little water and the washings likewise filtered into the graduated flask which is then immersed in a thermostat at 25°C and made up to the mark. A solution of 1.7 g of periodic acid (HIO_4, $2H_2O$ = H_5IO_6) in 45 ml of distilled water is also prepared; this is filtered into a 50 ml graduated flask and made up to the mark in a thermostat at 25°C.

The oxidative degradation of poly(vinyl alcohol) is followed at 25°C by viscosity measurements in an Ostwald viscometer (capillary diameter 0.4 mm). One proceeds as follows:

(1) 5 ml of the periodic acid solution are diluted to 10 ml in a graduated flask at 25°C using filtered distilled water. The flow time t_o of this solution is determined.

(2) In the same way, 5 ml of the poly(vinyl alcohol) solution are diluted to 10 ml with distilled water and the flow time t of this solution determined.

(3) Finally 5 ml of the poly(vinyl alcohol) solution are mixed with 5 ml of periodic acid solution and the flow time determined immediately and then at short time intervals until a constant value t_d is reached after a few minutes.

If it is assumed that the contribution of the periodic acid to the viscosity is negligible then the specific viscosity of the original poly(vinyl alcohol) is given by

$$(\eta_{sp})_o = (t - t_o)/t_o$$

and the specific viscosity of the final degraded polymer by

$$(\eta_{sp})_d = (t_d - t_o)/t_o$$

The limiting viscosity numbers are calculated by the equation of Schulz and Blaschke ($K_\eta = 0.27$; see Section 2.3.2.1) and hence the average molecular weights; from these one can estimate the number of cleavages of the original chain and hence the frequency of 1,2-diol groupings.

Example 5–19:
Hydrolytic degradation of an aliphatic polyester

1 g of a linear aliphatic polyester (e.g. a polyester of succinic acid and 1,6-hexanediol, see Example 4–02) are dissolved in dry tetrahydrofuran in a 100 ml graduated flask and made up to the mark at 30°C. 50 ml of this solution are mixed with 2 ml of 30% sulfuric acid and stored in a closed vessel, separate from the other 50 ml of solution, at 30°C (e.g. in the viscometer bath). The viscosities of the two solutions are determined immediately and then at hourly intervals in an Ostwald viscometer (capillary diameter 0.3 mm) at 30°C; for this purpose 3 ml of the appropri-

ate solution are pipetted each time from the flask into the viscometer. The two flasks are allowed to stand overnight at 30°C and the hourly viscosity measurements continued next day until the values remain essentially constant over 1 h, which will be the case after a total of 20 h. The viscosity of the sample without the sulfuric acid remains unchanged.

The specific viscosity (or simply the flow time) is plotted against reaction time. For the calculation of the specific viscosity the flow time of the solvent, t_o, must be determined for a mixture of 50 ml of tetrahydrofuran and 2 ml of 30% sulfuric acid; the viscosity of pure tetrahydrofuran is considerably raised by the addition of the acid. Finally 20 ml each of the hydrolyzed and unhydrolyzed solutions are dropped into 200 ml of methanol and the resulting precipitates compared; if the hydrolysis of the first sample is complete no precipitate will appear whatsoever.

Example 5–20:
Hydrolytic degradation of cellulose and separation of the hydrolysis products by chromatography

100 mg of cellulose (e.g. filter paper), that has been well shredded by hand, is mixed with 1 ml of 72% ice-cold sulfuric acid and, after pulping well with a glass rod, is kept overnight at 0°C. 1 ml of 25% sulfuric acid is added and the mixture held for 2 h at 50°C. After cooling, 40 ml of iced water is added and the mixture finally refluxed for 1 h.

To neutralize[1] the hydrolysis products the solution is passed through a column (of about 15 cm length and 1 cm width) packed with an anion exchanger (e.g. Amberlite IR–45 or the anion exchanger from Example 5–13). The column is subsequently washed with about 50 ml of distilled water; a drop of the final washings should give a negative test for sugar when applied to a piece of chromatographic paper and sprayed with aniline phthalate reagent after drying the paper. When this is the case the solution is evaporated in vacuum down to about 5 ml.

The hydrolysis products are separated by paper chromatography.[2] To improve the sharpness of separation the paper is impregnated with sodium dihydrogenphosphate; for this purpose a solution of 15.6 g (0.1 mol) of NaH_2PO_4, $2H_2O$ in 1000 ml water is prepared, 100 ml poured into a flat dish and the sheet of paper to be impregnated slowly drawn through the

[1] Neutralization with alkaline-earth carbonates is to be avoided at all costs, in order to prevent epimerization of glucose to mannose which is favoured by complex formation between mannose and alkaline-earth ions.

[2] *G. Hesse*, "Chromatographisches Praktikum", 2nd Ed., Akademische Verlagsgesellschaft, Frankfurt 1968.

solution. (The latter is quickly denuded of solute and must be renewed for each further sheet of paper). The paper is hung up to dry on a line, and is then cut into strips 10 cm wide and 60 cm long; a start line is drawn with a fine pencil 8 cm from the edge of the paper and the flow direction indicated with an arrow. With the aid of a micropipette 5 mm^3 of the following solutions are applied to the base line; (it is advisable to evaporate the solvent with a hot-air blower immediately after applying the samples to the paper to avoid undue spreading of the spots):

1. 5 mm^3 of hydrolysis product (corresponding to about 0.1 mg of solid material);
2. 5 mm^3 of a solution of 100 mg of glucose[1] in 5 ml of water (corresponding to 0.1 mg of solid material);
3. 5 mm^3 of a solution of 5 mg of xylose[1] in 5 ml of water (corresponding to 0.005 mg of solid material);
4. 5 mm^3 of a solution of 5 mg of mannose[1] in 5 ml of water (corresponding to 0.005 mg of solid material).

The chromatogram is developed by the descending method using a mixture of butanol/acetone/water (volume ratio 4:5:1) in a suitable glass tank. For this purpose sufficient eluant is placed in the tank to cover the bottom. The trough mounted in the upper part of the tank is filled half full, and the tank is then closed and allowed to stand for a few hours so that the air within it is saturated with the solvent. Only then is the paper strip placed in the tank with the upper edge dipping into the trough and held in place by a heavy glass rod so that the base line is about 3–5 cm below the edge of the trough. The tank is now closed and development allowed to proceed for 2–3 h, until the solvent front has reached the lower quarter of the paper. (The optimum time for development depends to some extent on the construction of the tank and trough, and must be found by experiment). The paper strips are then taken out and dried in the air. Next they are sprayed with aniline phthalate reagent and allowed to dry for 10 min in the air and for 10 min in the oven at 100°C; the sugars then become visible as violet-brown spots.

This very simple procedure yields, of course, only qualitative information concerning the composition of the hydrolysis products of cellulose. One may, however, estimate from the size and intensity of the resulting spots that the original cellulose consists of more than 90% glucose units and that xylose units are also detectable. Using more refined conditions, and with the assistance of photometric analysis of the paper chromato-

[1] E.g. "Kollektionen für chromatographische Vergleichszwecke" from Merck, Darmstadt.

grams under u.v.-radiation, similar procedures can yield quantitative results.[1]

Thin layer chromatographic separation of the resulting sugars is simpler and can yield quantitative results.[2] For this purpose glass plates are prepared with a layer of kieselguhr G (Merck) 250 μm thick, the kieselguhr being impregnated or mixed with a phosphate buffer of pH 5. About 1.5 mm from the lower edge of the plate 5 μl of the cellulose hydrolysis product are applied with the aid of a micropipette; 0.7% solutions of glucose, mannose, and xylose in 70% ethanol are also applied for comparison. A mixture of butyl acetate, ethanol, pyridine, and water (volume ratio 8:2:2:1) is used as eluant. Elution is carried out four times in a rectangular glass chamber, the plate being dried after each elution (separation distance about 16 cm, time about 60 min). The chromatogram is finally dried with the aid of a hot-air blower, sprayed with aniline phthalate reagent and placed in the drying oven at 105°C for 10 min. Approximate R_f values after four elutions and a separation distance of 16 cm: glucose 0.51, mannose 0.68, xylose 0.88.

Example 5–21:
Curing of a butadiene-styrene copolymer

Industrially the curing (vulcanization) of diene homopolymers and copolymers with elementary sulfur is carried out in a heated press at 100–140°C (hot curing); this cannot be done in a normal laboratory on account of the expensive apparatus required. However, the principle of curing can be illustrated by crosslinking a butadiene-styrene copolymer (Buna S) with disulfur dichloride (S_2Cl_2) at room temperature (cold curing):

(a) A small piece of a butadiene-styrene copolymer (see Example 3–47) is placed in a test tube and covered with disulfur dichloride; the stoppered sample is allowed to stand under nitrogen[3] for 1 h. The S_2Cl_2 is then poured off and benzene added; a sample that has not been treated with S_2Cl_2 is likewise covered with benzene. The solubility and swellability of the two samples are compared.

(b) The progress of the crosslinking during cold curing can be observed very nicely by the following experiment. 2 g of a butadiene-styrene copolymer are dissolved under nitrogen[3] in 100 ml of benzene in a

[1] See *G. Jayme* and *H. Knolle*, Fresenius' Z. Anal. Chem. *178* (1960/61) 84.
[2] *T. Krause* and *H. Teubner*, Holzforschung *27* (1973) 123.
[3] It is not absolutely necessary to work under nitrogen.

250 ml conical flask. 1 ml of disulfur dichloride is added to half this solution and vigorously shaken; the other half of the solution is likewise kept stoppered, but without the addition of S_2Cl_2. After 5 min the solution treated with S_2Cl_2 is already significantly more viscous than the reference solution; after 10 min gelation sets in and after 20 min a pudding-like mass is formed. After some hours the crosslinking is so far advanced, that as a consequence of the high crosslink-density a phase separation occurs and the solvent is partially exuded from the shrunken gel.

6 Subject Index

Bold page numbers refer to examples of preparations; also see the index of contents. For general properties of macromolecular substances see also under the heading "Polymer".